1st International Conference on Advanced Engineering and Technology (ICATECH 2018)

IOP Conference Series: Materials Science and Engineering
Volume 462

Surabaya, Indonesia
29 September 2018

ISBN: 978-1-5108-8094-8
ISSN: 1757-8981

Printed from e-media with permission by:

Curran Associates, Inc.
57 Morehouse Lane
Red Hook, NY 12571

CURRAN ASSOCIATES INC. proceedings .com

Some format issues inherent in the e-media version may also appear in this print version.

Printed by Curran Associates, Inc. (2019)

Institute of Physics
Dirac House, Temple Back
Bristol BS1 6BE UK

Phone: 44 1 17 929 7481
Fax: 44 1 17 920 0979

techtracking@iop.org

Additional copies of this publication are available from:

Curran Associates, Inc.
57 Morehouse Lane
Red Hook, NY 12571 USA
Phone: 845-758-0400
Fax: 845-758-2633
Email: curran@proceedings.com
Web: www.proceedings.com

Table of contents

Volume 462

The 1st International Conference on Advanced Engineering and Technology

29 September 2018, Surabaya, Indonesia

Accepted papers received: 14 November 2018
Published online: 8 January 2019

Preface

Papers

IOP Conf. Series: Materials Science and Engineering **462** (2019) 011001 doi:10.1088/1757-899X/462/1/011001

Preface

It is my great honor, and pleasure, to welcome you, in the name of the Adhi Tama Institute of Technology Surabaya, on the occasion of the 1st International Conference on Advanced Engineering and Technology (ICATECH 2018) held in Surabaya, Indonesia, September 29, 2018.

This seminar is the first international seminar which has the main purpose to bring researchers and academicians to share their knowledge and experience in Engineering, Design, Information System and Technology area. The conference serves as an excellent opportunity to meet each other and to exchange ideas with theme, Multidisciplinary Approach towards Sustainable Technology and Industry.

This proceeding contains selection papers from graduate students, faculty members, researchers, and academia from various universities and research institutions, and also professional associations and other related organization in infrastructure area. The selected papers are selected based on the paper quality and the relevancy to the theme.

A total of 52 manuscripts are selected to be submitted in IOP Conference Series: Materials Science and Engineering and to be presented in ICATECH 2018 seminar. Our deepest gratitude to all of our speakers, participants and contributors who have given the ICATECH 2018 their generous supports. Also to all the reviewers who helped us managing the papers so that all the manuscripts are well written. Many thanks are due to all our Organizing Committee members for their dedication and continuous efforts and hard work in preparing and organizing the seminar. We would like also to thank to all members of the Steering Committee, Scientific Committee and our distinguished international board of reviewers for all their support and advice.

Dr. Syamsuri, Adhi Tama Institute of Technology Surabaya
On behalf of ICATECH 2018 Organizing Committee
September 28, 2018

Peer review statement

All papers published in this volume of *IOP Conference Series: Materials Science and Engineering* have been peer reviewed through processes administered by the proceedings Editors. Reviews were conducted by expert referees to the professional and scientific standards expected of a proceedings journal published by IOP Publishing.

IOP Conf. Series: Materials Science and Engineering **462** (2019) 012001 doi:10.1088/1757-899X/462/1/012001

Study and Simulation of A Solar System for Drying Purpose in Rwanda

P D Uwitije[1], R Hantoro[1], M Y Nasri[2], G Nugroho[1]

[1]Department of Engineering Physics, Institut Teknologi Sepuluh Nopember, Surabaya 60111, Indonesia
[2]HVAC Department, University of Frères Mentouri Constantine 1, 25000 Constantine, Algérie

uwitipe@yahoo.fr

Abstract. This study aims to simulate and present the hourly thermal analysis of a forced convection solar dryer along with the parametric analysis of the drying process made using the drying characteristics of green pepper. Under the consideration of Rwanda weather condition on August 21st, the designed solar collector shows low performance before noon and high-performance afternoon. It generates output air temperatures above 50^{0}C from 12:00 until 19:00 with the highest value of 74^{0}C generated at 15:00. In the drying chamber, the green pepper drying time shows dependence on the air temperatures obtained from the solar collector and on the tray positioning number. The graphical results show that the products dried in the first tray dries quickly compared to those in the last trays, the effect of drying at high temperatures lead to the reduction in the drying time. The obtained results are confirmed by their similarities in curves trends with those of results diagrams in related literature studies.

1. Introduction

In the confrontation of the immense population increase and resource challenges facing humanity, sufficient provision in the food supply and renewable energy exploitation will take up vital importance. To cope up with current and future food supply, preservation techniques that optimise transport cost and reduction in food losses are required. An excellent and most applicable way to preserve products and to facilitate their transportation is the drying process. By reducing product moisture content, drying helps in the impediment of the microbial actions that are responsible for food spoilage and in the reduction of their weight. Unfortunately, mechanical thermal dryers are more energy intensive, have a high initial and fuel running cost and are the harmful impact on the environment. On the other hand, the traditionally used open sun drying is slow and fails in drying some particular products requiring intense energy. In Rwanda, the loss arising from products deterioration can be substantial especially for products such as fruits and vegetables deteriorating quickly. The use of solar drying in enclosed thermal systems becomes a very ideal and cost-effective alternative especially in the rural locations with non-existent grid connections [1]. Establishment of modelling techniques and simulation programs have been very useful in analysing and optimising the drying systems performance and in the forecasting of components temperatures and drying mechanism of different products [2]. Elaborated numerical studies for various solar dryers working in different locations have been carried out to study the system performance and drying process of different products. Dilip in India used a transient analytical model along with the drying characteristics of onions to evaluate the system's drying performance a solar crop dryer [3]. Using the drying characteristics of potatoes, Nasri and Belhamri in Algeria simulated an indirect forced convection solar dryer for drying agriculture products [4]. In their calculations, algorithm and code programs were established to solve the energy balance equation to calculate the hourly air temperatures of various components making the drying system. In Rwanda, the first study on drying mango slices in domestic solar dyer ascertained the promising application of reducing food spoilage [5]. The purpose of this study is to simulate a forced convectional solar dryer for drying the excess or non-sold out products fruits and vegetable in the local markets of Rwanda.

2. Materials and Methods

2.1. System Design

The drying system used in this present study is a typical indirect convection dryer designed to dry fruits and vegetables. It consists of two main parts, a double glazed conventional solar air heater and a drying chamber. The collector is tilted at 15° concerning the horizontal. On the top, it consists of two glass covers spaced by 15 mm with a thickness of 10 mm. In the middle, a galvanised sheet painted black of thickness 1 mm is used as an absorber whereas at its bottom it has polystyrene insulation of 4 cm thick. The drying chamber is a cubic box having a side length of 1 m walls. This consists of a 10 cm layer made up of brick and with external insulation of 4 cm thick layer of polystyrene. Inside the drying chamber, there are ten wired mesh racks separated by 10 cm in between on which 2.5 kg of products is laid. On the top, it is equipped with an extractor fan to ensure forced convection as shown in Figure 1. During the working time with bad weather conditions, the system is considered to have a backup air heater.

2.2. Theoretical Approach

In modelling the drying chamber, we have adopted the step by step mathematical model proposed by Danguent. This model accounts for the evolution of temperatures of all solar dryer elements in time and space. Much more it involves the effects of mass and system thermal inertia [4]. It consists of taking the drying chamber and cutting it into some fictitious slices in the direction of airflow. The obtained slice sections are delimited by two racks and the walls of the chamber as shown in Figure 2. The equations governing thermal and mass exchanges at the level of each tray section are written based on the energy and heat balance as follows:

Exchange at the external polystyrene wall face

$$\frac{m_{Pp}C_{Pp}}{4}\left(\frac{dT_{pe}}{dt}\right) = h_r.S_v(T_c - T_{pe}) + h_{am,pe}.S_v(T_{am} - T_{pe}) + k_p.S_v(T_p - T_{pe}) \quad (1)$$

Exchange through the wall between polystyrene and brick faces

$$\frac{m_{Pp}C_{Pp}}{4}\left(\frac{dT_p}{dt}\right) + k_pS_v(T_p - T_{pe}) = \frac{m_{Pb}C_{Pb}}{4}\left(\frac{dT_p}{dt}\right) + k_bS_v(T_p - T_{pi}) \quad (2)$$

Exchange at the internal surface of the brick wall

$$\frac{m_{Pb}C_{Pb}}{4}\left(\frac{dT_{pi}}{dt}\right) = k_bS_v(T_p - T_{pi}) + h_{ash,pi}.S_v(T_{ach} - T_{pi}) \quad (3)$$

The exchange between the product, internal wall face and the drying air at the internal wall face of the brick

$$\dot{m}_{ach}C_{Pair}(T^{\star}_{ach} - T_{ach}) = h_{ach,f}\,S_f\,(T_{ach} - T_f) + 4.\,h_{ach,pi}.S_v(T_{ach} - T_{pi}) \quad (4)$$

Exchange between the product and the drying air

$$m_fCp_{pf}\left(\frac{dT_f}{dt}\right) = h_{ach,f}\,S_f\,(T_{ach} - T_{pi}) - P_{ev} \quad (5)$$

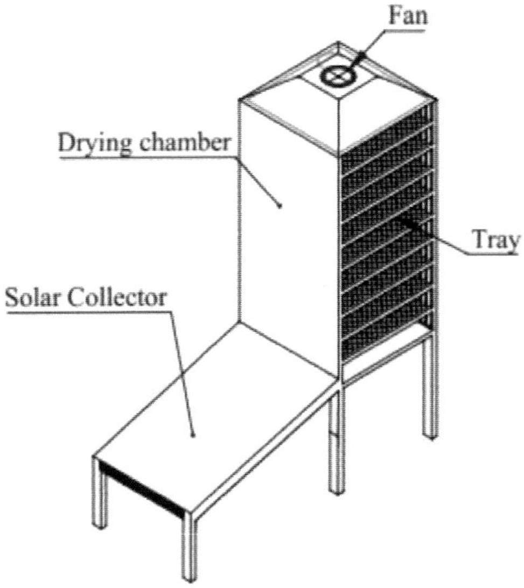

Figure 1. Diagram of the studied forced convection indirect solar dryer.

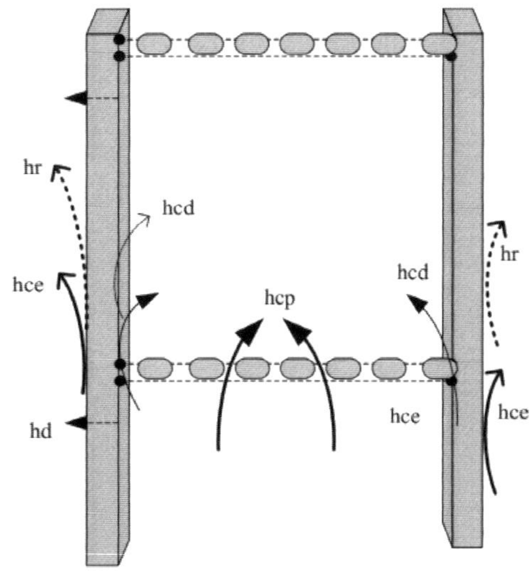

Figure 2. Energy balance on the tray inside the drying chamber.

2.3. Drying Kinetics

The kinetic model used to simulate the behaviour of the moisture transfer within the green pepper is the first order empirical model written as [6]:

$$\frac{dX}{dt} = -K(X - X_e) \tag{6}$$

With Xe representing the equilibrium moisture content for green pepper and K is a drying constant found in the experimental work for drying green pepper [7]. This constant K ensures the falling rate slope of the drying curve due to drying condition such as product diameter, drying temperature, air velocity and absolute humidity determined by:

$$K = K_0 D^{KD} T_{ach}^{KT} W_{ach}^{KW} U_{ach}^{KU} \tag{7}$$

Taking the moisture content $X = X_i$ at t=0 the solution is found for the moisture content at any time as in the following expression [6]:

$$\frac{X_t - X_e}{X_i - X_e} = e^{-Kt} \tag{8}$$

2.4. Numerical Approach

To solve numerically the set of equations obtained for the drying chamber, an implicit finite difference method is chosen and applied to discretise each equation in time and space. After, we are lead to resolve the set of obtained equations into a matrix system of $[A][T] = [B]$ in which the vector $[T]$ represents the vector of unknown quantities. Later a Matlab program is developed for the drying chamber in which an iterative Gauss-Seidel method is used to solve the system of equations. At the beginning of this simulation program, we assign at the beginning of the simulation arbitrary values at temperatures, which are equal to ambient temperatures, then heat exchange coefficients are calculated and injected into the system of equations above. After the resolution of the latter, the values of the temperatures calculated in this way are compared to the previous values. The calculations are considered complete when the desired precision is obtained. Otherwise, the calculated values are taken as arbitrary values, and the calculation is resumed to the desired precision.

3. Results and Discussion

The hourly distribution temperatures for different elements of the solar air heater and the output air temperatures are presented in Figure 3. The maximum fluid output temperatures above 50°C are generated from 12:00 up until 19:00 with the highest value of 74°C generated at 15:00 owing to maximum radiation and high ambient temperatures input at this time. It is observed that before 8:00, both temperatures do not increase much even though they are exposed to solar radiation for 3 hours. This is explained by the effect the received energy is used to warming up the system. Another remarkable observation is the high collector output temperature fluid at 19:00 despite sunlight absence. This is due to the logic of mass inertia for system cooling [4].

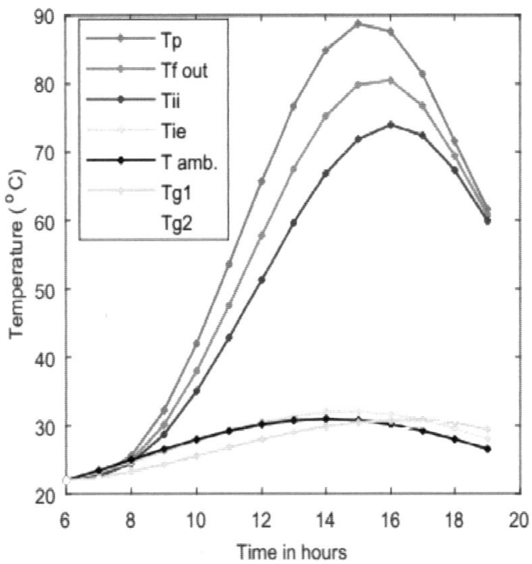

Figure 3. Variation in temperatures for different air collector elements

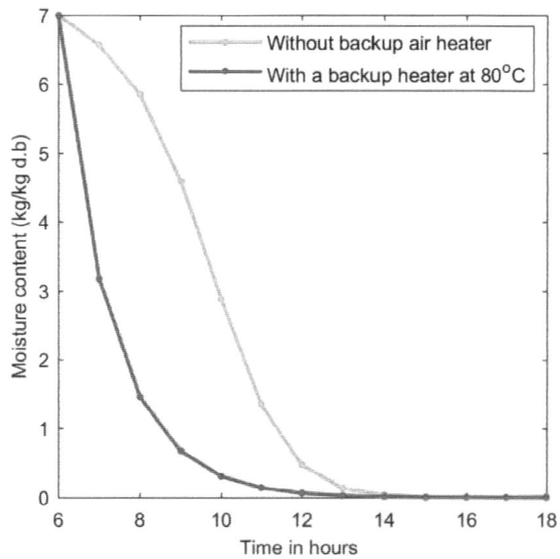

Figure 4. Effect of drying at constant and varying temperatures on the moisture content reduction

The performance of the drying chamber is evaluated by the decreasing product moisture content concerning the drying time. In Figure 4, the influence of drying with constant fluid temperatures at 80°C when the backup heater is added is shown in red curve whereas the influence of using varying air temperatures when there is no heater is depicted by the green curve. When drying at constant temperatures, product water removal is high at the beginning of the drying process owing to the quick evaporation of product moisture content in external layers of product. The moisture content reduction for the green curve represents the reduction due to collector air temperatures varying hourly depending on the radiation intensity reaching on the solar air heater. The reduction goes slow first in the morning but later is accelerated when much radiation intensity is gathered and after the ambient temperature has increased. Drying at a controlled constant temperature of 80°C insured by the solar collector with the addition of heating results in short drying of six hours from 6:00 am to 11:00 am compared to 8 hours needed when using solar air heater only. Similar results in curves shape are found for drying onions [6].

First, the effect of influential parameters is studied by inspecting the drying time at various drying temperature. In Figure 5 the effect of drying at various constant temperatures (40, 60 and 80°C) of air on the drying time of products in the last tenth tray is illustrated. Drying at 80°C occurred 3 hours earlier compared to the drying time at 60°C happening after ten hours whereas drying at 40°C requires more than fifteen hours of drying time. Similar findings in the graph trends are obtained when drying onions in the previous work [6]. At last, the performance of the drying chamber is evaluated by studying the reduction in product moisture content in different trays. The difference in drying time is illustrated in Figure 6 for trays 1 and 10. The drying time for products placed in the first tray is short compared to the product dried in the last tenth tray. The explanation might be based on the fact that the air temperature at the entrance of the drying chamber decreases as it moves from tray to tray. Another reason would be the decrease in the drying air capacity to take up water as it moves from tray to tray fixing more water vapour. The observed graphic trends are similar to the work of D. Jain in his work concerning a solar crop dryer [3].

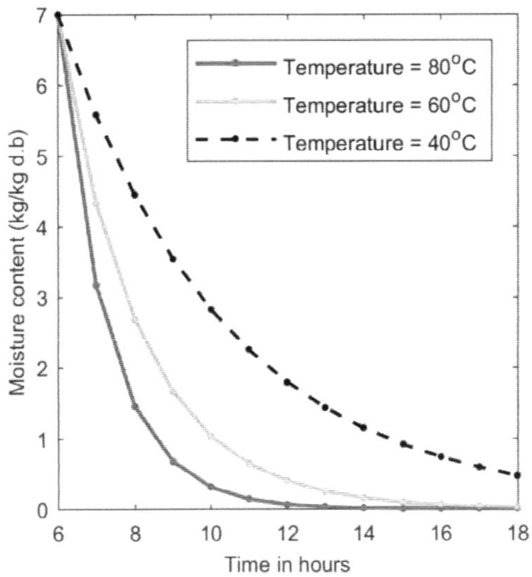

Figure 5. Effect of drying temperature on the moisture content reduction.

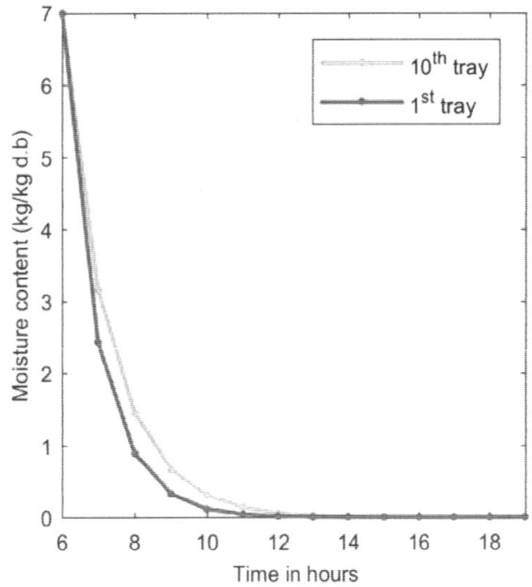

Figure 6. Variation in moisture content in the first and last tray.

4. General Conclusion

In this paper, we study the behaviour of an indirect forced convection solar dryer by the use of numerical modelling. The developed Matlab program fits perfectly for studying the thermal performance of the solar collector and drying chamber in Rwanda. To do this, we proceeded to the mathematical simulation of our dryer by developing and solving the respective mass and heat transfer mathematical models to reflect the physical reality of the working mechanism of a forced convection indirect solar dryer. The results showed that the designed solar collector generates high air temperatures capable of shortening the drying process and that using a backup heater increases the system performance, especially in the morning time. In the parametric study, the drying rate is much influenced by the drying temperature. It is also observed that products in different trays have different drying time. The graphical results obtained using the drying kinetics of green pepper presented similarities in the curve trends occurring in the diagrams of related work. This comparison allowed us to conclude that our simulations are valid. Using the developed code program will help in choosing appropriate measurements and in decision making during system design.

Nomenclature

C_p	:	Specific heat (J/kg K)
D	:	Characteristic dimension (m)
hr	:	Coefficient of heat exchange by radiation (W/m^2K)
j	:	Section number
K	:	Coefficient of heat transfer by conduction (W/m^2K)
$\dot{m}$	:	Mass flowrate (kg/s)
T	:	Temperature (K)
X	:	Moisture content (kg/kg d. b)

Subscripts

a	:	Ambient
ach	:	Heated air
b	:	Brick
f	:	Air fluid
f	:	Product
e	:	External
g_1	:	First glass cover
g_2	:	Second glass cover
ii	:	Internal part of the insulation
ie	:	External part of the insulation
p	:	Absorber plate
s	:	Sky
t	:	Time

5. References

[1] Yagnesh C B and Pravin R P 2018 A Comprehensive Review on solar dryer *Int. J. Ambient Energy* 2162–8246

[2] Om P, Vinod L, Anukul P, Anil K and Arbind K 2016 Review on various modelling techniques for the solar dryers *Renew. Sustain. Energy Rev.* 396–417–396–417

[3] Jain D 2007 Modeling the performance of the reversed absorber with packed bed thermal storage natural convection solar crop dryer *J. Food Eng.* 637–647–637–647

[4] Anon 2016 Simulation d'un sechoir solaire indirect á convection forcée pour les produits agroalimentaires *Sci. Technol. B* 57–62

[5] Hakizimana E, Emmanuel D and Charlote A 2017 Design and Optimization of Domestic Solar Dryer *SciencePG* 130–5

[6] Anon 2003 Design and simulation of a solar dryer for agriculture products *J. Food Eng.* 259–66

[7] Kiranoudis C T, B.Maroulis Z and Marinos-Kouris D 1992 Drying Kinetics of onion and green pepper *Dry. Technol.* 995–1011

IOP Conf. Series: Materials Science and Engineering **462** (2019) 012002 doi:10.1088/1757-899X/462/1/012002

Quality Control of Cigarettes Packaging using Convolutional Neural Network

A Nazar[1], M N P Nurwiyadi[1], M Syai'in[1], A Khumaidi[1], R Y Adhitya[1], N Rinanto[1], M K Hasin[1] and H A Widodo[1]

[1]Automation Engineering Study Program, Department of Marine Electrical Engineering, Shipbuilding Institute of Polytechnic Surabaya, Indonesia

Alfianuddin15nazar@gmail.com

Abstract. The tobacco industry is a source of incomes for our country. Currently, cigarette consumers in Indonesia are already very high. Quality is considered as an essential point for every industry to satisfy its customers. However, some problems often occur in the Industry. Including the decline in cigarettes quality. One of the reasons for the decline in the quality of cigarettes is the lack of selection of the sorting system on cigarette packaging. The lack of quality of the packaging system makes the quality of cigarettes decreased. In a tobacco company in East Java, the fetching of defective products is still done manually. Based on the problem, the author takes that study case as the background of this research. In this system, the packs of cigarettes will be detected and selected using image processing. On the image processing will be implemented as a method of convolution neural network (CNN) this algorithm has a function to detect texture from cigarette packaging. If there is non-conforming or defective cigarette packaging, the rejector will be active and discard the cigarette packaging and it will not enter into the next packing process. The result of training data from this study had a success rate of 96.06%. When using real-time data, the system is able to classify 148 data from 154 existing data correctly. The real-time system success rate is 94,59%.

1. Introduction

Quality control for the packing product by machine vision is a kind of inspection method which has a simple system and didn't take a high cost to be applied into an industrial system. The inspection of external product-packing quality by machine vision is a no-damage inspection method that is arising in recent years [1] In the industries of cigarettes in Indonesia, it have many inspection methods to check the cigarettes pack. The average company in Indonesia already use an automated system. But in several case, some industry still using manual inspection and this method is very subjective. With Convolutional Neural Network (CNN) that will be applied into the image processing, the quality control of the system will be good and the system will be automated. Basically, CNN is from Neural Network with many development inside it, and a Neural Network is commonly used artificial intelligence in many application [2-4]. CNN has been performed in many research. In 2017 J. Feng and his team have perform image processing defect identification of pipeline. The characteristics of general CNN have been well suited to deal with the problem on inspection pipeline. The proposed model can achieve higher accuracy and better robustness than the additional approaches [5]. CNN has recently succeeded in computer vision and pattern recognition. CNN have become the solution of an image object classification [6, 7] and object detection tasks [8, 9]. This method also goes into health research. The example is Automated ECG Classification [10], Fingerprint liveness [11], and classification of breast cancer [12]. Another research is to distinguish between natural and computer-generated images. The accuracy of the data is over 80% [13].

Convolutional Neural Network [14, 15] is a kind of Neural Network that uses for image processing. In this research, the output of CNN will classify that packaging of cigarettes where it's accepted or rejected. This classification will combine with sorting system of cigarettes. So the system will work automatically, This paper contains 5 section. Section II will explain the architecture and design of the research system. Part III is the data analysis. Section IV is the conclusions. Moreover, Section V is the references for this research.

2. Methodology

The purpose of this project is checking the cigarettes packaging automatically. This project will be applied to the prototype of a sorting machine that designed like a real plan. The explanation about Hardware, Software and the overall steps of this research can be explained as follows.

2.1Hardware Design

This project uses several Hardware. The hardware can be explained as follows.

1. Arduino Uno
   This device is used to process sensor and actuator programs that will be used on this system. Arduino software includes a variety library for a run any program and any different purpose [16].
2. Personal Computer
   This device is used to process the image with a convolution neural network method. Moreover, it will show the interface for this system
3. Logitech C170
   It used to take an image as input data for training, and this device will be a sensor for classifying the output.
4. E-18 D50NK Proximity Sensor
   This device is used to detect and count the object.
5. Motor DC 12V
   It's used to drive the conveyor for object movement
6. Pneumatic Hydraulic
   It's used for sorting cigarettes pack that have been classified by the sensor.

Figure 1 will show the diagram block of the system.

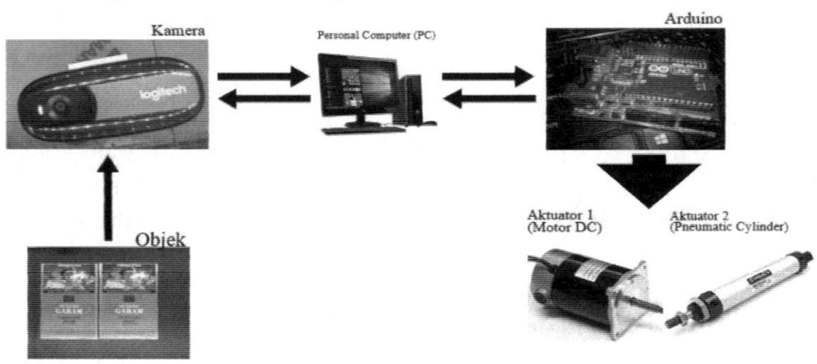

Figure 1. Hardware System Design

2.2System Design

For the first step of the system design, make the step of CNN first. Figure 2 will show the steps of CNN process. The first step of CNN Process is making dataset. This step is significant, because this dataset will be the data for training with Neural Network. The second process is converted to grayscale. The program of CNN is only read grayscale image. Therefore it must change the underlying model which has RGB, and it must be converted into Grayscale. Next, we will go to the primary process called Convolution. In the Convolution process all dataset will be processed automatically with the program, and in the last method, after Convolution is done, the dataset will be trained by using a neural network. After all the procedure, the system can classify the object.

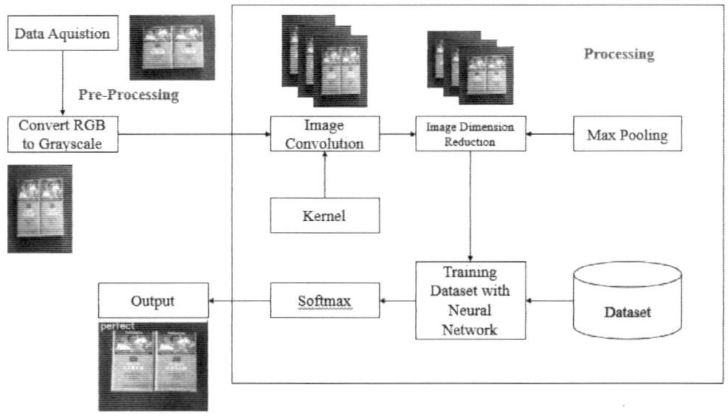

Figure 2. Process of CNN

There is 6 Step on this project, capture the purpose for the dataset, Convert the image from RGB to Grayscale, Image Convolution, Max-pooling process, Flatten Process, Training dataset with Neural Network, and Processing the Output. Before we create the steps of this project, we must create the architecture for the Image processing with CNN as shown in Figure 3.

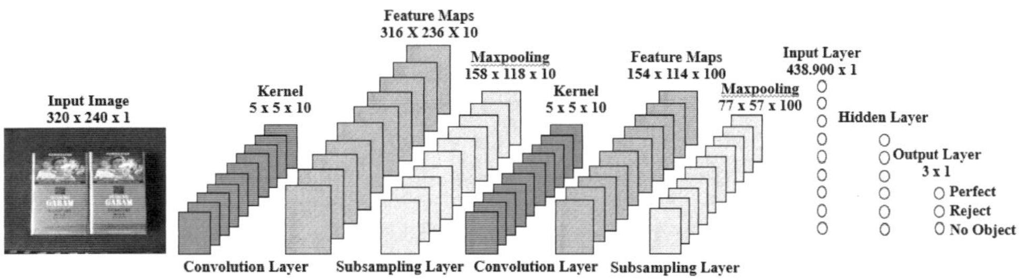

Figure 3. Convolution Neural Network Architecture

After we build the architecture of CNN, the step will be explained as follow.

2.2.1 Step 1 Capture the Object for Dataset. Image of the object is needed for Vector Input on Neural Network. So, we need to capture the object as much as possible. This project uses 320 x 240 resolution scale of the image. The system takes 3 classifications for the predicted output, the classification is perfect, reject, and no object. Every condition will take 100 datasets, so, the total is 300 for the dataset. Figure 4 will show several examples of the dataset.

Figure 4. Several Examples of Dataset

2.2.2 Step 2 Image Conversion from RGB to Grayscale. After capturing the object, we need to convert the image from RGB to Grayscale. This step is significant because CNN method is working in Grayscale Image [17].

2.2.3 Step 3 Image Convolution 1. In this Step, the grayscale input image will be combined with a Gaussian kernel. The function of this step is clear the noise. After combining input image array with Gaussian kernel array, this step will make a new variety called feature maps. Filter on this process is the step to repeat the image convolution until we got the best result. As shown in Fig. 3, the input image resolution is 320 x 240 x 1 pixels. That input pixel will be combined and processed with a Gaussian kernel which has 5 x 5 pixel with one stride and ten times of filters. This process will make a new array with 316 x 236 x 10. This process has the formula to get an array that will automatically process every single pixel on a dataset that we build. The Formula will be explained as follow.

$$Convolution = (A_n, B_n \times C_n, D_n) + \cdots\cdots\cdots + (A_{n+x}, B_{n+x} \times C_{n+x}, D_{n+x}) \quad (1)$$

After the Convolution is done, the next step is Max-pooling process. This is the process that will reduce the feature maps from 316 x 236 x 10 to 158 x 118 x 10. It was changed because this process will take 2 x 2 pixels of the feature maps with 2 stride that will slide to all area of the feature maps and choose the most prominent pixel.

2.2.4 Step 4 Image Convolution 2. This step is similar to Image Convolution 1. The difference is the resolution of the input image. Feature maps which are the result of the Image Convolution 1 will be an input image for this process that has a smaller decision. It would be 158 x 118 x 10 pixel. An array of this input image will be processed with array of kernel 5 x 5 x 10. It will produced a new array; the result is 154 x 114 x 100. The produced image will go to the max-pooling process which is produced new array and resolution of the input image. The new image is 77 x 57 x 100 pixel.

2.2.5 Step 5 Flatten Process. Flatten Process is the process of an image that will make pixel to be vector input for training step on Neural Network. As shown in Fig. 3, the vector input would be 438.900 data. The last data of Max-pooling process is 77 x 57 x 100 pixel. It will change array from 2 dimensions to 1 dimension. Moreover, the result is 438.900 data. This data will be processed with Neural Network. In this case, this research is using several hidden layers with a different number of neurons. The selected number of neurons is a neuron that produces the best data for classifying the output. The output is divided into three classes (binary classification), there are :

- (1 0 0) = No Object
- (0 1 0) = Perfect
- (0 0 1) = Reject

2.2.6 Step 6 Fully Connected. After the vector input has been created, the process will continue to the training process with the python program. Moreover, the last step is Fully Connected. This step is used to classify the output of the image. The production is Perfect, Reject, and No Object. The production will be used to sort the cigarettes pack.

3. Test Result

The training process has several experiments. It uses much analysis to optimize the best data of an image. In this research, the number of used neurons are 24 neurons and 100 epoch for training the data.

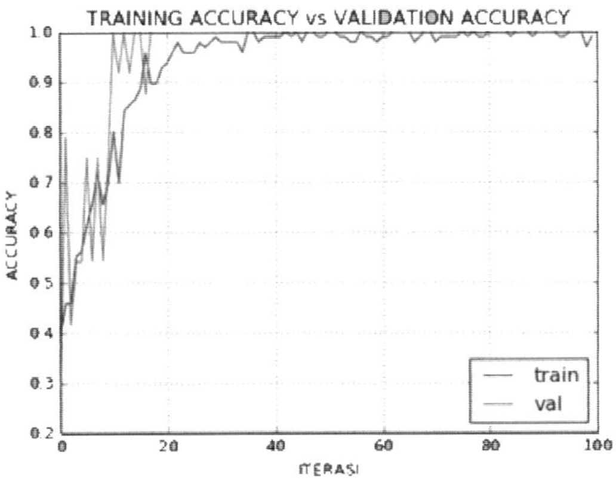

Figure 5. Mean Square Error Graphic

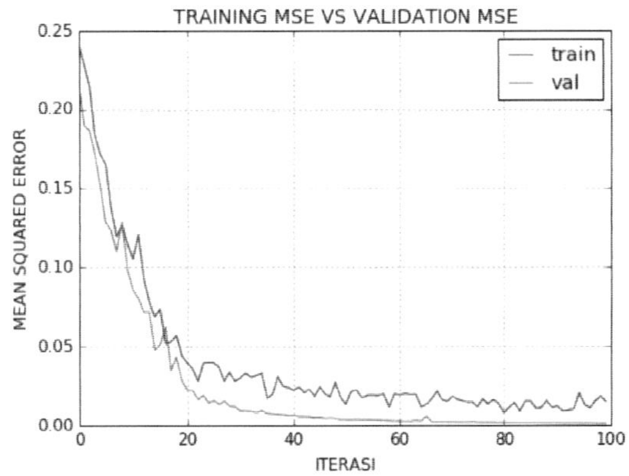

Figure 6. Accuracy Graphic

Figure 5, shows the performance of Mean Square Error. MSE training on the data begins at 0.23. When the iteration runs, MSE in Training data has decreased and its approaching to zero. In the 40th iteration, MSE in Training data starts to be constant at 0.04 - 0.02. While for Testing data begins at 0.21. When the iteration begins, the Testing data is decreased. When it reaches the 40th iteration, the Testing data starts to be constant at 0.01.
From the **Error! Reference source not found.**, the graphics show the line of the graph is stable when the iterations are over 40. The data is able to classify with the accuracy up to 100%. For this graphic, it has the formula to count it manually, and it will be shown as follows.

$$P = \frac{\sum train}{n_i} \times 100\% \qquad (2)$$

As explained in the discussion of neuron analysis, Training data starts at 0.39 or 39%. In the 40th iteration, Training data values are stable at 0.98-1 or 98% - 100%. From the graph, the average training data is 96.06%, it means that from 96 training data, 90 data can be detected correctly. For the results of the Testing data, the first iteration has a value of 0.29 or 29%. In 20th iteration, the data is very stable at 1 or 100%. The Testing data obtained an average amount of 95.62%. Means, from 24 testing data samples, the system is able to detect 23 data precisely. Figure 7 will be shown the classification of CNN

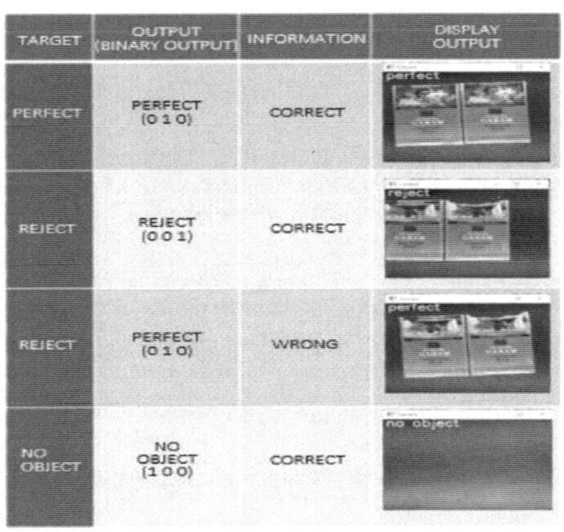

Figure 7. Classification of CNN Result

4. Conclusion

Based on the experimental result of CNN architecture were almost perfect. From the results of architectural CNN analysis, kernel results are obtained according to the architecture; this is using 5 x 5 kernels with a total of 10 filters. The convolution process in this architecture occurs two times, and it produces 438,900 Input Vectors for the training process with the neural network. From this research, the number of Hidden layers that are suitable for this application is 24 hidden layers. If the method uses the existing data, then the accuracy percentage is 95.62% from 24 data tested. While in real time testing, the obtained testing percentage is equal to 94.59%.

5. References

[1] Hu F, He X, Niu T, editors. Study on the Key Image Processing Technology in the Inspection of Packing Quality for Small-Pack Cigarettes. Second Workshop on Digital Media and its Application in Museum & Heritages (DMAMH 2007); 2007 10-12 Dec. 2007.

[2] Adhitya RY, Ramadhan MA, Kautsar S, Rinanto N, Sarena ST, Munadhif I, et al., editors. Comparison methods of Fuzzy Logic Control and Feed Forward Neural Network in automatic operating temperature and humidity control system (Oyster Mushroom Farm House) using microcontroller. 2016 International Symposium on Electronics and Smart Devices (ISESD); 2016 29-30 Nov. 2016.

[3] Hanggara PO, Syai'in M, Paradisa PF, Arifin MZ, Sarena ST, Syaiin M, et al., editors. Prototype control and monitoring system safety device from leakage ammonia at marine loading arm with comparison of Neural Network (NN) and Extreme Learning Machine (ELM) method. 2017 International Symposium on Electronics and Smart Devices (ISESD); 2017 17-19 Oct. 2017.

[4] Putra RY, Kautsar S, Adhitya RY, Syai'in M, Rinanto N, Munadhif I, et al., editors. Neural network implementation for invers kinematic model of arm drawing robot. 2016 International Symposium on Electronics and Smart Devices (ISESD); 2016 29-30 Nov. 2016.

[5] Feng J, Li F, Lu S, Liu J, Ma D. Injurious or Noninjurious Defect Identification From MFL Images in Pipeline Inspection Using Convolutional Neural Network. IEEE Transactions on Instrumentation and Measurement. 2017;66(7):1883-92.

[6] Krizhevsky A, Sutskever I, Hinton GE. ImageNet classification with deep convolutional neural networks. Proceedings of the 25th International Conference on Neural Information Processing Systems - Volume 1; Lake Tahoe, Nevada. 2999257: Curran Associates Inc.; 2012. p. 1097-105.

[7] Szegedy C, Liu W, Jia Y, Sermanet P, Reed S, Anguelov D, et al. Going deeper with convolutions2015. 1-9 p.

[8] Girshick R, Donahue J, Darrell T, Malik J. Rich Feature Hierarchies for Accurate Object Detection and Semantic Segmentation. Proceedings of the 2014 IEEE Conference on Computer Vision and Pattern Recognition. 2679851: IEEE Computer Society; 2014. p. 580-7.

[9] Ferrari A, Lombardi S, Signoroni A. Bacterial colony counting with Convolutional Neural Networks in Digital Microbiology Imaging. Pattern Recognition. 2017;61:629-40.

[10] Zhai X, Tin C. Automated ECG Classification Using Dual Heartbeat Coupling Based on Convolutional Neural Network. IEEE Access. 2018;6:27465-72.

[11] Jung HY, Heo YS. Fingerprint liveness map construction using convolutional neural network. Electronics Letters. 2018;54(9):564-6.

[12] Bardou D, Zhang K, Ahmad SM. Classification of Breast Cancer Based on Histology Images Using Convolutional Neural Networks. IEEE Access. 2018;6:24680-93.

[13] Quan W, Wang K, Yan D, Zhang X. Distinguishing Between Natural and Computer-Generated Images Using Convolutional Neural Networks. IEEE Transactions on Information Forensics and Security. 2018;13(11):2772-87.
[14] Abdel-Hamid O, Mohamed A, Jiang H, Deng L, Penn G, Yu D. Convolutional Neural Networks for Speech Recognition. IEEE/ACM Transactions on Audio, Speech, and Language Processing. 2014;22(10):1533-45.
[15] Khumaidi A, Yuniarno EM, Purnomo MH, editors. Welding defect classification based on convolution neural network (CNN) and Gaussian kernel. 2017 International Seminar on Intelligent Technology and Its Applications (ISITIA); 2017 28-29 Aug. 2017.
[16] Martínez-Santos JC, Acevedo-Patino O, Contreras-Ortiz SH. Influence of Arduino on the Development of Advanced Microcontrollers Courses. IEEE Revista Iberoamericana de Tecnologias del Aprendizaje. 2017;12(4):208-17.
[17] A Khumaidi IM, Giliang Prilian Arinata, Z.M.A. Putra. Segmentasi dan Estimasi Jarak Bola dengan Robot Menggunakan Stereo Vision. 5th Indonesian Symposium on Robotic Systems and Control 2017 (ISRSC 2017); Bandung: Universitas Pendidikan Indonesia; 2017. p. 140-4.

IOP Conf. Series: Materials Science and Engineering **462** (2019) 012003 doi:10.1088/1757-899X/462/1/012003

Assessment to Mechanical Material Properties of Natural and Metakaolin based Geopolymer Stabilized Soil

L L Lestari[1,2,3], R A A Soemitro[2], M Hattab[3], J J Ekaputri[2], and D D Warnana[2]

[1]Department of Civil Engineering, Institut Teknologi Adhi Tama Surabaya, Surabaya 60117, Indonesia
[2]Faculty of Civil, Marine, and Geo Engineering, Institut Teknologi Sepuluh Nopember, Surabaya 60111, Indonesia
[3]Laboratoire de Physique et Mécanique des Matériaux, CNRS FRE 3236, Université de Lorraine, Ile du Saulcy, 57045 Metz Cedex1, France

laraslaila.lestari@itats.ac.id

Abstract. This study discussed the mechanical properties of natural and metakaolin based geopolymer stabilized soil. The mechanical soil properties is an important soil properties that determine the soil strength parameter and its behaviour. It is proposed to do an assessment on 2 types of soil: natural (remolded) soil and metakaolin based geopolymer (MKG) stabilised soil of Bengawan Solo River Embankment soil. The soil sample are treated on oedometer test in order to know soil compressibility and permeability. The result shows that the implication of MKG for natural (remolded) soil only have an influence in reducing swelling. Also for the permeability, it is found that despite the high porosity (index high vacuum), its permeability is smaller than the permeability of other mixtures.

1. Introduction

Initially, the existing soil embankment of Bengawan Solo River continuously experienced an erosion, so an artificial embankment was built to protect the village from flooding. In order to upgrading this protection, sediment was removed from the riverbed and spread on top of the existing embankment to dry. This soil material was compacted in 40 cm using light compaction plant to a dry density 80% - 85% of the standard proctor optimum, on the dry side of optimum [1]. The soil type along Bengawan Solo River is dominated by the alluvium sediment soil which is in the form of silty soil. The silty soil known as soft soil which has low shear strength and is vulnerable to water content changes.

It is proposed the chemical stabilization to improve the physical and mechanical soil properties. Chemical stabilisation usually occupies lime, Ordinary Portland Cement (OPC), and fly ash to increase the soil solid grain volume with the chemical reaction between stabilizer and soils. However, the utilisation of OPC, lime, etc. are energy intensive and emit large quantity of CO_2. Geopolymer as a soil stabilizer has a high strength, low energy consumption, low cost, excellent adhesion to aggregates, and low CO2 emissions during synthesis [2]. [3] and [4] also stated that polymer has lower shrinkage volume than OPC, and more mechanical strength with short setting time. Hence, the utilisation of geopolymer (in this case is metakaolin based geopolymer, MKG) as a next-generation of soil stabilizers besides the traditional stabilizers offers an effective yet promising alternative [2].

For further analysis on soil strength and behaviour of natural and treated soils, this research studied the experimental study of mechanical soil properties such as compressibility and permeability derived from oedometer test.

2. Material Properties and Experimental Methods

2.1. Material Properties

The soil material for this research derived from the surficial soil (-0.50 m from original ground level) of Bengawan Solo river embankment. The soil fraction dried in an open air and sieved (No. 40) with soil properties shown in Table 1.

Table 1. Natural soil properties.

Physical Properties	Cassagrande		Fall Cone
Liquid Limit (w_L, %)	44.05		45.50
Plasticity Limit (w_P, %)	24.62		17.00
Plasticity Index (IP)	19.43		28.50
G_s		2.69	
Shinkage Limit (w_{SL}, %)		16.45	
	Granulometry		
Clay (%)		45.2	
Silt (%)		44.33	
Sand (%)		10.47	
Gravel (%)		0	

The soil which passed sieve no. 40 (425 μm) is classified as inorganic clays of low to medium plasticity, silty slay (CL) according to Unified Soil Classification System (USCS). Some basic properties are shown in Table 1.

2.2. Experimental Methods

The oedometer test was designed to determine compressibility soil subject to a vertical load as well as determine the preconsolidation pressure. It consists in subjecting the sample to a series charges applied and to measure soil compaction over time in each of these charges. Soil samples are placed inside a rigid cylindrical mold with a diameter D = 50 mm with no radial deformation loading $\varepsilon_2 = \varepsilon_3 = 0$. The test is deducted in a dimensionally stable cell and greased and placed between two stones porous to allow water to flow, the load is applied in steps, the compressive loads is exerted vertically on the specimen size D = 50 mm, h = 20 mm below a piston. The deformation is recorded by reading a displacement sensor or a dial hence the time versus displacement curves could be plotted in real time. The materials were prepared initially in slurry state $w_i = 1.5\ w_L$ and settled for 24 hours before it is remolded into the cell to ensure the homogenization. After that, the materials were settled for 24 hours after remolded under the piston. The piston acted as a preconsolidation load which has the mass of 2.7 kPa.

3. Soil Behaviour – Results

3.1. Compressibility

The comparison of natural (remoulded) soil (B.Solo) and MKG Treated Soil (B.Soil_Tr (MKG)) could be seen in Figure 1. The compression curve of MKG Treated Soil tend to have the same pattern as natural (remoulded) soil with bigger value of compression index (C_c). Opposed to the value of swelling index (C_s) which is decreased. The trend shows that the implementation of MKG have a major influence in reducing the swelling.

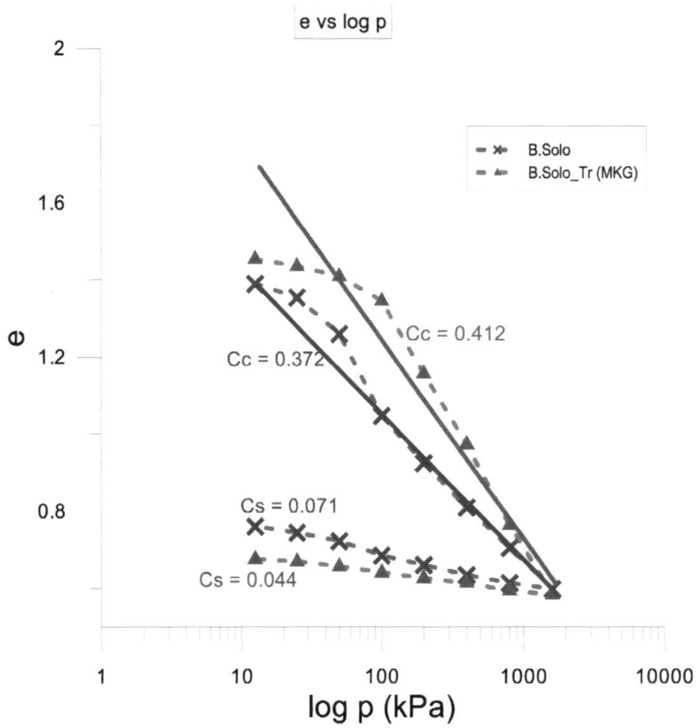

Figure 1. Oedometer path of natural (remolded) soil (B.Solo) and MKG treated soil (B.solo_Tr (MKG)).

For natural remoulded soil, the result shown in the Fig 1. The value of Cc = 0.372 and the swelling index is 0.071 with the ratio of Cc / Cs = 5.2. The curve in Fig 1. is tend to bend after 100 kPa. The bend is followed by two tendencies; first, for low stress domain of remoulded Bengawan Solo soil samples (σ'_v = 12.5 kPa – 100 kPa) shows ideal curve. It expresses a process of progressive degradation of the cementation between the grains. Second, after the sample reached high stress domain $\sigma'_v > 100$ kPa, the curve changes drastically and gives a high value of compression index. It corresponds to a cementation breaking between grains of material. The soil sample observed by Scanner Electron Microscopic (SEM) in order to convince the grain structure of soil in powder form (Figure 2.).

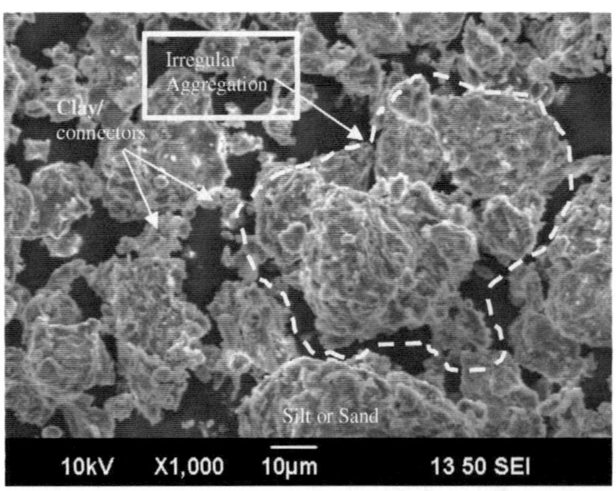

Figure 2. SEM Images of natural soil in powder form (the magnification 1000x).

3.2. Coefficient of Consolidation and Permeability

From oedometer test, we could draw the deformation vertical vs square time curve. Casagrande method is applied to determine t_{50} and t_{100}, and the method of Taylor to determine t_{90}. The coefficient of consolidation is defined by the equation:

$$C_v = \frac{T_v d^2}{t} \tag{1}$$

Where T_V is a time factor depending on degree of consolidation, t is the time required to achieve the degree consolidation and d is the distance of drainage, in this test it is equal to half the thickness the sample because the drain is allowed to both sides.
When oedometer test, coefficient of consolidation, C_v, can be determined by the Casagrande method or the method of Taylor. This study presents two methods, first is the method of Taylor (eq. 2.) and second is the method of Casagrande (eq. 3).

(Taylor Method) $$C_v = \frac{0.848d^2}{t_{90}} \tag{2}$$

(Casagrande Method) $$C_v = \frac{0.197d^2}{t_{50}} \tag{3}$$

where t_{90} and t_{50} are the times needed to achieve a degree of consolidation of 90% and 50% respectively.

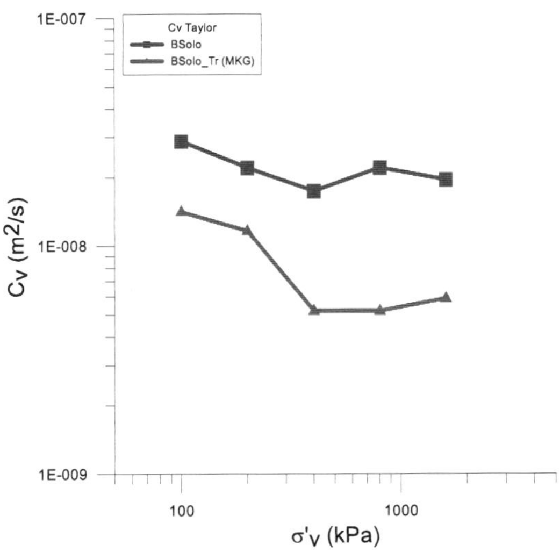

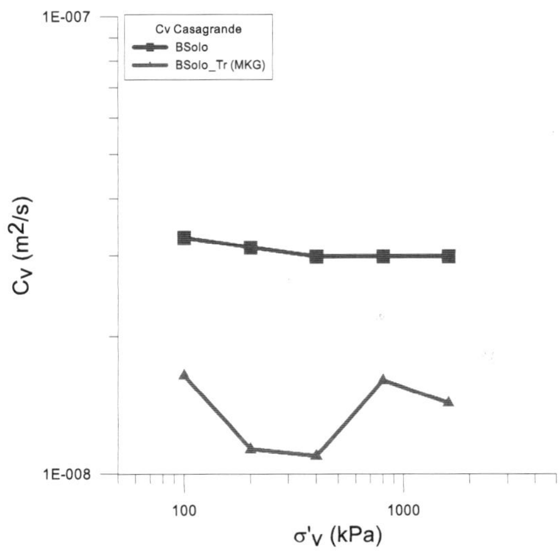

Figure 3. Variation of coefficient consolidation in function of σ'_v.

Figure 3. show changes in the coefficient of consolidation (Taylor method) according to the effective vertical stress. For Natural Soil, starting from 400 kPa point, the C_v coefficient decreases slightly. On contrary, the observed change is related to the characteristic of highly plastic clays. The values of C_v for these materials result in permeability variations during the test, they are calculated by eq. 4:

$$K_v = \frac{C_v \gamma_w}{E_{oed}} \tag{4}$$

Where E_{oed} is the oedometric module $E_{oed} = \frac{d\sigma'_v}{d\varepsilon_v}$ and γ_w is the specific weight of water.

Figure 4a shows variations in the permeability of the materials according to effective vertical stress. The permeability of the natural remolded soil (Bengawan Solo soil) and MKG treated mixture fall more

rapidly from a stress of 400 kPa. This reduction in permeability from these stress levels might explained by the performance of C_c slopes during the oedometer test.
This can be interpreted by the variation of the microstructure: in the field of low-stress compression is due to compression of large inter-aggregate pores or inter-domain. This occurs with the expulsion of water molecules characterized by low physicochemical interaction clay-water.
During passage to the field of high stress, some other mechanism intervenes related to the interaction between the adsorbed water and the surface of the clay inside the domains. Eliminating this water becomes more difficult [5].

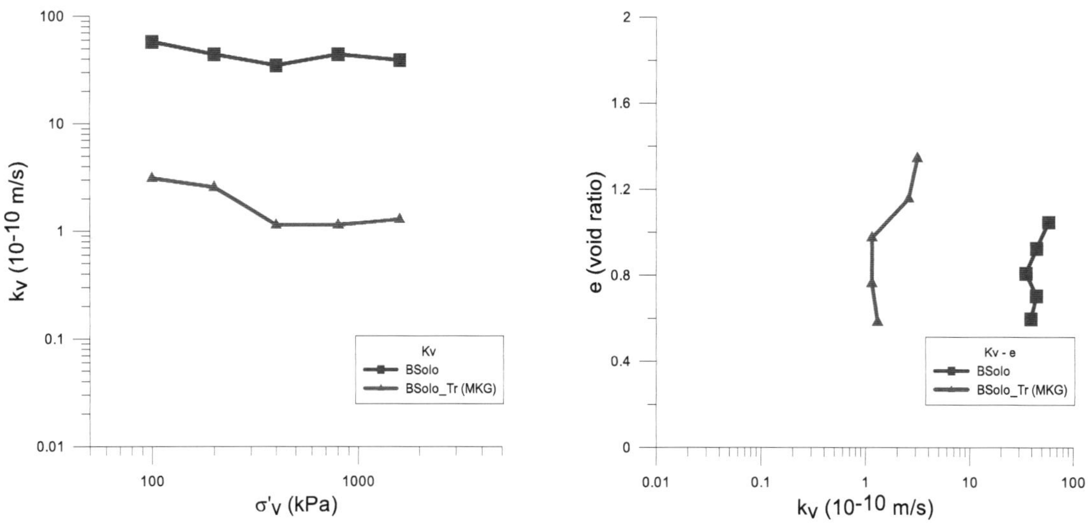

Figure 4. Variation of permeability in function of σ'$_v$.

The permeability of natural remoulded soil sample almost linearly decrease during the test but increases for Kaolin K13 treated. In Figure 4b., it could be seen that natural remoulded soil show changes in void ratio depending on the permeability: it is found that despite the high porosity (index high vacuum), its permeability is smaller than the permeability of other mixtures. This result is coherent with previous study [5].

4. Conclusions

From the assessment of material properties of natural and MKG treated soil. it can be conclude that:

- The compression curve shows that the implication of MKG for Bengawan Solo Silty Soil only have an influence in reducing swelling. This happen due to the chemical composition (Si/Al ratio) and treatment of the samples. In literature, mechanical behavior of samples could be augment if the soil cured in certain temperature and time.
- Natural remoulded soil compression curve is bending. It is followed by two tendencies; in the low stress domain, it expresses a process of progressive degradation of the cementation between the grains. In high stress domain σ'v > 100 kPa, the curve changes drastically and gives a high value of compression index. It corresponds to a cementation breaking between grains of material.
- The permeability of the natural remoulded soil and MKG treated soil mixture fall more rapidly from a stress of 400 kPa. This reduction in permeability from these stress levels might explained by the performance of C_c slopes during the oedometer test. It is found that despite the high porosity (index high vacuum), its permeability is smaller than the permeability of other mixtures.
-

5. References

[1] G. El Mountassir, M. Sánchez, R. A. A. Soemitro, and E. Romero, "Behaviour of compacted silt used to construct flood embankment," *Proc. ICE - Geotech. Eng.*, vol. 164, no. 3, pp. 195–210, 2011.

[2] M. Zhang, H. Guo, T. El-Korchi, G. Zhang, and M. Tao, "Experimental feasibility study of geopolymer as the next-generation soil stabilizer," *Constr. Build. Mater.*, vol. 47, pp. 1468–1478, 2013.

[3] D. Hardjito and B. V. Rangan, "Development and properties of low-calcium fly ash-based geopolymer concrete," *Res. Rep. GC*, p. 94, 2005.

[4] J. J. . Ekaputri, K. Maekawa, and T. Ishida, "the Use of Geopolymerization Process for Boron Fixation in Fly Ash," *RILEM. First Int. Conf. Adv. Chem. Mater.*, pp. 225–232, 2010.

[5] T. Hammad, "THESE Tammam HAMMAD Grade de docteur de l ' Ecole Centrale de Paris Spécialité : Géotechnique Laboratoire d ' accueil : MSS – MAT CNRS UMR N ° 8579 COMPORTEMENT DES SEDIMENTS MARINS DE GRANDE PROFONDEUR : APPROCHE MULTI - ECHELLE," 2010.

Acknowledgement

This research work is fully supported by the following research funding:

- JICA Joint Research under Predict ITS - Phase 2 (Technical Cooperation Project for Research and Education Development on ICT in ITS Phase 2), in 2012. "Ground Inclinometer Development Based on Variations of Soil Water Content and Lateral Displacement for Slope Stability Field Monitoring"
- International Research Collaboration (managed by LPPM – ITS and Directorate General of Higher Education Indonesia, DIKTI), in 2014 and 2015. "Assessment to the River Embankment Stability with Respect to Rainfall and Seismic – Induced River Embankment Failures" – 2014 "Assessment to River Conditions Affected by River Profile, River Water Fluctuation and Sediment Rate" - 2015
- PEER (Partnerships for Engagement in Research) Science cycle 3 project by USAID-NSF, in 2014. "Sediment Transport Evaluation on Bengawan Solo River (downstream and estuary) to Minimize Sedimentation and Flood Combining Effect on Nearby Infrastructure"

Micro-Structural Characterization of the bond strength capacity of adhesive material in the alternative of cold-formed steel frame system

I Komara[1,2], E Wahyuni[2], P Suprobo[2], K Taşkin[3]

[1]Department of civil engineering, Institut Teknologi Adhi Tama Surabaya, Jalan Arief Rachman Hakim, Surabaya 60117, Indonesia
[2]Department of Civil Engineering, Institut Teknologi Sepuluh Nopember (ITS), Jalan Raya ITS, Surabaya 60111, Indonesia
[3]Department of Civil Engineering, Anadolu University, Eskişehir 26555, Turkey

indrakomara@itats.ac.id

Abstract. The introduction of adhesive material in the structural connection of cold-formed steel (CFS) as alternative connection material has a great deal on the strength capacity of bonded joints. This paper evaluates the microstructural characterisation on CFS connection. Two different types of adhesive material, A-SK, and A-3M are taken. The experiment is made to get mechanical properties and is followed by classifying incidence x-ray diffraction (XRD) and scanning electron microscopy (SEM) to study the behaviour inside the adherence. The specimen was penetrated to classify layer surfaces, to present the structural component and to get their morphology. The result is that both materials have similar bonded joints strength capacity which meets the required standard. Quartz and calcite on A-SK are mainly located on the innermost surface separately, while A-3M mostly mixed close to the base steel. In that matter, they have been beneficial for knowing the wide variety of structural characterisation by their elemental composition phases.

1. Introduction

The use of cold-formed steel (CFS) for the function of structure element has been established lately due to several advantages over timber, hot-rolled steel and reinforced concrete, such as ease of construction, valuable cost, excellent strength capacity to the weight ratio of the section, high in flexibility and less environmental issues [1]. However, one disadvantage of using CFS is difficulty in improving the capacity and reliable connection due to thin cross sections and a small area of contact. For the CFS system, the screw connection is typically used. In application, when some screws are increased the fracture mechanism expecting to occurred in this connection. This application is received growing attention as many as a failure of the CFS system leads to significant structural failure as well as potential fatalities [2].

There are several alternatives for designing the CFS system connection such as friction welding and mechanical joints, but by far the most suitable method that gives a similar capacity is artificial joint by using adhesive material [3-5]. The adhesive has some enticing benefits for joining components. They do not interrupt the section surface, they spread reaction loads more widely and evenly, have a higher natural sealing capability and in some point lead to lighter designs with active coherency [6]. Fact, the adhesive material is a complex combination of some chemical and mechanical effects taking place within little surface depth.

With these problems and benefits, experimental research was conducted by comparing test of two of the most popular adhesive material that is implied in Indonesia, first is better known as super glues, water-based and second is well known as retrofitting paste, apparently like cement paste. Both of them are popular for a general using adhesion with no mixing and have quick cure times. This research aimed to compare the bond strength and behaviour of the relative performance of widely available adhesives bonding common substrates. Since surface bonding is an essential element of adhesive use, a significant key to this test in assessing the capacity of each adhesive. Also, speciality experimental methods were tested to determine whether they performed better than general purpose adhesives in their intended application, CFS system.

Staging of the experimental study was conducted correctly. It is maintained to keep the total number of test to a functional level; tensile testing was carried in two phases, where the first phase of a trial process to provide an acceptable result for the second phase. In a first phase, six classes of adhesive from two manufacturers, A-SK and A-3M, were tensile lap coupon tested with a variety of surface preparation methods and common material substrates. Because both adhesive materials are sensitive to moisture and temperature, each combination was also placed in an environmental chamber at elevated humidity. Further, the results of the first phase, a single parameter of adhesive was chosen as the focus of the second phase, in which competitive offerings from different manufacturers were tested in a coupon tensile test. Standardised mould dimension placed there. Subsequently, the structural characterisation test was taken from a cut dimension of coupon tensile test attached to the substrate. They were sized and specified to be analysed in microscopy, in which settings for diffraction analysis and examined the imaging analysis.

2. Materials

2.1. Adhesive Selection

Two selected adhesive material, A-SK, and A-3M were considered to be analysed for the first reason is to optimise the performance to be bonded, in this case using the CFS section. The second reason is to address the needs of joint condition, such us: fit-up, environmental and type of loading. Two adhesive materials were selected over the familiar brand which both the desired adhesives and primer were readily available at a local constructional supply, while the other products would require to be ordered.

In details for framing construction such roof truss framing system. The A-SK also offers several benefits such as low odour adhesive which is ideal for any setting where harsh fumes or flammability are an issue, serves peel strength and high shear. It is also affirmed for impact resistance, 10 minutes work life and 20 minutes handling strength, bonds oily metals with minimal surface prep and bonds most plastics, composites, and metals. Furthermore, Acrylic adhesives or A-3M often referred to as resin or epoxy glue, are contained two-part adhesive material that offers capacity in strength and durability with the fast installation process, although it implies strong odour of the mechanical and chemical process which mostly heating the area of adhesion.

In addition, A-3M provide several benefits in the installation process, i.e., practice with a faster-curing speed, have no issue when attaching to the oily surface or no-clean prepared bonding layer and have a high ability to bond many materials such as glass, metals, and plastics. However, recommended applications are implied in the substrate of metals and plastics. Those are customarily taken place for industries or manufacturing.

2.2. CFS selection

CFS as a substrate for bonding based primarily on their function and popularity. The CFS was obvious steel choices, as was lightweight steel to represent a structural element.

3. Methodology

The primary objective was to apply the most consistent processes one could reasonably expect production processes and also to give an alternative adhesive material that achieved a well-bonded in support of the capacity of the CFS connection. It is a significant distinction in all joint testing, but mainly when dealing with adhesive, subtle changes in preparation can have a substantial impact on the result and because the influencing parameters are difficult to quantify. As an example, while the test procedure in this test conforms in principle to ASTM D638-14, the standard protocol for tensile lap coupon test, ASTM D638-14 need that coupons be shaped from the sheet stock to eliminate the potential burrs or twist that are possible in sheet metal shearing or forming. It will minimise the bond gap and maximise adhesion performance.

3.1. XRD Measurement

XRD experimental measurements were taken at Anadolu University, Turkey with a Rigaku MiniFlex 600 diffractometer. It comprehensively equipped with the help of a cobalt X-ray tube, a beam collimator and an XRD laser that concentrated on the fine specimen. The sharp specimen cut in several small sizes which was taken from CFS connection. The substantial cutting was rushed and cleaned then analysed by using a Lynx Eye linear detector with a diameter 500 μm. To support the observation, XRD generator was controlled reducibly from 40 kV and 30 mA.

A fully featured XRD machine was used with a step width 0.03° utilising a counting time 3 s/step over a 2θ range of 10-100°. This method performed qualifiedly and quantitively of polycrystalline materials. First, in qualitative measurement, unknown substances were identified by appealing the diffraction results of every phase. Other, in quantitative analysis, solid mixtures were characterised to assign crystalline compounds.

(a) (b)

Figure 1. Experimental study of diffraction analysis;
(a) XRD specimen, (b) Diffractometer

3.2. SEM Measurement

SEM was measured with a Phenom Pro-X from Anadolu University. The sample structures were physically examined and magnified the size of observation determined their elemental composition. This observation included viewing three-dimensional images of microscopic structures. This term has exposed the identification of chemical elements.

These observations were divided into several magnifications. Every magnification was enhanced to see various characterisation. In detail, the sample was set in glutinous holder then titling and cooling inside the workbench of mapping line measurement. The siding was also considered for imaging an even greater diversity. Samples were Imaging in light optional with 20 – 135x magnification range and

80 – 100.000x magnification range of electron optical with 12x maximum digital zoom. Then, specified acceleration voltages were also taken from 10 kV to 15 kV with an adjustable range from 9.8 kV and 15.2 kV for imaging and analysis mode, respectively.

(a) (b)

Figure 2. The mapping line of scan analysis measurement;
(a) SEM measurement specimens; Coating process and inserted sample to mould, (b) Reading process; running and magnifying process

4. Data Analysis and result

4.1. XRD analysis

Conventional XRD analysis of solid adhesive material obtained by cutting the CFS connection side-to-side as a part of the connection. It may examine the rigorous surface interaction of the long-range order produced as a consequence of very short-range interactions.

As a result, A-SK layer informed a high content of quartz and calcite and low silicon and carbon materials. It may expect that the surface of the segment, is more abundant in quartz and calcite, rather than another content. All of this seems to confirm that A-SK has no additional element that affects concentration on its composition. Those specimens exhibited under the peaks at $2\theta = 20°$-, 80°, which is illustrated in Figure 2. It was well-explained that the more defined peaks of A-SK, demonstrate the region between $2\theta = 20°$- 30°.

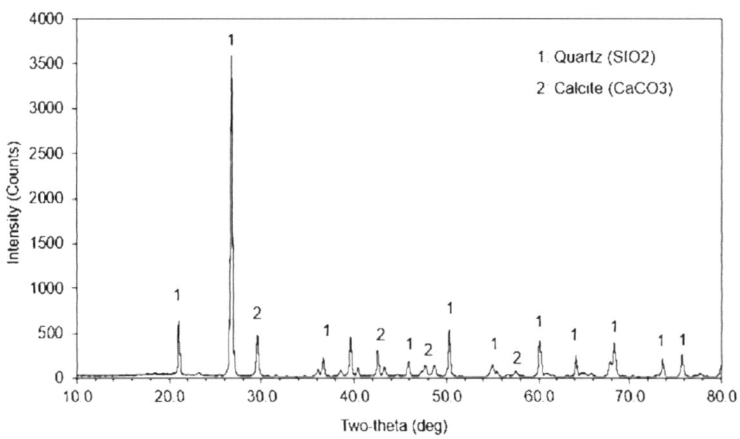

Figure 3. Decomposition peaks of A-SK using peak fitting program

Furthermore, to determine the crystallinity percentage of semi-crystalline adhesion, a well-defined mixture of the amorphous and crystalline region should be assumed. Figure 3 shows the decomposition peaks of A-SK; through peak fitting program, the A-SK crystallinity percentage was around 46%.

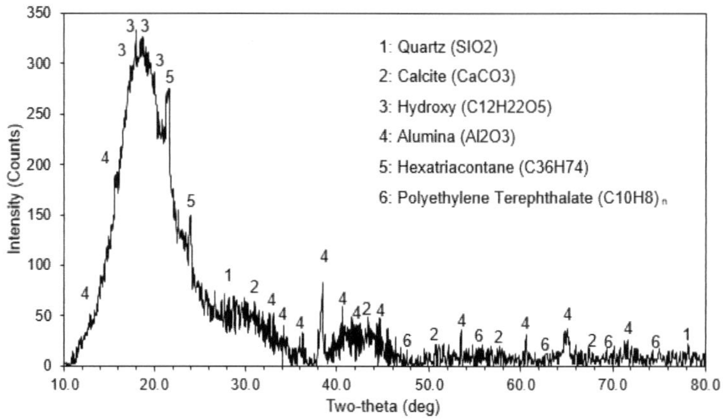

Figure 4. Decomposition peaks of A-3M using peak fitting program

4.2. SEM Analysis

Another objective is to achieve a morphological characterisation of two different phases of the adhesive material. It was taken to know the proportion of material which could help the strength and fracture behaviour. This research tested to indirectly identify the morphologies by its composition, shaped pattern and unity mixture of each part. This method was made from a penetration process by using a magnifying laser. Preceding research had newly measured an effluence depth of around 200 nm to describe the best condition of steel-to-steel connection [14-15].

The appearance of A-3M surface was reasonably uniform based on the clear diffracting result as expressed in Figure 5. It gains information of hydroxyl and hexatriacontane as a primary phase, respectively. Calcite was added but in impoverished part and the alumina seen filling the small peak of the portion. Quartz was also given considered percentage while the polyethylene terephthalate perchance examined negligibly. This analysis acquired by smoothing the solid layer that performs in steel to steel connection. Its connection indicated high contents of quartz and calcite, and gain some elements from CFS, i.e., magnesium and calcium were also classified, 1.5% and 2.2% respectively. The results of A-SK were not similar (seen in Figure 5). The results diffraction was excellent with only detected the majority quartz and calcite, signify that the outermost layer surface is loaded in the silicon and oxygen.

Clarifying images of scan is hard to be taken due to a small size pointed material. First, a comprehensive early scan of the solid surface was conducted. Second, identify a dot scan by highlighting the most prepared surface of the morphologies. Bear in mind, the selected surface, the typically small size of the observed zone, does not represent to contain all the given morphologies as XRD did but should lid those that are most prodigal. Fourth, Find the spot then Magnify the resolution which increased until the different part of layer morphologies could be naturally observed, spotting the area of the morphological feature, e.g., quartz, calcite, and another particle, where the spectrum would subsequently be studied. The 256 times magnification of amplification were increased gradually with different frequency. Then, this spot was performed at the previously identified area of the morphological feature by 10000x magnification.

A3M contained a little portion of calcite; it is proofed that no similar crystalline shape to ASK. Calcite in ASK itself is relatively indicated as a significant proportion analogously with XRD analysis. A particular dispersion was studied in the (cm-1) wavelength shifts of the characteristic peaks of the studies above.

Thus, deal with the evaluations obtained by XRD, it looks like to be confirmed that each adhesive material is preferentially contained calcite and silicon even with different proportion which seems represent the globular formation of various morphologies. The sharp one is classified as calcite. The structure located furthest presented a rock-like with scarping two major colour appearance in ASK, while another was fine paste with unshaped of thin reminiscent of the goethite crystals illustrated in A3M.

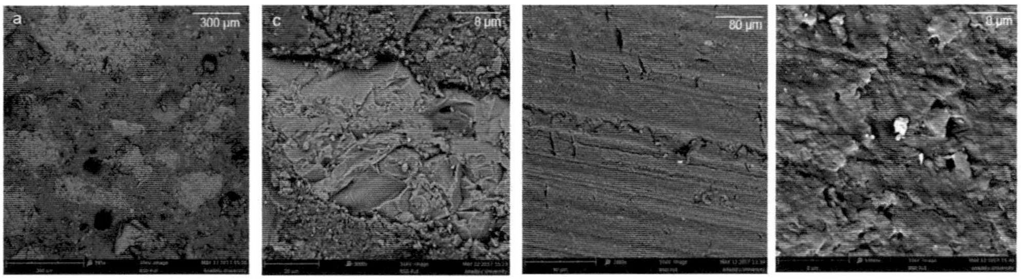

(a) (b) (c) (d)

Figure 5. SEM images of A-SK and A-3M;
(a) A-SK 256x magnification, (b) A-SK 10000x magnification, (c) A-3M 256x magnification, (d) A-3M 10000x magnification

In addition, the A3M surface showed a very same firm dark reddish-purple colour and its appearance was almost similar in all the marked area. Still, in these surfaces, A-3M is more disparate, informing different classification from one marked area to another. Figure 5 above illustrate, the different morphologies scanned in several marked regions of the surface including vary magnification. Different from A3M, ASK explained light colour, a bit grey to dark and its appearance was not uniform. Some holes also presented in this layer. Contrariwise, other observed areas were characterized by an extensive background of blackish colour, possibly silicon, mottled with the whites' globular formation of different morphologies. The sharp one is classified as calcite.

5. Conclusion

Several number properties affect bonding quality besides joints formed with the adhesive and adherents. Two aspects that important as a bonded parameter are the strength capacity and composition of the adhesive material. From this study, it successfully obtained that both adhesive materials fitted as an acceptable alternative material for CFS system connection. It was proved by the bond strength capacity to raise the standard of its specification. On the other hand, structural characterisation on A-SK and A-3M occurred different result, while A-3M given highly diverse morphology through peak fitting program the A-SK crystallinity percentage was around 46%. The same distribution of phase contents to the previous surface of both adhesive, A-SK data is considerably accepted since only two implied element which will be easy to be modified on the field.

6. References

[1] A. H. Cagatay, Lateral load behaviour of cold-formed steel wall panels (Atilim University, Ankara, 2013)

[2] I. Komara, E. Wahyuni, P. Suprobo, T. Kivanc. Assessing the Tensile Capacity of Cold-Formed Steel Connections using Self-Drilling Screws and Adhesive Materials. IJASEIT. 8, **2**, 397-404 (2018)

[3] I. Komara, E. Wahyuni, P. Suprobo, A study on Cold-formed Steel Frame Connection: A review, IPTEK The journal for technology and science, 28, **3**, 83-89 (2017)

[4] I. Komara, T. Kivanc, E. Wahyuni, P. Suprobo. The experimental study of cold-formed steel truss connections capacity: Screw and adhesive material, Journal of current construction issue, 3, **2**, 229-24 (2017)

[5] I. Komara, E. Wahyuni, P. Suprobo. Studi numerik perilaku sambungan baut dan adhesive pada struktur rangka atap baja ringan. Proc. Konferensi nasional paskasarjana teknik sipil, Insitut Teknologi Bandung, **4**, 29-38 (2016)

[6] LeBono, J., Barton, L., Birkett, M. Low temperature tensile lap-shear testing of adhesively bonded polythelene pipe. International journal of adhesion and adhesive. **74**, 57-63 (2017)

[7] Lula, E., Leite, T., Alves, C., Santana, I., Almeida, A., Costa, J. Parameters that influence micro tensile bond testing of adhesive systems. RGO. **62**, 65-70 (2014)

[8] Armentani, E., Laiso, M., Caputo, F., Sepe, R. Numerical FEM Evaluation for the structural behaviour of hybrid (bonded/bolted) single-lap composite joint. P. Struc. Integ. **8**, 137-153 (2018)

[9] Kalina, T., Sedlacek, F., Krystek, J. Determination of the influence of adherent surface on the adhesive bond strength. MATEC w. conf. **157**, 05012 (2018)

[10] Fuente, D., Alcántara, J., Chico, B., Diaz, I., Jiminez, J.A., Morcillo, M. Characterisation of rust surfaces formed on mild steel exposed tomarine atmospheres using XRD and SEM/Micro-Raman techniques. Corrosion Sc. **110**, 253-264 (2016)

[11] Ohtsuka, Y., Tsubouchi, N., Kikuchi, T., & Hashimoto, H. Recent progress in Japan on hot gas cleanup of hydrogen chloride, hydrogen sulfide and ammonia in coal-derived fuel gas. Powder Technology, 190, **3**, 340-347 (2009)

[12] AISI. North American Specification for the design of Cold-Formed Steel Structural Members. American Iron and Steel Institute and Canadian Standards Association. AISI S100-16. Massachusetts, Canada: CSA Group (2016)

[13] I. Komara, T. Kivanc, E. Wahyuni, P. Suprobo. Experiment on cold-formed steel c-section joint with screw and adhesive material, MATTER: IJST, 3, **2**, 51-63 (2017)

[14] ASTM. Standard Test Method for Tensile Properties of Plastics. ASTM D638-**14**. ASTM International. West Conshohocken. PA. (2014)

[15] ASTM. Standard Test Methods for Tension Testing of Metallic Materials. ASTM E8 / E8M-**16a**. ASTM International. West Conshohocken, PA, (2016)

Acknowledgements

The author implied a gratefully acknowledges the financial support for this study from Indonesia Endowment Fund for Education, the ministry of finance of the Republic of Indonesia under LPDP scholarship. The authors would also like to express their gratitude to the Ministry of Research, Technology and Higher Education of the Republic of Indonesia for the support and facilities provided. The last, the author like to thank the Anadolu university for allowing the micro-structural investigation. A special acknowledgement must be given to Levent Koroglu (Eskisehir, Turkey) for Micro measurements and analysis.

IOP Conf. Series: Materials Science and Engineering **462** (2019) 012005 doi:10.1088/1757-899X/462/1/012005

Analysis of Location Tracker Devices (GPS Microcontroller STM 32) on The Position of Solar-powered Electric Bicycles

Syamsuri[1], H S Maulana[1], D Fachrudin[1]

[1]Institut Teknologi Adhi Tama Surabaya, Department of Mechanical Engineering, Jl. Arief Rachman Hakim 100, 60117 Surabaya, Indonesia

syamsuri@itats.ac.id

Abstract. Solar electric bicycles aim for remote areas, where there is no electricity coverage. To find out the position of the solar electric bicycles, therefore research on solar electric bicycles was made with the addition of an STM 32 microcontroller. This research was conduct to determine the speed and holding time of data retrieval during the data delivery from location to serve. This research was carried out by analysing the variations of holding 15 minutes with a delay of 3 seconds: holding 30 minutes with a delay of 2 seconds: holding 1 hour with a delay of 30 seconds. The research results showed that the difference between GPS (Global position system) and the delay was 4%. From the result of this study, conclude that the use of GPS microcontroller STM32 was good enough.

1. Introduction

Energy consumption continues to increase, causing all countries to compete to create alternative energy sources. Every discovery of new things about energy should qualify that generate a large enough amount of energy but minimises the cost and no negative impact on the environment. Therefore, the search for new energy is directed to solar energy by using solar cell panels that can convert solar energy into electrical energy [1]. One form of application of the use of solar cells usually found in vehicles [2] like an electric bicycle using solar cells. Electric bicycles use solar cells to be a solution for solar energy utilisation, where this vehicle has high mobility and does not cause pollution. To make it easier for users to find out the location of an electric bicycle using solar cells, a Global Position System (GPS) is installed on the bicycle.

Position tracker system makes easy to search or find some object, especially for tracking the position of solar-powered electric bicycles that have been developed by adding an STM32 GPS microcontroller in this study. Research on tracking positions has been widely carried out, one of them is research conducted by [3]. On that research, a GPS is placed with a variety of specific locations; then the GPS data is accessed from a different location using short message system (SMS). The results of this study indicate that the GPS system has a fairly high level of accuracy with an average shift value of 16.25 m. Research on GPS on moving objects has been examined by [4] and [5], both researchers use Geographic Information System (GIS) to show geographic information on the map so that the coordinates of the position data from that location are sent to the user via SMS. The results of both studies show that the system used can work very well. In research conducted by [4], it shows that the average time to send location data through SMS requires around 17.8 seconds. The use of GPS and GSM network (by SMS data) on a moving object has been widely used and applied not only to vehicles but also on other moving objects [6-8]. The use of GPS is also very helpful for users because it has the wide range of capacity as reported by researchers [9, 10].

This study aims to determine the effect of the data transmission period to the delay time data received from a solar electric bicycle that uses GPS microcontroller STM32 and comparing the results of the speed on the speedometer and the speed on the GPS. In addition to the mileage on the odometer and mileage tool on GPS microcontroller STM32 also discussed.

2. Research Method

To determine the effect of the data transmission period to the delay time data received and comparison of speed on the speedometer with the mileage on the odometer and mileage on the GPS microcontroller STM32 on electric bicycles using solar cells. Then, in this study using a folding bicycle modified by adding a solar panel to the top of the bicycle as shown in Figure 1. Three solar panels with a capacity of 50 Watt peak are assembled to produce 150 Watt peak. This power is used to fill a 48 Volt capacity battery which is used to drive an electric bicycle as well as the power supply for the GPS microcontroller STM32 circuit. The circuit scheme of the GPS system is shown in Figure 2. In this framework, the battery acts as the main power for charging the GPS device. An STM32 is used as a CPU in this circuit, max 32 as a component for input/output, and GSM MODUL is used to send data to serve. In the GPS circuit, there is one line connected to the antenna cable, the antenna on the GPS is tasked with finding signals so that GPS can continue to get signals from satellites. In this study, the retrieval data is only taken 5 data at every holding intervals to represent the accuracy of the timer is the holding time of 15 minutes, 30 minutes and 60 minutes. The three variations of holding time are chosen to determine the significant delay time that occurs between the data transmission time and the data receiving time. As was the case in previous studies, but on that study using holding time of 1 minute and 10 minutes to determine the delay time that occurred [4].

Figure 1. Electric bicycle with three solar panels.

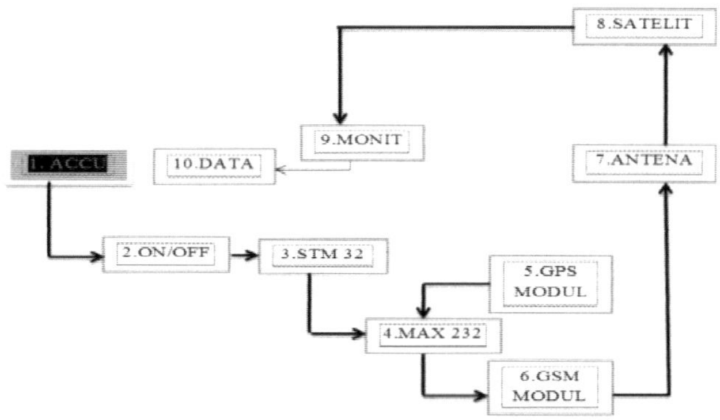

Figure 2. The circuit scheme of the GPS system.

3. Result and Discussion

3.1. Delay Time Data Reception on Sending GPS Data to The Server

To determine the effect of data transmission time to the reception delay time, then the 5 data is taken at each time interval to represent the accuracy of the timer which is the holding time of 15 minutes, 30 minutes and 60 minutes. The collection of experimental data samples was carried out in Ronggojalu village, Masangan Wetan, Sukodono District, Sidoarjo. Figure 3 shows the results of tracking from a solar cell electric bicycle. It shows that the tracking results are in accordance with the path that is traversed.

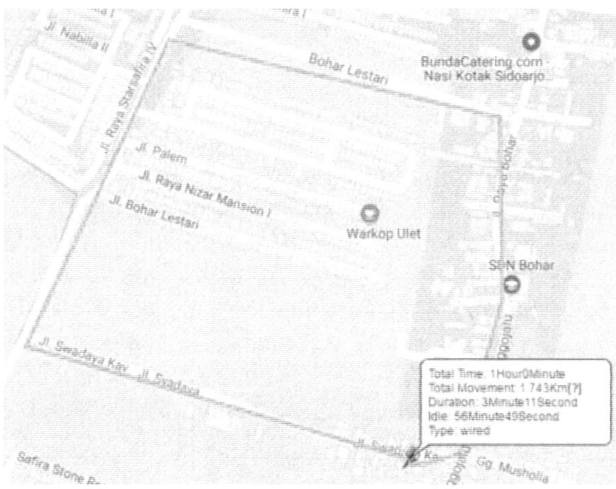

Figure 3. The results of tracking the position of the electric bicycle with solar cell power.

In the experiment with holding time of 15-minutes, it shows that at number 1 the delay is 0 seconds, while in the number 2 between sending data and receiving data in the monitor there is a time delay of 2 seconds as shown in Table 1. In general, from this test, the average delay time between sending and receiving data is not too long, ranges from 3 seconds / 5 tests = 0.6 seconds at each holding time. Experiments with holding time of 30-minute indicate that experiments number 3 and 4 have a delay time 1-second as shown in table 2. With these conditions, generally at holding time 30 minutes has an average delay time of 2 seconds / 5 test = 0.4 sec testing at each holding time. Whereas in the experiment with holding time of 60 minutes shows the delay time is quite long in experiments number 2 and 4 which is 7 seconds and 35 seconds, as shown in table 3. This causes the average value to increase to 42 seconds slightly: 5 tests = 8.4 seconds at each holding time. All data can be seen in Tables 1, 2 and 3.

Table 1. Test data with 15-minute holding time.

No	Date	Sending time	Receive time	Longitude	Latitude	Delay
1	2017-08-27	16:25:34	16:25:34	112.697774	-7.392929	0 second
2	2017-08-27	16:40:34	16:40:36	112.697774	-7.392948	2 second
3	2017-08-27	16:55:36	16:55:36	112.697794	-7.392931	0 second
4	2017-08-27	17:10:36	17:10:37	112.697794	-7.392931	1 second
5	2017-08-27	17:25:37	17:25:37	112.697825	-7.392907	0 second

Table 2. Test data with 30-minute holding time.

No	**Date**	**Sending time**	**Receive time**	**longitude**	**latitude**	**Delay**
1	2017-08-27	17:40:37	17:40:37	112.697805	-7.392928	0 second
2	2017-08-27	18:10:37	18:10:37	112.697805	-7.392911	0 second
3	2017-08-27	18:40:37	18:40:38	112.697805	-7.392900	1 second
4	2017-08-27	19:10:38	19:10:39	112.697805	-7.392921	1 second
5	2017-08-27	19:40:39	19:40:39	112.697835	-7.392900	0 second

Table 3. Test data with 60-minute holding time.

No	**Date**	**Sending time**	**Receive time**	**longitude**	**latitude**	**Delay**
1	2017-08-27	20:10:37	20:10:37	112.697856	-7.392900	0 second
2	2017-08-27	21:10:37	21:10:44	112.697805	-7.392907	7 second
3	2017-08-27	22:10:44	22:10:44	112.697805	-7.392933	0 second
4	2017-08-27	23:10:44	23:11:19	112.697866	-7.392938	35 second
5	2017-08-28	00:10:19	00:10:19	112.697815	-7.392927	0 second

In general, from Tables 1, 2 and 3, almost all the holding time taken by the data is real-time data when sending data. This means that in general the delay time is very small this is caused because the GPS signal on the STM 32 is very good. This might change because the presence of an antenna blocks the incoming signal on the GPS to send to the server. Delay time is the difference between sending GPS data with the time to receive the server data. In the three tables, it was found that there was an accumulation of high delay time with a duration of 35 seconds in table 3 of the 4th data, which in that data used holding time of 60 minutes. This is due to a barrier that interferes with the GPS signal to send data. From the third result of the holding time variation that is tested, the best delay time and feasible to use is at holding time of 30 minutes.

3.2. Comparison of Speed on GPS with Speed on Speedometer

The test carried out this time intended to determine the difference in speed displayed on GPS with speed displayed on the speedometer installed on a solar cell power electric bicycle. The results of these experiments show that GPS and speedometer show the speed measurements generally have the same value, as shown in Table 4.

Table 4. Speed testing data with GPS and Speedometer.

No	Date	Sending time	Receive time	longitude	latitude	Speed on GPS	Speed on Speedometer
1	2017-08-22	04:06:25	04:06:25	112.769382	-7.343834	9.8 km/hour	9.5 km/hour
2	2017-08-22	04:06:30	04:06:30	112.769566	-7.343875	5.0 km/hour	5.0 km/hour
3	2017-08-22	04:06:33	04:06:33	112.769669	-7.343895	8.8 km/hour	8.7 km/hour
4	2017-08-22	04:06:37	04:06:37	112.769833	-7.343942	9.0 km/hour	8.9 km/hour

From the data obtained the speed difference shown on the GPS on the monitor screen with a speedometer can be calculated as follows:
Number 1:

$$\frac{(speed\_GPS - speed\_speedometer)}{speed\_GPS} \times 100\% = \frac{9.8 - 9.5}{9.8} \times 100\% = 3\%$$

With the same formula, the difference in speed is shown in GPS and Speedometer is obtained in experiment Number 2 = 0%, Number 3 = 1%, and Number 4 = 1%. With these results, the difference in speed shown by GPS and Speedometer is around 4%, indicating that the use of GPS with STM32 microcontroller is quite accurate and good.

3.3. Comparison of GPS Mileage with The Odometer Mileage

The testing data below is the result of testing of a solar electric bicycle. This test aims to compare between the use of GPS and the odometer to determine the distance of each. The measurement data can be seen in Table 5 and Figure 4.

Table 5. Mileage data using GPS.

No	Speed (km/hour)	Distance (km)
1	9.3	10.310
2	9.5	10.588
3	10.2	10.883
4	6.7	11.204
5	9.2	11.327
6	13.6	11.675
7	8.5	12.141
8	8.1	12.428
9	12.5	12.756
10	12	12.868

Figure 4. The odometer distance is at the start and end conditions.

Speed data retrieval on GPS cannot be stable due to the influence on the signal so that it affects the transmission of speed data that appears on the monitor screen. So what is done is to calculate the mileage from a bicycle, for example, the data above using GPS in table 5 shows the initial distance is 10,310 km, and the final distance is 12,868 km. Thus the distance is 2,558 km. While on the odometer shows the number of starting bicycle runs is 48324.8 km and ends at a distance of 48327.3 km. It is seen that the odometer shows the bike mileage of 2.5 km. Then the percentage of error reading between GPS and odometer can be calculated by using the equation below,

$$\frac{(\text{distance } GPS - \text{distance } odometer)}{\text{distance } GPS} \times 100\% = \frac{2.558 - 2.500}{2.558} \times 100\% = 2.2\%$$

then the difference in reading error between GPS and odometer is 2.2%. It can be ascertained that this GPS device is quite accurate and good.

4. Conclusions

From data retrieval and data analysis from location tracker devices with GPS microcontroller STM32 to the position of solar electric bikes can be concluded that the test results show that the average delay in receiving data in holding time of 15 minutes is about 0.6 seconds, holding time of 30 minutes is about 0,4 second and holding time of 1 hour is about 8.4 seconds. This because there is an accumulation of delays, from the results of the holding time that gets the best delay time and is worthy of use is holding time of 30 minutes. The results also showed that the GPS ratio with a speedometer was around 4% and the GPS ratio with an odometer was around 2.2%. This shows that the use of GPS microcontroller STM 32 is quite good. This result is expected to be used as a reference for the development of location tracker devices, especially for other transportation devices.

5. References

[1] Devabhaktuni V, Alam M, Shekara Sreenadh Reddy Depuru S, Green RC, Nims D, Near C. Solar energy: Trends and enabling technologies. *Renewable and Sustainable Energy Reviews*. 2013;**19**:555-64.

[2] Rizzo G. Automotive Applications of Solar Energy. *IFAC Proceedings Volumes*. 2010;**43**(7):174-85.

[3] Budiawan T. *MOBILE TRACKING GPS (GLOBAL POSITIONING SYSTEM) MELALUI MEDIA SMS (SHORT MESSAGE SERVICE)*. Semarang: University Diponegoro; 2011.

[4] Hanifah R, Isnanto RR, Christyono Y. Simulasi Sistem Informasi Geografis (SIG) Pemantauan Posisi Kendaraan Via SMS Gateway. *Transmisi*. 2012:5.

[5] Abidin DZ. SISTEM ONLINE UNTUK PELACAKAN PAKET MENGGUNAKAN GPS *Jurnal Processor*. 2011;**6**(2):13-24.

[6] Deblauwe N, Ruppel P, editors. *Combining GPS and GSM Cell-ID positioning for Proactive Location-based Services*. 2007 Fourth Annual International Conference on Mobile and Ubiquitous Systems: Networking & Services (MobiQuitous); 2007 6-10 Aug. 2007.

[7] Meneses F, Moreira A, editors. *Using GSM CellID Positioning for Place Discovering*. 2006 Pervasive Health Conference and Workshops; 2006 29 Nov.-1 Dec. 2006.

[8] Tian Z, Yang J, Zhang J, editors. *Location-based Services Applied to an Electric Wheelchair Based on the GPS and GSM Networks*. 2009 International Workshop on Intelligent Systems and Applications; 2009 23-24 May 2009.

[9] Lai Y, Lin J, Yeh Y, Lai C, Weng H. A tracking system using location prediction and dynamic threshold for minimizing SMS delivery. *Journal of Communications and Networks*. 2013;**15**(1):54-60.

[10] Le-Tien T, Phung-The V, editors. *Routing and Tracking System for Mobile Vehicles in Large Area*. 2010 Fifth IEEE International Symposium on Electronic Design, Test & Applications; 2010 13-15 Jan. 2010.

Acknowledgement

We would like to thank Direktorat Riset dan Pengabdian Masyarakat (DRPM) from Kementerian Ristek Dikti for funding this research through the skim "Penelitian Unggulan Perguruan Tinggi (PUPT)" flagship research program.

The Production of Activated Carbon from Indonesian Mangrove Charcoal

A Budianto[1], E Kusdarini[2], S S W Effendi[1] and M Aziz[1]

[1]Department of Chemical Engineering, Institut Teknologi Adhi Tama Surabaya (ITATS), Jalan Arief Rahman Hakim 100, Surabaya 60117, Indonesia; email:
[2]Department of Mining Engineering, Institut Adhi Tama Surabaya (ITATS), Jalan Arief Rahman Hakim 100, Surabaya 60117, Indonesia; email: esti@itats.ac.id

budichemical@itats.ac.id

Abstract. Activated Carbon is an important material for purification, colour removal, adsorbent, and catalyst in the chemical industry. One of the iodine industries in East Java use active carbon as adsorbent that captures iodine from the solution. Indonesia still imports activated carbon despite the numerous raw materials of activated carbon in Indonesia. This research aims to determine the effect of the concentration of phosphoric acid (activating agent) and activation temperature on the production of activated carbon from mangrove charcoal. The activating agent used was phosphoric acid with 0.4-1.0 M and 2 hour immersion. The activation temperature was 650-725°C. The activated carbon of the research results was tested for its proximate level, iodine number, and specific surface area. The results showed that water content ranged from 0.11-0.38%, iodine number was around 769.075-1,019.087 mg/g. The highest iodine number was 1,019.87 mg/g and the surface area of 354.977 m^2/g that occurred at 725 °C with activator concentration of 1 M. Proximate analysis, iodine number, and specific surface area of activated carbon from mangrove charcoal have met Indonesian National Standardization (SNI 06-3730-1995).

1. Introduction

Indonesia is a country with the most number of islands in the world which makes Indonesia have the longest coastal area in the world. This fact makes the coastal area as important area. Two regions in East Java that actively cultivate mangroves are Tuban and Banyuwangi. Dry wood from these mangroves is widely used for firewood and charcoal, some of which are not even utilized. While at the same time, wood waste is very important for industry, especially for charcoal briquettes [1]. This utilization can still be increased to activated charcoal which is needed by various chemical industries and water treatment.

Changes in charcoal or wood into activated carbon require further processing. Some researchers have tried to use charcoal, sawdust, and agriculture waste for activated carbon. They examined the production of activated carbon from mahogany sawdust. The activation process was carried out using sodium carbonate and caustic soda, however, the absorption capacity of benzene still did not meet the requirement. The absorption capacity of this benzene was 14.5% [2]. Alpian et, al. tried to produce activated carbon from gelam wood, the result showed that the activated carbon produced from gelam wood has met SNI 06-3730-1995 standard [3]. Orange peel and Sapelli sawdust were made into activated carbon using microwave with rapid heating [4][5]. Eragrostis Plana Nees leaves were made into activated carbon with the help of microwave as well [6]. Masthura and Zulkarnain examined the formation of pores in the heating of mangrove wood and coconut shell. This research showed that the most pore formation of mangrove wood and coconut shell were respectively at the temperature of 500 °C and 900 °C [7]. However, this research still has not shown the complete success of activated charcoal production from mangrove wood.

This research modifies the production method of activated charcoal from wood. The modified method was employed to transform mangrove charcoal to activated charcoal that complies with Indonesian National Standardization (SNI). The novelty of this research is the process of making activated carbon from mangrove charcoal using two methods of alternating activation, namely thermal and chemical. It is expected that the research results will be a reference for producing commercial activated carbon in the future.

2. Problem and data presentation

2.1 Problem

The production of activated carbon from mangrove charcoal is still a main attraction, especially in economic terms. Some researchers have made activated carbon from raw wood but the results still cannot be used to produce for commercial purposes. In 2018, Mastura and Zulkarnain have started a research on the effect of activation temperature on pore formation, but the research data were incomplete in which were not equipped with iodine absorption and SNI standards [7]. Accurate research is required since the commercial production of activated carbon from mangrove charcoal needs accurate data on procedure, proper activation temperature, activation time, chemical concentration for activation, and activated carbon specification.

2.2 Aim

This research aims to determine the effect of phosphoric acid concentration and activation temperature on the product proximate, iodine absorption, the addition of surface area after activation, and the best operating conditions for producing activated carbon to the product specifications compared to SNI 06-3730-1995.

3. Methodology

3.1. Data collection

This research was conducted in ITATS Chemical Engineering and mangrove charcoal was bought in the nearest traditional market in Surabaya. Mangrove charcoal was activated in two stages. The first stage was soaking the charcoal in phosphoric acid with certain concentration for 8 hours. Then, the charcoal was rinsed twice with distilled water at a temperature of 85°C until neutral pH was obtained. The dried mangrove charcoal was then ground and sieved to pass the 40 mesh filter. The ground mangrove charcoal was thermally activated for 2 hours at the selected temperature. Then, the activated carbon was cooled down in a desiccator and dried at 105°C for 6 hours followed by an analysis. Activated carbon analysis covered specific surface area with BET method, iodine absorption, and proximate analysis.

4. Data Analysis and result

4.1 Water content

Water content is one of the requirements of activated carbon. High water content may reduce the quality of activated carbon. According to SNI 06-3730-1995, the standard water content of activated carbon is less than 15%. Water content of mangrove charcoal activated carbon can be seen in Figure 1.

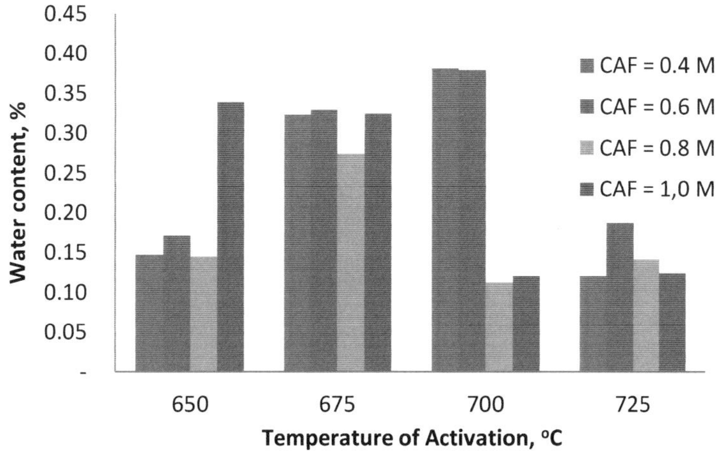

Figure 1. Water content of mangrove charcoal activated carbon at various temperatures

Figure 1. is the result of water content analysis of activated carbon at various temperature and phosphoric acid concentration. The water content of mangrove charcoal activated carbon ranged from 0.11-0.38% indicating that the water content of this activated carbon has met the SNI 06-3730-1995. This water content was better than activated carbon from bituminous and mahogany coal [2][8]. Besides due to thermal activation, the low water content was also caused by drying procedure at the end of the process.

4.2 Ash content

Ash content is one of the important active carbon requirements. If the ash content is high, it will reduce the absorption of this activated carbon. SNI 06-3730-1995 requires a maximum of 100% ash content in the activated carbon. Data on the ash content of mangrove charcoal activated carbon with thermal activation at various temperature and phosphate acid chemical activation are shown in the Figure 2. The maximum ash content of this activated carbon ranged from 6.8-10 % which has met Indonesian National Standardization (SNI 06-3730-1995), in which the maximum ash content of activated carbon is 10%.

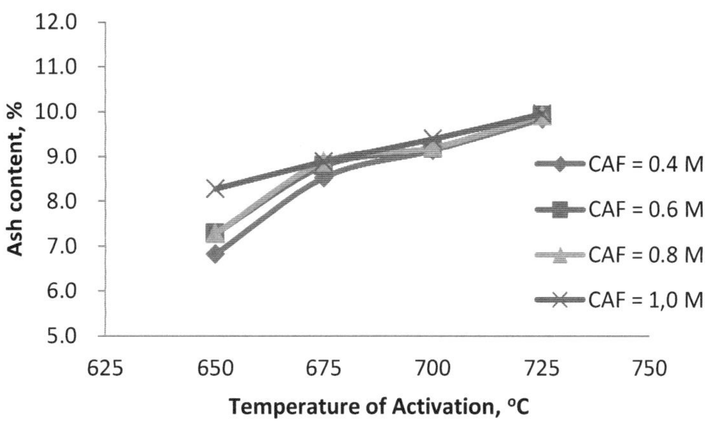

Figure 2. Ash content of mangrove charcoal activated carbon with thermal activation at various temperature and phosphate acid chemical activation

The increase of activation temperature from 650°*C* to 725°*C* may increase ash content from 6.8-9.8%. This happens due to the partial increase of the volatile matter temperature will leave activated carbon. The ash content in this study was better than in the activated carbon from bituminous coal [8][9]. High ash content may cover the pores of activated carbon.

4.3Volatile Matter

Volatile matter is a parameter describing the number of materials decomposed from activated carbon in certain temperature range. The maximum volatile matter allowed by SNI 06-3730-1995 is 25%. The data on activated carbon volatile matter of mangrove charcoal are presented in Figure 3. The volatile matter content of this activated carbon ranged from 19.8-23.4%. This number shows that all variables of activation temperature and phosphate acid concentration have formed activated carbon that met the requirement of volatile matter.

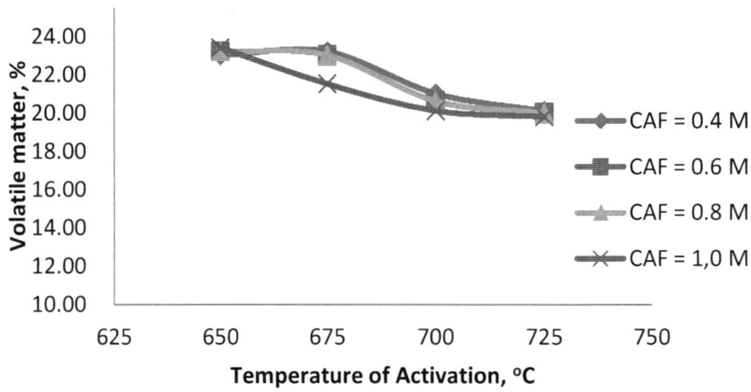

Figure 3. Volatile matter content in mangrove charcoal activated carbon with chemical and thermal activation at various activation temperatures and phosphoric acid concentration (CAF)

4.4 Fixed Carbon

Fixed carbon is a parameter describing the amount of carbon bound from activated carbon at certain temperature range. Fixed carbon content required by SNI 06-3730-1995 is 65%. The data on the analysis results of mangrove charcoal activated carbon can be seen in Figure 4. The fixed carbon content of this activated carbon ranged from 68.13%-70.25% showing that all activation temperature ranges and phosphate acid concentration used have met the requirement of fixed carbon. The highest fixed carbon content of mangrove charcoal activated carbon was 70.51% occurred at operating condition of 725°C thermal activation and phosphate acid concentration of 1.0 M. Fixed carbon in this study was higher than bituminous coal [8].

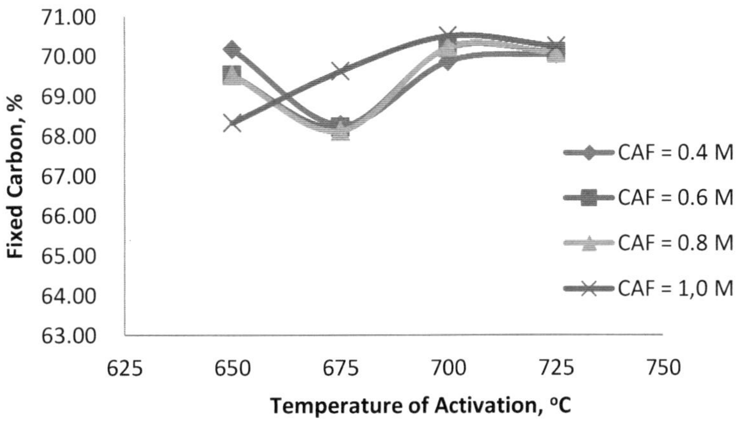

Figure 4. Fixed carbon content of mangrove charcoal activated carbon with chemical and thermal activation at various temperatures and phosphate acid concentration

4.5 Iodines Number

Iodine number is a number showing the ability of 1 g activated carbon in absorbing iodine in mg. This iodine number is an absolute requirement of SNI that must be fulfilled by activated carbon. The iodine number required by SNI 06-3730-1995 is 750 mg/g. Data on iodine number of activated carbon is presented in Figure 6. This figure shows that iodine number of activated carbon from mangrove charcoal activated by phosphate acid with 0.4-1.0 M concentration and activation temperature ranged from 650-725°C is around 769–1019 mg/g. This showed that all operation conditions chosen in the experiment of producing mangrove charcoal activated carbon have met SNI requirement in terms of iodine number. The increase of temperature from 650°C to 725°C may increase iodine number up to 19-24 percent. The highest iodine number of mangrove charcoal activated carbon is 1,019 mg/g occurred at 725°C operation condition and phosphate acid concentration of 1.0 M. Iodine absorption in this study was lower than activated carbon from mahogany [2].

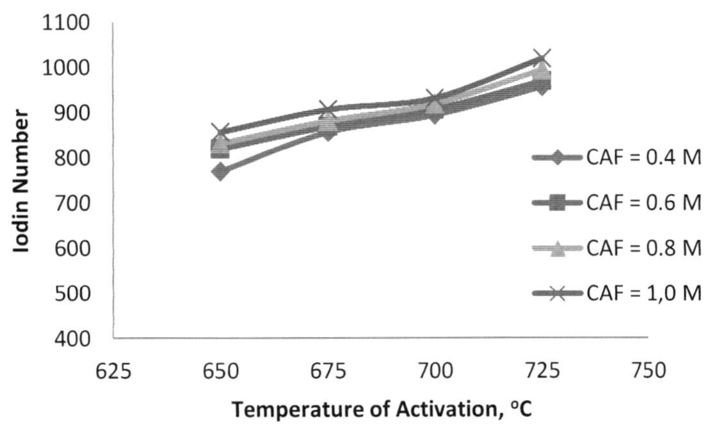

Figure 5. Iodine number of mangrove charcoal activated carbon activated using chemical and thermal activation at various activation temperatures and phosphate acid concentration

4.6 *Specific Surface Area*

Specific surface area of mangrove charcoal before activation was 136.345 m2/g. Activation of 1.0 M phosphate acid and heating at 725°C produced activated carbon with surface area of 354.977 m²/g. The activation process has succeeded in increasing specific surface area by 160%. The specific surface area of this activated carbon was better than activated carbon from cacao shell, but lower than those from coconut shell [10],[11],[12]. The specific surface area of this activated carbon is sufficient to be commercial activated carbon.

5. Conclusion

Activation temperature and phosphate acid concentration affected proximate properties, specific surface area, and iodine number of mangrove charcoal activated carbon. Water content, ash content, volatile matter content, tethered carbon content, iodine number, and specific surface area of mangrove charcoal activated carbon in this research have met the SNI–06-3730-1995 activated carbon standard. The highest iodine number was 1,019 mg/g occurred at 725°C and phosphate acid concentration of 1.0 M. The specific surface area of this mangrove activated carbon was 354.977 m^2/g. The best operating conditions for producing activated carbon from mangrove charcoal was at 725°C with phosphate acid concentration of 1,0 M.

6. References

[1] A Saleh, I Novianty, S Murni, A Nurrahma 2017 Analisis kualitas briket serbuk gergaji kayu dengan penambahan tempurung kelapa sebagai bahan bakar alternatif, Al Kimia volume 5 No. 1, pp 21-33.

[2] R Pujiarti, JPG Sutapa 2005 Mutu arang aktif dari limbah kayu Mahoni (Swietenia macrophylla King) sebagai bahan penjernih air, Jurnal Ilmu dan Teknologi Kayu Tropis. Volume 3 No.2 pp. 33-36.

[3] Alpian, TA Prayitno JPG Sutapa, and Budiadi 2010 *Kualitas Arang Aktif Kayu Gelam dan Aplikasinya untuk Meningkatkan Kualitas Air,* Jurnal Ilmu dan Teknologi Kayu Tropis. Volume 8 No.2 pp. 155-168.

[4] SS Lam, RK Liew,YM Wong, PNY Yek, N L M, CL Lee, and HA Chase 2017 Microwave-assisted pyrolysis with chemical activation, an innovative method to convert orange peel into activated carbon with improved properties as dye adsorbent, journal of cleaner production, No. 162, pp 1376-1387.

[5] PS Thue, GSD Reis, EC Lima, JM Sieliechi, GL Dotto, AGN Wamba, SLP Dias, FA Pavan 2017 Activated carbon obtained from sapelli wood sawdust by microwave heating for o-cresol adsorption, Res Chem Intermed, Nomor 43, pp.1063–1087.

[6] MR Cunha, EC Lima, NFGM Cimirro1, PS Thue, SLP Dias, MA Gelesky, GL Dotto, GSD Reis, FA Pavan 2018 Conversion of Eragrostis plana Nees leaves to activated carbon by microwave-assisted pyrolysis for the removal of organic emerging contaminants from aqueous solutions, Environmental Science and Pollution Research, Volume 25 Issue 23, pp. 23315–23327

[7] Masthura, P Zulkarnain 2018 Characterization of Activated Carbon Microstructure of Coconut Shell and Mangrove Wood, *Elkawnie: Journal of Islamic Science and Technology Vol. 4, No.1,*pp 45-54.

[8] E Kusdarini, A Budianto, D Ghafarunnisa 2017 Produksi Karbon Aktif dari Batubara Bituminus dengan Aktivasi Tunggal H_3PO_4, Kombinasi H_3PO_4-NH_4HCO_3, dan Termal, Reaktor, Volume 17 no 2, pp 74-80.

[9] M Rahim and OS Indriyani 2010 Pembuatan Karbon Aktif dari Batubara Peringkat Rendah, Jurnal Teknologi Media Perspektif, Volume 10 No. 2, pp. 60-114.

[10] A Budianto, Romiarto and Fitrianingtyas 2016 Pemanfaatan Limbah Kakao (Theobroma Cacao L) sebagai Karbon Aktif dengan Aktifator Termal dan Kimia, Prosiding Seminar Nasional Sains dan Teknologi Terapan IV, pp. 208-213.

[11] SPD Kaman, IAW Tan and LLP Lim 2017 Palm oil mill effluent treatment using coconut shell – based activated carbon: Adsorption equilibrium and isotherm, Proceeding of MATEC Web of Conferences 87, 03009, pp. 1-6.
[12] IAW Tan, MO Abdullah, LLPB Lim, THC Yeo 2017, Surface Modification and Characterization of Coconut Shell-Based Activated Carbon Subjected to Acidic and Alkaline Treatments, Journal of Applied Science & Process Engineering Vol. 4, No. 2, pp. 186-194

Acknowledgements
ITATS research team would like to thank the head of YPTS and ITATS as well as the head of Chemical Engineering Department for the permission given to use laboratory facilities for this research.

Formulating a Plan Model for Controlling Water Pollution in Kali Surabaya Based on Obedience Analysis of IPLC Implementation

Yulfiah[1]
[1]Master of Environmental Engineering, Institute of Technology of Adhi Tama Surabaya, Indonesia
yulfiah@itats.ac.id

Abstract. Indication of pollution is worrisome occurring in the water of Kali Surabaya which becomes the main source of Perusahaan Daerah Air Minum (PDAM) or Municipal Waterworks. Therefore, research on it is necessarily conducted by (a) analyzing the obedience of IPLC (Ijin Pembuangan Limbah Cair or Liquid Waste Disposal Permit) implementation, (b) identifying the obstacles to employ IPLC, and (c) formulating a plan model for controlling pollution and recovering the water quality of Kali Surabaya based on the results of both analysis and identification. A comparative descriptive method was used in this research as it gave systematical, factual, and accurate description about facts and research object characteristics. Comparative study was then carried out to investigate the relationship among the existing facts. This research formulated the plan model for controlling pollution and recovering the water quality of Kali Surabaya by using active participation from industrial community.

1. Introduction

Ijin Pembuangan Limbah Cair (IPLC) or Liquid Waste Disposal Permit is a set of environmental regulation that must be owned by every activity involving liquid waste disposal into water. Every industry must meet the requirements of *IPLC* before disposing liquid waste into water. It must have adequate *Instalasi Pengolahan Air Limbah (IPAL)* or Liquid Waste Management Instalation and standardized waste water quality[4].

Kali Surabaya is part of Brantas watershed which supplies the main water for *PDAM* Gresik and Surabaya. It serves more than 5 millions people. However, along the river bank of Kali Surabaya becomes a disposal place of liquid and solid wastes resulted from people and industrial activities. There are more than 300 industries along Kali Surabaya. Industrial waste disposal has caused serious water pollution at some parts of Kali Surabaya river bank including Driyorejo and Warugunung Districts. Therefore, controlling pollution resulted from industrial activities is required as it serves as an effort for improving the water quality of Kali Surabaya.

The initial phases of polution control program of Kali Surabaya were (a) analyzing the obedience during IPLC implementation, (b) identifying the obstacles to implement IPLC, and (c) designing a plan model for controlling pollution and recovering the water quality of Kali Surabaya based on the results of analysis and identification.

2. Theoritical Review

River water pollutant can be classified in three things: (1) different locations of pollutant sources namely spot source and non-spot source; (2) history of river water formation; and (3) pollutant types such as pollutants of organic, pesticide, volatile, neutral, and acid [5].

The dispersion of river water pollutant from the spot source will decrease along with the movement of pollutant away from the spot source [6-8]. It will keep going until its content level becomes very low or harmless due to the ability of river self-purification. Self-purification will continue as long as it does not exceed the limit of river water ability in *swapentahiran* or natural water purification process. The purification process depends on time, distance, pollutant type, and river physical condition [2]. Meanwhile, Hammer (1986) argues that self-purification is determined by factors of flow quantity, time, movement to downstream, water temperature, and aeration [3]. Purification process involves the mechanisms of filtration, sorption, chemical process, decomposition, and dilution [9-10].

3. Method

Comparative descriptive method was employed in this research. It can give systematical, factual, and accurate description on facts and research object characteristics. Comparative study was then carried out for investigating the relationship among the existing facts.

Research variables included: (a) the qualities of river water and waste water, (b) the arrangement of IPLC conduct, (c) the obstacles encountered by industry during IPLC conduct, and (d) factors needed in controlling pollution and recovering the water quality of Kali Surabaya.

The target of this research were industrial areas of Driyorejo Warugunung at Driyorejo District in Gresik Regency and Waru Gunung District in Surabaya city. The industries which became the target were selected based on the data analysis of IPLC ownership, industrial waste quality monitoring, and environmental cases. In other words, the targeted industries were classified into Group I (industries with IPLC in good environmental performance), Group II (industries with IPLC in bad environmental performance), and Group III (industries without IPLC).

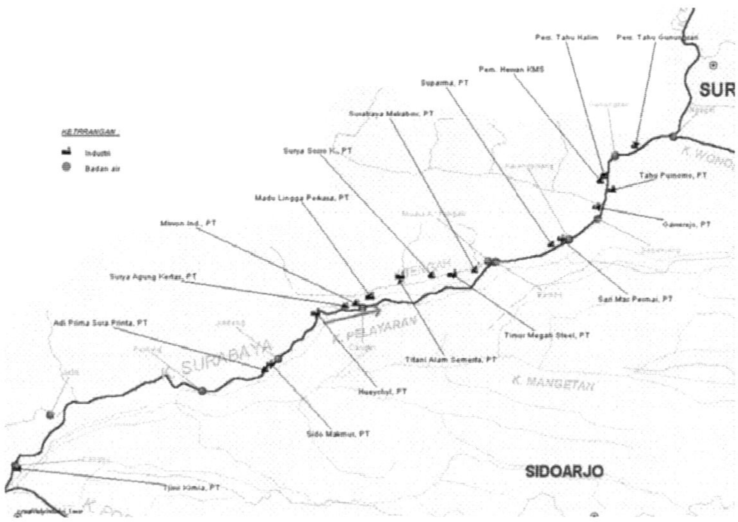

Figure 1. Sampling Point for Waste Water and River Water

4. Result and Discussion

The result of analysis on waste water quality, particularly in the targeted industries of Groups I and II showed that all tested parameters did not exceed standardized waste water quality. In short, all targeted industries in Groups I and II had obeyed IPLC implementation. Meanwhile, the result of analysis in Group III reported that there were 3 industries obeyed IPLC provisions but 1 industry disobeyed. All tested water parameters of the three industries in Group III exceeded standardized waste water quality. Only one industry met the threshold of waste water quality standard. Obstacles encountered by industries during IPLC implementation were as follows:

1. Institution giving business permit did not give sufficient monitoring.
2. The caretakers of business had low awareness in controlling pollution resulted from their industrial waste (self control).
3. People had low participation and care in monitoring industrial waste or in other words, they much depended on the government in terms of monitoring.
4. The implementation of material delict article of UU No. 23 in 1997 was weak as there was found many obstacles in the field justification
5. The caretakers of business had missperception as they thought that the cost of waste management and control/environmental pollution burdened the production cost.
6. Industries did not give transparancy in environmental management information.

The strategical assumption in the plan model for controlling pollution and recovering water quality of Kali Surabaya is that all components of community must have standardized environmental management activity. Therefore, information transparency in environmental management activity, fund allocation, and socialization to build people's awareness to actively participate in environmental management are required. The key element in the program of controlling pollution and recovering water quality of Kali Surabaya is the empowerment program through partnership, awareness and attitude development, monitoring and controlling. The strategy modeling is illustrated in Figure 2.

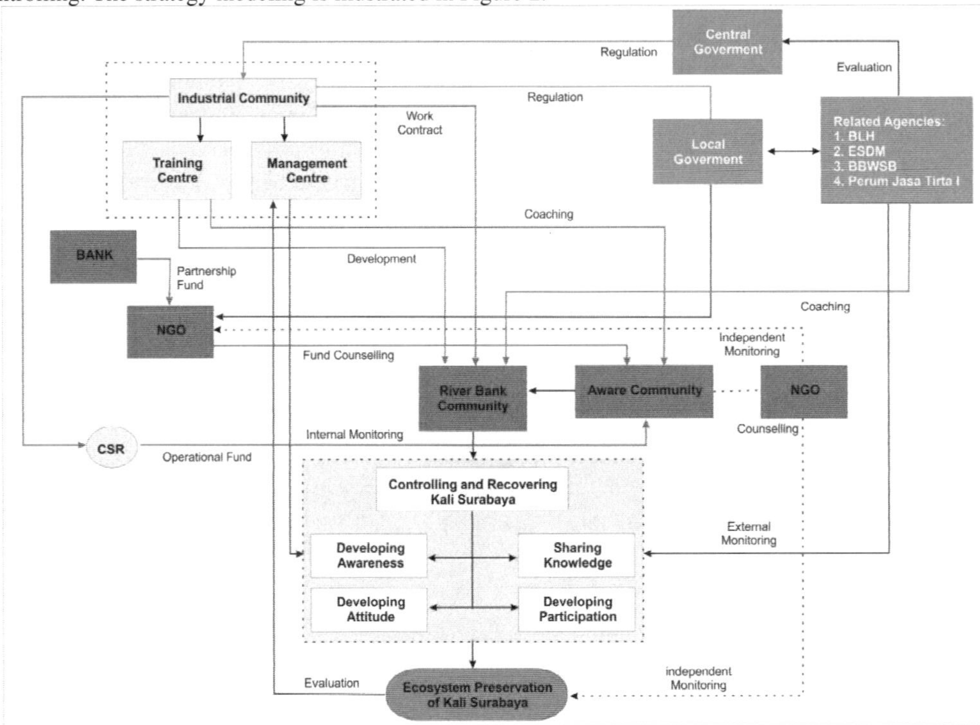

Figure 2. Plan Model for Pollution Control and Water Quality Recovery of Kali Surabaya Based on Active Participation of Industrial Community

5. Conclusion

The conceptual plan model for controlling pollution and recovering water quality of Kali Surabaya consists of two sub-models i.e. a partnership model on environmental management and a model for controlling pollution and recovering water quality of Kali Surabaya. Both models are supported by people empowerment. The implications of policy are deciding task and responsibility through local regulation supported by community empowerment in environmental management and optimizing fund for environmental management by creating stipulation based on the agreement among parties who give special attention to problems of Kali Surabaya.

6. References

[1] Appelo, C.A.J. Postma, D. 1993. *Geochemistry, Groundwater and Pollution*. A.A Balkema Publishers. Rotterdam.

[2] Domenico Patrick, A. Schwartz Franklin, W. 1990. *Physical and Chemical Hydrology*. John Wiley & Sons. New York.

[3] Hammer, M.J., 1986, *Water and Wastewater Technology,* John Wiley & Sons Inc., New York.

[4] Orlob, G.T., 1986, *Mathematical Modeling of Water Quality : Streams, Lake, and Reservaoirs*. John Willey and Son, New York.

[5] Ward, R.C. Robinson, M. 1990. *Principles of Hydrology*. Mc. Graw Hill International Editions. London.

[6] J. Wang, X.D. Liu, J. Lu, "Urban River Pollution Control and Remediation," Procedia Environmental Sciences, 13 (2012), pp. 1856 - 1862.

[7] Paul Whitehead, Gianbattista Bussi, Mohammed Abed Hossain, Michaela Dolk, Partho Das, Sean Comber, Rebecca Peters, Katrina J. Charles, Rob Hope, Sarwar Hossain," Restoring Water Quality in the polluted Turag-Tongi-Balu River System, Dhaka: Modelling Nutrient and Total Coliform Intervention Strategies," Science of the Total Environment, 2018, pp 631 - 632.

[8] Sayyed Ali Noorhosseini, Mohammad Sadegh Allahyari, Christos A. Damalas, Sina Siavash Moghaddam, "Public Environmental Awareness of Water Pollution from Urban Growth: The

Case of Zarjub and Goharrud Rivers in Rasht, Iran," Science of the Total Environment, 2017, pp 599 - 600.

[9] David N Lerner, Alison Holt, "How Should We Manage Urban River Corridors ?," presented at the 18th Biennial Conference of International Society for Ecological Modelling, Procedia Environmental Sciences, 13 (2012), pp. 721 – 729.

[10] Hui XU, Duozhi LV, Yinghua FAN, "A Pragmatic Framework for Urban River System Plan in Plain River Network Area of China," 2012 International Conference on Modern Hydraulic Engineering, Procedia Engineering, 28 (2012), pp. 494 – 500.

The Configuration of Engine-Sail Catamaran Fishing Vessel

P I Santosa[1]

[1]Lecturer of Naval Architect Department, FTMK ITATS

mpic.pramudya@gmail.com

Abstract. In the field of shipping, consumption of fossil fuel is quite large, especially as fuel to drive ship-engines. The high fuel price is not at all beneficial to the shipping industry and fishermen as users of ship-engines. The use of fuel for ship-engines is not only uneconomical but also environmentally unfriendly. The more expensive fuel is anticipated by the reuse of sail in the form of Engine-sail Vessels (ESV) as Fishing Vessels. This paper presents a study on the efficient use of fuel in ESV as a driver that does not utilize fuel to develop of environmentally friendly fishing vessels. There is a potential savings in the use of fuel 70% when compared with conventional ship-engines only.

1. Introduction

The mission of a fishing boat is to catch fish from the sea to get fish that meet the quality in appropriate ways and deliver the fish to land or to another vessel for further processing. Today, the catch of these fishing vessels supplies the daily food needs and sustains the food security of millions of people in the world [1].

From these fishing activities have an impact on the increase of air pollution levels (such as CO_2, SO_2 and NO_X) in the atmosphere, especially on fish boats that use diesel engines fuel by fossil fuels. The impact of these activities is one of the most important problems in the world to look for solutions that fish ships operating in the world become environmentally friendly fishing vessels [2].

In general, the operation of a fishing boat is always associated with economic and environmental issues. Economic factor is the cost of fuel, while the environmental factor is related to the level of pollution produced. Economic problems and strong environmental pressures force ship designers and owners to create more efficient vessels to minimize the use of ship propulsion. Reduced magnitude of ship propulsion (and fuel requirements) can be fulfilled since the ship design stage by creating more efficient hull design and propulsion systems as well as ship operational activities including ship operations, such as Engine-sail Vessels (ESV) [3].

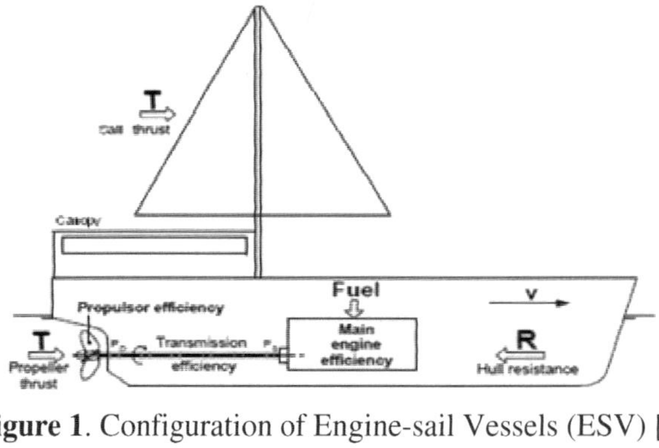

Figure 1. Configuration of Engine-sail Vessels (ESV) [1]

Figure 1 shows the configuration of the ship with the engine-sail driver. The concept of energy conversion in the ship's configuration is to convert fossil energy (fuel) and wind energy into the required thrust force of the ship through the propeller (engine) and sail. The thrust (*T*) force generated from one or more of the ship's propulsion sources operates simultaneously or together known as the hybrid system [1].

Experimental model test on a symmetrical catamaran has been carried out and reported in [3]. The experimental model test was carried out at ITS towing tank as shown in Figure 2 and the principle particular of the model and full ship was given in Table 1.

Figure 2. Experimental model test

Table 1. Principle particular

Parameter	*Catamaran*	*Demihull*
LWL (m)	14.5	14.5
B (m)	7.118	1.318
H (m)	1.44	1.44
D (m)	0.694	0.694
C_B	0.434	0.434
Displ. (ton)	11.8	5.9

Table 2 showed the results of resistance test which indicated that the wider the separation between the hulls (S/L), the lower the total resistance.

Table 2. Result of resistance test [3]

Run No.	V (knots)	Fr	Catamaran Resistances (kN)		
			S/L = 0.2	S/L = 0.3	S/L = 0.4
1	5.788	0.250	1.821	1.659	1.659
2	6.218	0.268	2.141	1.851	2.061
3	6.677	0.288	2.443	2.239	2.348
4	7.051	0.304	2.852	2.678	2.947
5	7.560	0.326	3.460	3.568	3.547
6	8.032	0.347	4.467	3.954	3.766
7	8.384	0.362	4.844	4.345	4.341
8	8.818	0.380	5.149	4.790	4.662
9	9.233	0.398	5.807	5.592	5.515
10	9.813	0.423	7.101	6.448	6.138

Data's from Table 2 is further used as a basis for designing the concept of catamaran fishing boat with a combination of engine-sail [1][8].

2. Literature Review

2.1 Ship Moving Theory

The vessel may move forward due to a sufficient thrust to resist ship resistance at a certain service speed [4]. Based on the above concept, then requirement of ship can move.

$$T \geq R_T \text{ or } T - R_T \geq 0 \tag{1}$$

where T is Thrust (kN) and R_T is Resistance (kN).

2.2 Ship Resistance

The total vessel resistance (R_T) is calculated according to equation (2) where ρ is the sea water mass, C_T is the total resistance coefficient, WSA is the wet surface area, and V is the velocity of the vessel [4].

$$R_T = \frac{1}{2} \rho \, C_T \, (WSA) \, V^2 \tag{2}$$

2.3 Thrust

Thrust (T) is the energy or force required to drive the vessel and can be expressed as equation (3) [4]

$$T = R_T / (1 - t) \tag{3}$$

where t is thrust deduction factor. For double screw [3], the t is defined as

$$t = k_R \cdot wt \tag{4}$$

where k_R is 0.5 for thin rudder. The wt could be found using

$$wt = -0.0458 + 0.3745\, C_B^2 + 0.1590\, D_w - 0.8635\, Fr + 1.4773\, Fr^2$$

where $$D_W = \frac{B}{\nabla^{1/3}} \sqrt{\frac{\nabla^{1/3}}{D}}$$

To move itself, the thrust (T) force generated through the propeller and the sail must be greater than the existing total vessel resistance which is mathematically expressed in equation (5) [4].

$$T\, propeller + T\, sail > R_T \tag{5}$$

where $T\, propeller$ is the propeller thrust which could be defined as

$$T\, propeller = K_T . \rho . n^2 . D^4 \tag{6}$$

where K_T is the thrust coefficient, ρ is the salt water density, n is the propeller Rpm, and D is the propeller diameter. The sail thrust defined as

$$T\, sail = q \, . \, As \tag{7}$$

where q is the dynamic wind pressure. The q could be defined as

$$q = ½ \times \rho \times £ \times V_w^2 \; (\text{ton/m}^2)$$

where ρ is the air mass density which is equal to γ/g. γ is the weight per unit volume which is equal to 1.2265 t/m^3, g is equal to 9.81 ms^{-2}, £ is the wind pressure coefficient, V_w is the wind speed, and As is the sail area.

2.4 Powering

Engine conventional is the prime mover of the vessel which works by converting the fuel energy to rotate the blades thereby producing sufficient thrust to resist ship resistance at certain service speeds. One of the most fundamental methods of power sharing in this conventional driving force is to distinguish between the effective power (P_E) required to drive the ship and power delivered (P_D) on the ship propulsion unit [4]. The formulations used according to [4] are as follows

$$\text{Effective power (PE)} = R_T \times V_s \tag{8}$$

Delivered power (PD) = PE/ Hd (9)

Quasi propulsive coefficient (ηD) = ηP. ηH. ηR (10)

Service power (Ps) = PD/ ηT (11)

where ηT is 0.98 (with gearbox) or 0.95 (without gearbox).

Installed power (*PI*) = Ps + Margin (12)

where Margins (*roughness, fouling, weather*) is ranged by 15 – 20% depending on the ship route.

Sail is one of the propeller props without a propeller on the ship that can work due to the wind force (catch the wind) on the surface of the sail, resulting in a drifting force on the ship at a certain speed. The Sail is one of the propulsion devices on the ship. As with other propulsion devices such as propellers, the sail is attempted to produce an optimal thrust force, in order to produce maximum velocity of the ship. The forces on the Sailing Ship, as the ship moves due to the thrust of the propeller or sail there will be a lift that will lift the hull from the water. In addition, obstacles caused by the aerodynamic resistance of the sail are the forces acting on the sailboat, [5].

Determination of Sail Area almost as a comparison of sail area (*As*) with wetted surface area (*WSA*) is between 2.0 and 2.5. Comparison of sail area (As) with wetted surface area (*WSA*) known as sail ratio (*SR*), [6]. There is another way according to [7], where the determination of *SR* depends on the LWL of the ship by using the graph shown in Figure 3.

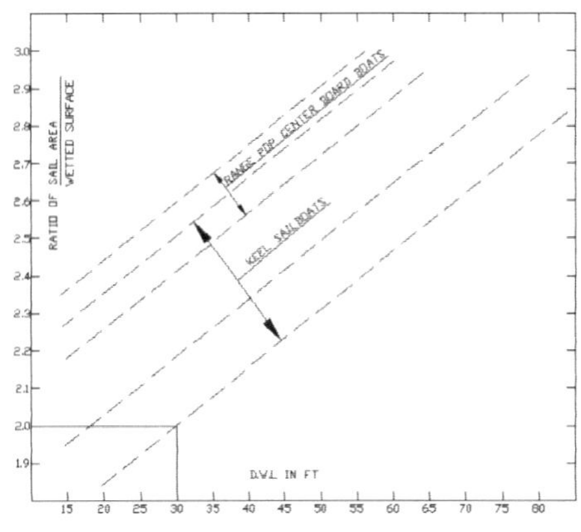

Figure 3. Graph of *SR-LWL* relationship which can be used to determine Sail Area with 15 - 80 feet or 5 - 25 m LWL limitation.

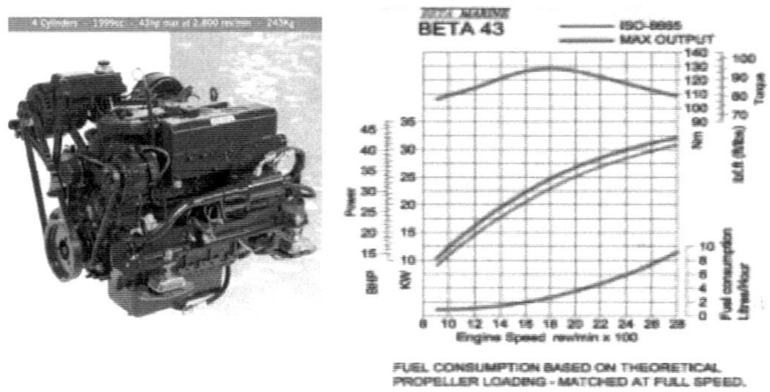

Type: Beta Marine, Power engine : 43 hp max at 2,800 rev/min, Fuel consumption : 10 Lt /hr (at continuous Ratting), Engine Dimension : L=978 mm, B=622 mm, H=740 mm, weight=243 kg

Figure 4. Specification of BETA43 engine series

2.5 Fishing Vessel

In general, the normal voyage profile of fishing vessels according to [9] are: (1) the ship departs and operates in the port (*departure from port*), (2) the ship goes to the location of the fishing ground (*outward bound*), (3) the ship arrives at the location fishing ground and fishing (*on fishing ground*), (4) the ship leaves the location of the fishing ground to the port (*homeward bound*), and (5) the ship arrives at the port and docked at the port (*arrival at port*).

In its operation, a fishing vessel must be completely safe (*very seaworthy indeed*). Even in bad weather, the ship must work. All work on the fishing boat must be done quickly, starting from the process of catching until the processing of the catch is a function of time. The slow catching process causes the fish to run all (migration), while the sluggish processing of the catch causes the fish to be damaged [1].

3. Method and Result

This research is a continuing research that has been done by [3]. All material data and information are using the results of previous research to support scientific/academic and its application.

Based on the results in Table 2, it can be calculated and thus provided the effective power (PE) of 32.4 kW. With the *quasi-propulsive coefficient* (ηD) of 0.664, then the *delivered power* (*PD*) will be of 50.21 kW and with the *transmission losses* (ηT) without gearbox of 0.98 will give *service power* (*Ps*) of 51.2 kW. Finally, if the *total efficiency* (*PE / PI*) is 54%, it yields the *installed power* (*PI*) of 60 kW. The engine specifications used is 2 x 43 HP or 2 x 30 kW as shown in Figure 4.

Furthermore, by using Figure 3 it can be found that the boat with 14.5 m LWL will have approximately 125 $m^2$of sail area. Meanwhile, the calculation of *thrust* (*T*) is obtained from Table 2, which is 6.685 kN and with thrust deduction factor (t) of 0.038. The Propeller thrust (*Tp*) is obtained at the service speed (*Vs*) of 9.8 knots while the thrust force (*Ts*) is calculated using equation (7) and reached at *wind speed* (*Vw*) of 19.2 knots. A hybrid curve chart was further derived from the above calculation and shown in Figure 6. The curve can be used to estimate the efficiency of ESV.

4. Discussion

It can be found from the tank test work as reported in [3] that at the service speed (Vs) of 9.81 knots, the total resistance (R_T) is 6.14 kN, thus the installed engine power (PI) required is 60 kW or 2 x 30 kW to adjust the shape of catamaran. The specifications of engine was shown in Figure 4 which shows the characteristics of the BETA43 engine connecting the engine running (rpm), engine power (P), fuel consumption (FC) and engine torque (Q). The maximum engine speed is 2800 rpm with fuel consumption of about 10 litres/hour and 110 Nm engine torque. Furthermore, Figure 5 can be used as a reference to calculate the efficiency of fishing vessel.

By using equation(4), it can be determined that the magnitude of thrust deduction factor (t) is 0.038, hence the thrust of ship (T) is 6.685 kN, whilst the total ship resistance is 6.138 kN. It means that as the thrust is greater than total ship resistance thus the boat will move, sea also equation (1).

As the fishing vessel is designed to use engine and sail drive (*ESV*), the generation of thrust to move the ship comes from propeller and sail. When the ship moves, the propeller and sail can work individually/separately or together as a hybrid system both with service speed (Vs) of 9.813 knots.

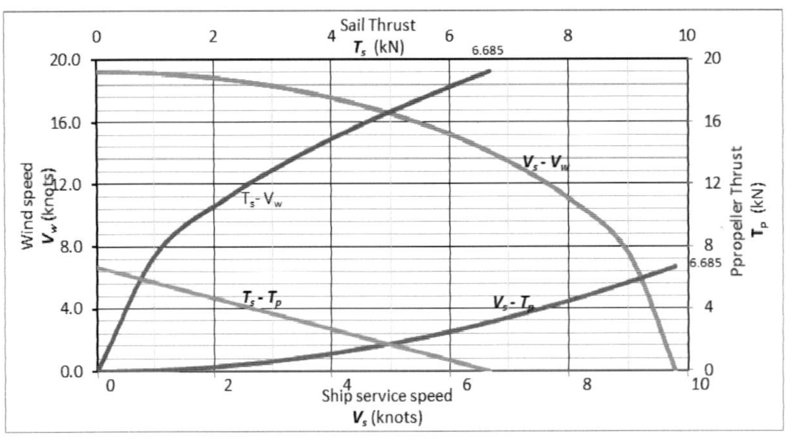

Figure 5. Hybrid curve chart [1]

The relationship between propeller thrust (T_P), see equation (3) and sail thrust (T_S), see Equation (7) is shown in Figure 6 called T_P–T_S curve. Meanwhile, the relationship between wind speed (V_W) with the sail thrust (T_S) is shown in Figure 6 called the V_W–T_S curve. Likewise, the relationship between ship service speed (V_S) and propeller thrust (T_P) is also shown in Figure 6, namely the V_S–T_P curve. The relationship between wind speed (V_W) and ship service speed (V_S) is called the V_W–V_S curve.

Table 3. Remark of hybrid curve chart

No	Number of Axis	Relationship Between
1	Axis [1] is T_s Sail Thrust (kN),	$T_s - T_p$Curve use ordinate [1] and abscise [2]
2	Axis [2] is T_p Propeller Thrust (kN)	V_s - V_wCurve use ordinate [3] and abscise [4]
3	Axis [3] is V_s Ship speed (knots)	$V_s - T_p$Curve use ordinate [3] and abscise [2]
4	Axis [4] is V_w Wind speed (knots)	T_s - V_wCurve use ordinate [1] and abscise [4]

Table 3 shows the number of axis and relationship of *Hybrid Curve Chart* as shows in Figure 6. If the economic value of the use of wind energy is equated with the economic value of fossil energy use, then to calculate the ESV efficiency and how the hybrid system at ESV will be explained in detail in the following case study. In order to test the validity of the present approach as summarized in Figure 6, 2 case studies are given, Case Study 1 and Case Study 2.

4.1 Case study 1 (CS1)

It is assumed that a fishing vessel uses hybrid driving force (engine, sail) and operated at a service speed (V_S) of 9.8 knots, whilst the sea breeze blows (V_W) with a speed of 12 knots. Furthermore, estimate thrust caused by T_S screen/sail, thrust generated by T_P propeller, the propeller contribution to V_S hence the vessel can be operated at the service speed of 9.8 knots, and the fuel consumption (F_C).

The answers is described as follows:

For completion of CS1 using Figure. 6 *Hybrid Curve Chart*, Fig. 8 *Rpm-Vs Chart,* and Fig. 9 *Rpm-FC Chart.* Choice value of 12 kts *Vw* Wind speed in the axis [4] and then generate horizontal line [a] through pass the *Ts – Vw* curve and the *Vs- Vw* curve, next from the intersection it drags the vertical line [b] and [c] From the intersection of the line with the curve *Ts-Tp* drag a horizontal line [d] , *(see green line in Figure 6*).

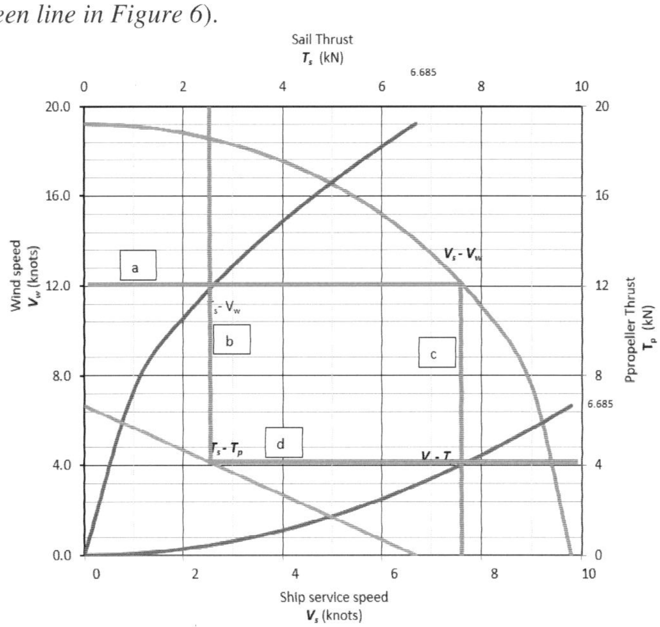

Figure 6. Application of CS1 use *Hybrid Curve Chart*

The results of *Hybrid Curve Chart* readings obtained:

1) Line [b] with the axis [1] is obtained 2.4 kN *Ts*
2) Line [c] with the axis [3] is obtained 7.6 knots *Vs*
3) Line [d] with the axis [2] is obtained 4.3 kN *Tp*

Furthermore, use Figure 7 to find when *the rotation of engine* abt. 1300 Rpm obtained 7.67 / 2 = 3.835 knots *Vs*. (*See red line*), and then use Figure 8 to find when the rotation of engine about 1300 Rpm obtained magnitudes of *fuel consumption* (*FC*) is abt. 1.8 liter/hour/engine or 3.6 liter/hour (*see red line*).

Results of the completion of CS1:

When 12 knots wind speed (*Vw)* blows on the sail generates 2.4 kN sail thrust (*Ts*) and automatically propeller thrust (*Tp*) is abt. 4.285 kN with the contribution of servicing ship speed*Vs* of 7.67 kts and 3.6 ltr / hour fuel consumption (*FC*).

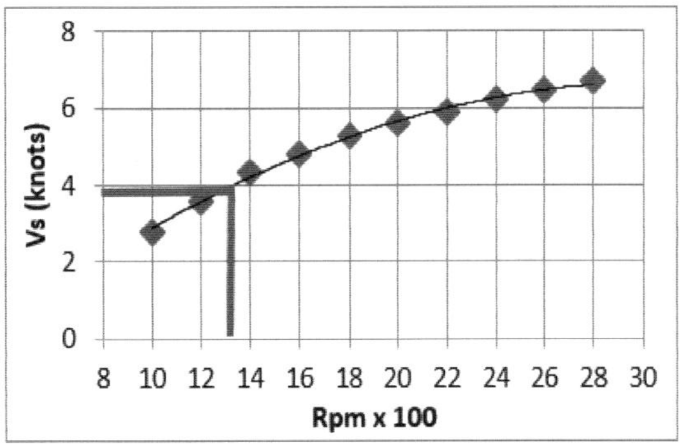

Figure 7. *Rpm-Vs Chart*

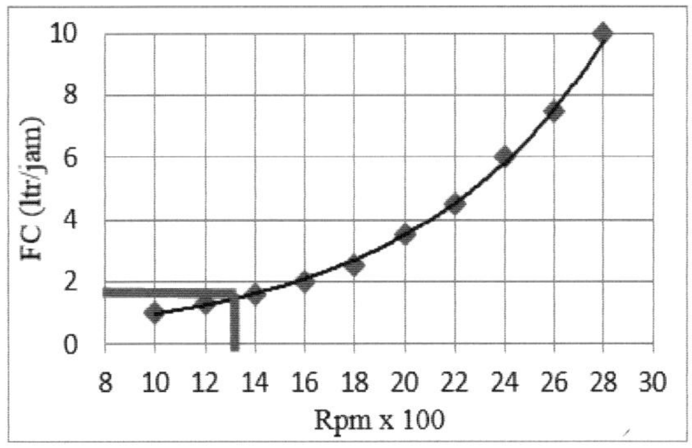

Figure 8. *Rpm-FC Chart*

4.2 Case Study 2 (CS2)
When ESV is operated the sea breeze blows (V_W) with the varying speeds of 0.0, 4.0, 8.0, 16.0, and 19.2 knots. By using *Hybrid Curve*, *Rpm-*V_S, and *Rpm-*F_C, estimate: thrust of sail (T_S), thrust of propeller (T_P), contribution of propeller to T_S thus the vessel can be operated at the service speed (*Vs*) of about 9.813 knots, and the fuel consumption (F_C). In the same way as in CS1, the following results obtained are: (see Figure 9 and 10)

Table 4. The hybrid system work at the service speed (*Vs*) 9.813 knots

Sail		*Propeller*		*FC/Engine (Liter/hour)*	*Prime mover work*
Vw (Knots)	*Ts (kN)*	*Tp (kN)*	*Engine Rpm*		
0.0	0	6.685	1700	4.5	Engine
4.0	0.5	6.185	1550	3.8	Engine-sail
8.0	1.2	5.485	1480	3.7	Engine-sail
16.0	5.4	1.285	1000	2	Engine-sail
19.2	6.685	0	0	0	Sail

Table 4 shows the Hybrid System (ESV) work, the propeller and sail can work individually/separately or together as a hybrid system when the ship moves.

Figure 9 shows the relationship between *wind speed (Vw)* and *fuel consumption (FC)*. Figure 10 also shows the efficiency/savings of fuel usage in ESV operations which alternately and fully utilize wind energy.

Service speed (Vs) on fishing boats is a major requirement because fishing vessels must arrive at the fishing ground as quickly as possible so as not to lose the right time to catch fish. This fishing vessel sailed with operational speed of around 9.8 knots. At this speed the ship will experience a drag force

(*RT*) of 6,423 kN with the need for thrust (*T*) of 6.685 kN. To meet the needs of the thrust force is supplied from propeller propulsion (*Tp*) with a 2x30 kW engine and sail (*Ts*) with an area (*As*) 125 m^2.

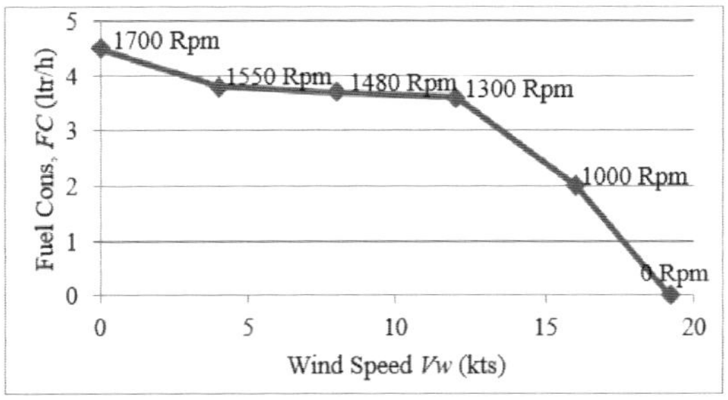

Figure 9. Result of CS2 use *Vw - FC Chart*

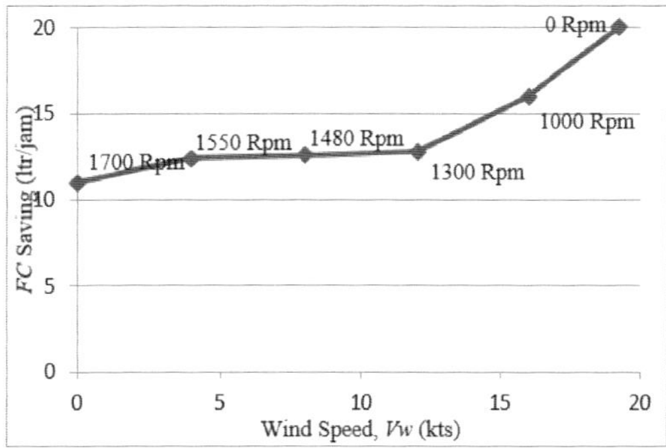

Figure 10. Result of CS2 use *Vw-FC Saving Chart*

The optional of service speed is adjusted to the voyage profile of the fishing vessel itself. When the ship operates in the port area (*departure from port*) in an empty load state, the ship will only use a speed of about 2 to 3 knots. At this time the use of engines and sail is highly recommended. When the ship goes to the location of the fishing ground (*outward bound*) in an empty load state and after the ship is on the high seas it will use full service speed of 9.813 knots, because the fishing vessel must arrive at fishing ground according to the planned time (on time) by using the engine and sail simultaneously. When the ship arrives at the location of the fishing ground in an empty load state and then doing the fishing operation on the fishing ground will tend to use a speed of about 4 to 7 knots because the fishing equipment set up requires a rather fast time and if it is not quickly worried the fish will run all. At this time the use of machines and sail simultaneously / alternately is highly recommended depending on the situation. When the ship leaves the fishing ground location (*homeward bound*) to the port in full load condition with the catch, fulfilment of thrust force is generated from the use of the engine and sail simultaneously to reach speed service 9.813 knots. And when the ship arrives at the port (*arrival at Port*), the ship only uses its speed of about 2 to 3 knots by using the engine and sail alternately. The use of wind energy to drive ESV with service speed of Vs 9.813 knots requires a *Vw* wind speed of 19.2 knots to produce a thrust force of T of 6.685 kN.

5. Conclusion

Application of hybrid technology is very useful when applied to catamaran fishing vessels. The development of hybrid vessel gives a promising expectation in order to reduce the use of fossil fuels. It has been found that the use of sail in combination with the operation of engine (ESV) was to be very useful. Economic review indicates the potency of cost saving and could be save about 70% fuels consumption. The system can reduce the emission of greenhouses gases. There is shown a good promise that can lead to significant savings in fuel consumption and hence reduce emission of greenhouse gases. The present work apparently portraits of study into the development of more energy efficient and less polluted fishing vessel.

6. References

[1] Pramudya IS, IKAP Utama, Wasis DA, 2017, Kapal Ikan Katamaran : Tinjauan Aspek Kebutuhan Energi, Fungsi dan Performa Kapal yang Ramah Lingkungan, Disertasi –MT093350.
[2] Korol I. and Latorre, R., 2010, Development of Eco-Friendly Fishing Vessel An ecological vehicle powered by renewable energy, EVER Monaco.
[3] Utama IKAP., Setyawan, D, Jamaluddin, A, Sugiarso, A., 2010, Development of Catamaran Fishing Vessel, IPTEK, Journal for Technology and Science, Vol.21, No.4.

[4] Molland AF., Turnock SR., Hudson DA., 2011, Ship Resistance and Propulsion – Practical Estimation of Ship Propulsive Power, ISBN:978-0-521-76052-2 Hardback, CUP, USA.
[5] Marchaj, 2003, C A, *Sail Performance, Techniques to Maximize Sail Power*, Revised Edition, ACl, London.
[6] Larson L., 2007, Principles of Yacht design, 3rd ed.,ISBN-13: 9780071487696, Mc Graw Hill
[7] Kinney, F.S., 1977, Skene's Element of Yacth Design, ISBN: 071361823x, ACB Ltd, London.
[8] P.I. Santosa, I.K.A.P Utama and W.D Aryawan, 2016, *Power Estimation of HCFV Based on Voyage Profile*, International Journal of Science and Research (IJSR), ISSN (Online): 2319-7064.
[9] Hind, JA., 1982, Stability and Trim of Fishing Vessels - second edition, Fishing News Book Ltd, Farnham, Surrey, England.

IOP Conf. Series: Materials Science and Engineering **462** (2019) 012009 doi:10.1088/1757-899X/462/1/012009

Identification of Flip Folder Model on Folding Machine

W S Pambudi[1], E A Zuliari[1], R A Firmansyah[1], Y A Prabowo[1], A Syaifurrizal[1]

[1]Electrical Engineering Department, Institut Teknologi Adhi Tama Surabaya, Indonesia

wahyusp@itats.ac.id

Abstract. Folding Machine is an indispensable tool in the small-scale laundry industry that aims to save time and energy. The main component of this equipment is a flip folder that will be controlled in the form of deflection of the angle position in order to operate the folded clothes. In the design of the controller, what needs to be done is the identification of the plant to determine the characteristics of the system. Modelling section in this paper is using relationships between input variables; input components are often difficult to measure the parameters of the plant. Therefore, this research will identify the system based on the offline input and output data measurement by using Extended Least Square (ELS) and Auto Regressive Exogenous (ARX) model approach. Based on the results of the identification, the second order damped plant equation can be approximated by first order reduction model with the characteristic transient response that having a time constant (τ) of 0.1706 seconds, the rise time of 0.500 seconds, the settling time of 0.511 seconds and the delay time of 0.1178 seconds.

1. Introduction

Several processes of cleaning clothes that can be run by the machine are the process of washing and drying, while the process for folding clothes still using the manual way. It is necessary to apply an automated technology to be able to do this folding process. One of the automation technologies that can be applied to the process of folding is by using the automatic folding machine. This tool uses the angle position deflection control mechanism of the flip. The working principle of the fold on the folding machine resembles a robot arm manipulator 1 DOF (Degree of Freedom). Several flip pieces will be arranged in such a way that can make the process of folding clothes [1–2].

A prototype folding machine has been designed using a simple mechanism with a semi-automatic system for clothing folding. This design uses three flips that are moved with DC motors on each flip. As the controlling unit to operate automatically use, the microcomputer (EV3 Brick) is connected with rotary encoder sensor; mechanical folding machine design is using LEGO mind storm EV3 robot module [2]. The use of this robot module aims to simplify the design because all the devices are easy to arrange. In order to the angular position of the flip to conform to the target, the used refractive control is proportional-integral-derivative (PID). PID is used to keep the deflection from the preserved flip angle according to its reference value [3–4]. The design of the PID controller is based on the model equation of the DC motor that contained in the LEGO mind storm EV3 robot module [3]. The given PID controller does not provide sufficient results on the performance of this folding machine prototype.

In each controller design, the modelling form of the plant is the most fundamental part to know the characteristic properties of the plant [5]. By knowing the characteristic properties of the plant, it can design the right controller and produce good performance [6]. In the folding machine, the plant that becomes the main purpose of control is the folding folder that is driven by the DC motor EMG30.

Modelling to obtain the plant model is using the relationship between the inputs, input components are difficult to obtain the value of some parameters. Under that matter, in this research will be applied a method of flip folder model identification model with EMG30 DC motor through output measurement offline data. Then do the testing and analysis on approach model to get the appropriate parameter.

2. Folding Machine

Folding machine is a tool that used in the process of folding clothes. This tool consists of 5 fold folders to be arranged in such a way that will form a fold of clothing. Each flip folder has a different area and angle of deflection. The arrangement of the flip can be seen in Figure 1, as well as in Figure 2. Shows the flip position when the deflection angle is 30°. In each fold folder as the actuator component of the drive is a DC motor that is controlled by a microcontroller component. The rotary encoder sensor is used to measure the deflection angle that formed by the fold of the folder. The working mechanism of the folding folder will move each other alternately when the clothes are finished in the iron, and they are on the worktable; then the flip will start moving from start flip B. C, A, D and E.

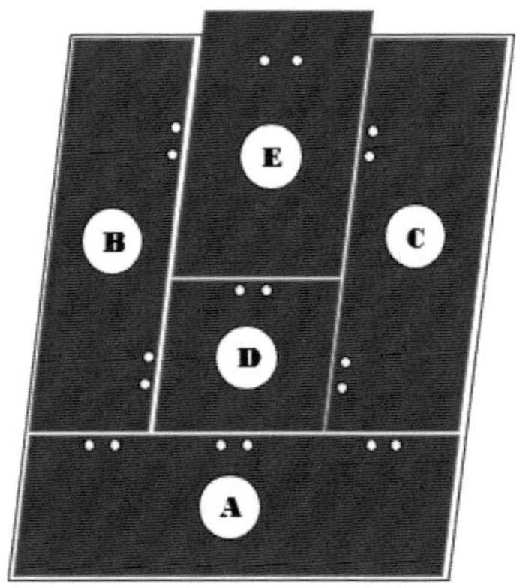

Figure 1. Compilation of flip folders folding machine.

Figure 2. Folding machine flip folder position.

3. Model Identification

3.1. Extended Least Square

The method of identification of plant parameters is done by going through the process of measuring input data and output in an offline condition. The input signal u (k) is given dynamically by giving a random signal to the system; the output y (k) of the system response will be recorded along with the input signal. Plant modelling is used the model form of the stochastic system with ARX model approach (Auto-Regressive Exogenous) [7]. The desired plant model has a 2nd order level, so the ARX plant model equation is as follows:

$$\frac{y(k)}{u(k)} = \frac{b_0 q^{-1} + b_1 q^{-2}}{1 + a_1 q^{-1} + a_2 q^{-2}} \quad (1)$$

In different equations, it can be represented as

$$y(k) = -a_1 u(k-1) - a_2 u(k-2) + b_0 u(k-1) + b_1 u(k-2) \quad (2)$$

moreover, can be simplified to be

$$y(k) = \varphi^T(k-1)\theta \quad (3)$$

with

$$\varphi^T(k-1) = \begin{bmatrix} -y(k-1) \\ -y(k-1) \\ u(k-1) \\ u(k-2) \end{bmatrix} \theta = \begin{bmatrix} a_1 \\ a_2 \\ b_0 \\ b_1 \end{bmatrix} \quad (4)$$

where $\varphi^T(k-1)$ is the input and output measurement data and θ is the plant parameter.

The identification process is needed to obtain the parameter value of the plant (θ). By using the Extended Least Square method, this method is a recursive method through a quadratic gradient error formulation with a criteria function by involving all errors [7]. The parameter values of the plant will continue to be calculated until the error meets the minimum criteria as shown in equation (5) and (6).

$$J_{min} = \frac{1}{2} e^2 \quad (5)$$

$$\hat{\theta}(k) = \theta(k-1) + F(k)\{y(k) - \varphi^T \theta(k-1)\}\varphi(k-1) \quad (6)$$

$$F(k+1) = F(k) - \frac{F(k)\varphi(k-1)\varphi^T(k-1)F(k)}{1 + \varphi^T(k-1)\varphi(k-1)} \quad (7)$$

3.2. Root Mean Square Error

The model that will be obtained through the process of identification of plant parameters by using the method of Extended Least Square will be very diverse by the input signal given to the system. To find out the equation of model that close to the actual condition with the real plant, then this research is used RMSE (Root Mean Square Error). RMSE is the average root value of the square error which shows how big the error deviation value from the zero value. It can be written mathematically as equation (8).

$$\text{RMSE} = \sqrt{\frac{\sum_i^n (y_i - \hat{y}_i)^2}{n}} \quad (8)$$

where y_i is the identification data on the i-iteration, $\hat{y}_i$ is the approximate model data in the i-iteration and n is the sum of all data.

4. Result and Discussion

In model identification, the assignment is given a random input signal in the form of flip deflection angle position with the deviation of 0 to 150°. This condition is due to a mechanical design that limits the deflection limit of the flip. The identification process is done through three experimental variations to get mixed results, so it can be analysed to get the closest model. Furthermore, the identification of Extended Least Square is represented by the equation of transfer function model from each experiment. As a validation method, the RMSE is used to know how close the model with the real plant. The result of the identification process is obtained by the plant model that is shown in Table 1.

Table 1. Model identification result

Trial Number	Model	RMSE
1	$\frac{16.41\,s + 4.167x10^{-4}}{s^2 + 1324\,s + 4.191x10^{-4}}$	5,48
2	$\frac{18.28\,s + 4.605x10^{-4}}{s^2 + 1286\,s + 4.622x10^{-4}}$	5,28
3	$\frac{13.95\,s + 3.475x10^{-4}}{s^2 + 1224\,s + 3.481x10^{-4}}$	6,13

Table 1 shows the second experiment has the smallest RMSE value of 5.28. Therefore from the identification of flip folding machine plant model, the closest model is located in the second experiment. To know the response characteristics of the plant model that has been obtained, then it was tested by a step input signal for the system that composed by the open loop and closed loop and giving a reference of 1 rad angle. The result of the open loop and closed loop response is shown in Figure 3 and Figure 4.

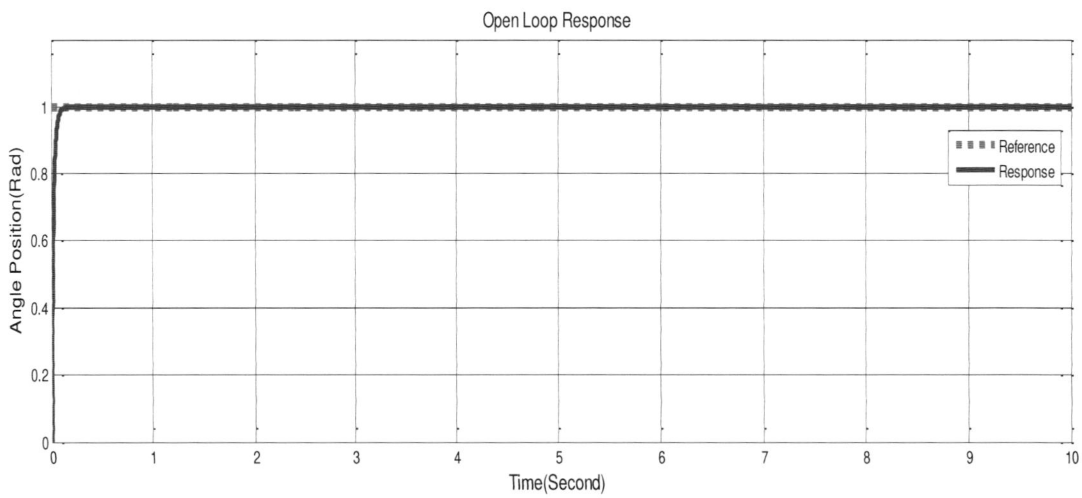

Figure 3. Open loop response.

Figure 3 shows the obtained open-loop response has the same characteristic properties with the 1st order system. It means the flip folder plant has a different real root value, so the system is second order and it is overdamped with the damping coefficient $\xi > 1$. This system can be approximated by the reduction model of the 1st order system. The characteristic of the transient response of this plant has a time constant (τ) = 0.1706 seconds, the rise time of 0.500 seconds, the settling time of 0.511 seconds and the delay time of 0.1178 seconds.

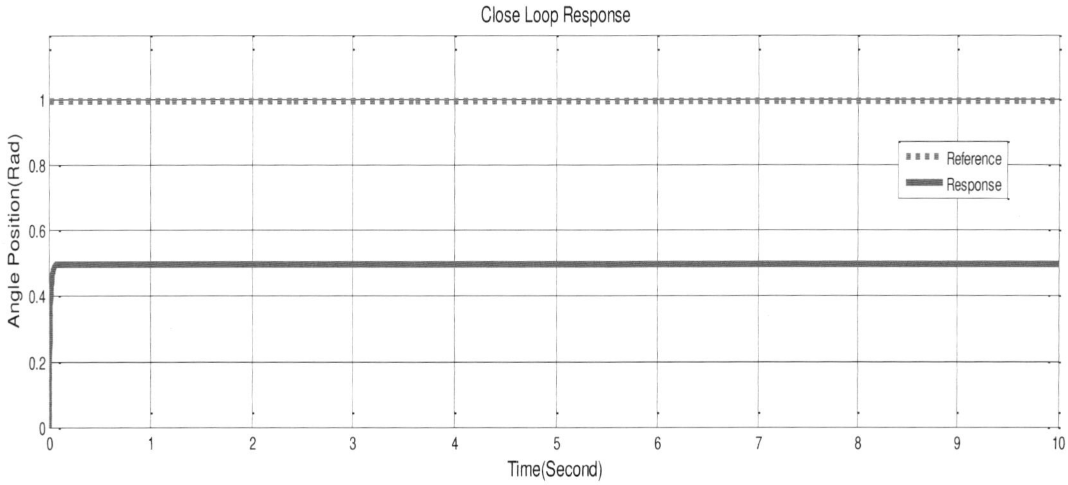

Figure 4. Closed loop response.

The analysis of the ability of the system in processing the error signal is prepared in a closed loop. Figure 4 shows the response system cannot reach the steady-state condition by the given reference with the steady-state error that occurs is equal to 50%. To overcome these conditions, the system is tested by providing a controller in order to reduce the value of the error. The Proportional Integral Derivative (PID) controller is the easiest method to be applied and most widely used on a conventional controller. To find out the results, then the PID controller is given to the system by using the trial and error method. The experiment is using the value of Kp = 1.5, Ki = 50 and Kd = 0.001. The PID results are shown in Figures 5 and 6.

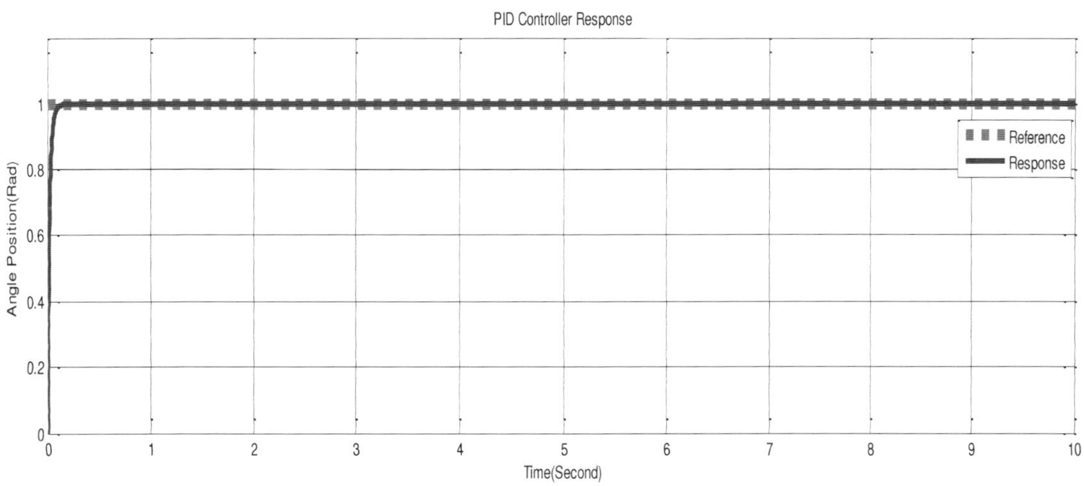

Figure 5. PID Controller Response.

The response results in Figures 5 and 6 shows the PID controller can keep the reference value as desired. Steady-state conditions can be obtained in 0.2763 with the transient response characteristics that having 0% steady state error, the time constant (τ) = 0.1742 seconds, rise time of 0.512 seconds, settling time of 0.522 seconds and delay time of 0.12 seconds.

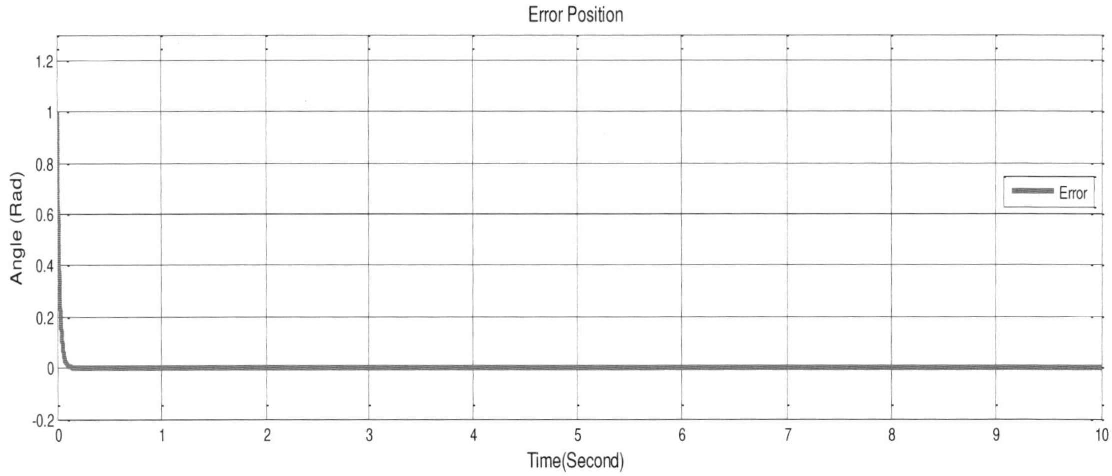

Figure 6. Error position response.

5. Conclusion

Based on the previous discussion it can be concluded that the process of identification of the plant by using Extended Least Square has produced the 2nd order equation. The properties of plant characteristics that have been obtained have some different real roots, so the plant is overdamped, and the 1st order reduction model can approximate it. The results of the PID controller show the controller performance which can reduce the position error steady state of 0%.

6. Reference

[1] Wahyu S Pambudi, Jan P B A S Pelawi. Simulasi Folding Machine Dengan PID, P, PI, PD and Fuzzy–PD (Proportional Differential), *Jurnal Sains dan Teknologi, Poiteknik Negeri Tanah Laut Kalimantan Selatan*. 2015; 1(1): **25-33**.

[2] Wahyu S Pambudi, and Titiek Suheta. "Implementation of Fuzzy-PD for Folding Machine Prototype Using LEGO EV3." *TELKOMNIKA*, 2018; 16(4): **1625-1632**.

[3] Wahyu S Pambudi, Titiek Suheta, dan Jamal Muhammad. "Aplikasi Sistem Kontrol Sudut Dengan Metode PID pada Arm Flip Folding Machine Menggunakan Lego Mindstorm EV3." *Seminar Nasional Sains dan Teknologi Terapan V (SNTEKPAN V) Institut Teknologi Adhi Tama Surabaya*, 2017: **B105-B110**.

[4] Khaled Sailan, Klaus-Dieter Kuhnert. "DC Motor Angular Position Control using PID Controller for the purpose of controlling the Hydraulic Pump". *International Conference on Control, Engineering & Information Technology (CEIT'13)*. Sousse. 2013; 1: **22-26**.

[5] Tang, Wei-Jie, Liu Liu, and Qian Wang. "DC Motor Speed Control Based on System Identification and PID Auto Tuning",*36th Chinese Control Conference. Dalian: IEEE*, 2017. **6420**.

[6] Liceaga-Castro, Jesus U., Irma I. Siller-Alcalá, Jorge Jaimes-Ponce, and Roberto Alcántara Ramírez. "Series DC Motor Modeling and Identification". *International Conference on Control, Artificial Intelligence, Robotics & Optimization. Mexico: IEEE*, 2017. **248**.

[7] Yuliyanto Agung Prabowo. "Perancangan dan Implementasi Kontroler PID Adaptive Untuk Trajectory Planning Gerakan Longitudinal Pada Sistem Navigasi UAV(Unmanned Aerial Vehicle)". Surabaya: Institut Teknologi Sepuluh Nopember, 2012.

IOP Conf. Series: Materials Science and Engineering **462** (2019) 012010 doi:10.1088/1757-899X/462/1/012010

Mooring Experimental Study of Motion Response for Pendulum Wave Energy Converters

I S Arief[1], I K A P Utama[2], R Hantoro[3], J Prananda[1], T R Arvisa[1], R F Kusuma[1]

[1]Dept. of Marine Engineering, Institut Teknologi Sepuluh Nopember, Indonesia
[2]Dept. of Naval Architecture, Institut Teknologi Sepuluh Nopember, Indonesia
[3]Dept. of Engineering Physics, Institut Teknologi Sepuluh Nopember, Indonesia

irfansya@its.ac.id

Abstract. Wave Energy Converters (WECs) is the type of Ocean energy need continual motion while operation for producing steady electricity. The mooring configuration as the answer for making sustainable WECs during operational life on the rough sea. The authors are researching trimaran pontoon of pendulum WECs type with anchors and four cables of mooring. The purpose of this research to determine the heave, pitch, and roll response from the pontoon. In this research, there are variations of wave height and wave periods. Experiment located at hydraulics laboratory in regular wave condition. The scale-up of the pontoon is (1:10) compared to real dimension. IMU (Inertial Measurement Unit) sensor used in this experiment to detect pontoon motion. Pontoon has the optimum condition at wave periods at 0.6s and height 0.4 m. Heave motion best displacement is 0.372m, Roll motion value is 1.425 degree and Pitch is 4.6164 degree. The pontoon needs fairlead for placed the mooring cables.

1. Introduction

People use energy as a primary necessity in everyday life. As the growth from economic life, the energy needs will be increased. Each year the demand for electrical energy escalates up significantly. According to BP (British Petroleum) data from 1995 to 2015, energy demand increased annually by 2.2% [1]. Indonesia's energy needs by 2020 are projected to exceed 250 TWh per year [2]. Every year, Indonesia has to build a new power plant to fulfil the electricity demand.

Power plants in Indonesia are still dominant sourced from fossil fuels (coal, gas, oil). According to statistical data in 2015 by the Directorate General of Electricity, more than 80% of power plants in Indonesia supply from fossil fuel [3]. In order to reduce the dependence of power plants on fossil fuels, The Ministry of Energy and Mineral Resources (ESDM) has a target 23% the total electricity supply will be produced from Renewable Energy by 2025 [4].

Ocean energy is the type of renewable energy which captured the energy from the ocean. Ocean energy classified into several types, tidal range energy, tidal stream energy, wave energy, and ocean thermal energy conversion (OTEC) [5]. INOCEAN (Indonesian Ocean Energy Association) in 2011 has ratified Indonesia's ocean energy potential. The wave energy has the practical potential of 1.2 GW, tidal energy of 4.8 GW [6]. Indonesia’s shoreline is number 4 longest in the world after Russia (95,181 km) [7]. It shows that Indonesia has huge potential for capturing wave energy.

Pendulum system is type Wave Energy Converters (WECs), its motion based on the movement of ocean waves, the waves will make rotational movement on the pendulum. The pendulum will be coupled with the generator and convert to electricity [8]. The factor of the dimensions and the total mass of pontoon will affect the total power will be generated by generator [9]. One of the most important aspect to convert electricity successfully in WECs is Foundation and Mooring, its secured The WECs device to the bed of the ocean [10]. Researchers had done several previous studies of motion response on the pontoon. However, anchoring and mooring system still needs to be proved for making the good device in extreme wave condition. Catenary mooring system study was presented in [11], Paredes had been tested using three cables. The experiment results had good performance in the rough sea waves test at the laboratory. The motion of heave, pitch and surge also have maximum amplitude value.

The purpose of this paper is to investigate the motion response of trimaran pontoon in regular wave conditions. When we found the motion, it will predict the pendulum oscillation. The authors did a variation test in wave period and wave height, we added mooring cable in portside and starboard of the pontoon. For finding the motion, we used the IMU (Inertial Measurement Unit) Sensor to detect heave, pitch, and roll motion. Strouhal equation had been used to show the scale-up of the pontoon. The pontoon we used scaled in (1:10).

2. Theory and Method

2.1 Experimental Case Study

The experiment located in the Hydraulics Laboratory of Institut Teknologi Sepuluh Nopember. Regular wave condition was conducted, two variations of wave height (H) at 0.2 m and 0.4 m. The three variations of the period established with 0.4 s, 0.6 s, 0.8 s. As we look the Laboratory in Figure 1, The laboratory could also generate the irregular waves.

Figure 1. Hydraulics Laboratory of Institut Teknologi Sepuluh Nopember.

The trimaran pontoon scaled (1:10) to the real dimension, the length of the pontoon is 49.5 mm; the width is 32 cm. The total mass of pontoon is 2.248 kg, the centre of gravity located at 2.2617 cm below the centre of the large cylinder. The moment inertia of the trimaran pontoon is (0.0109 0.0319 0.0308) Nm. The shape of the trimaran pontoon similar to *jukung* (Indonesian fisherman boat) shown in Figure 2. The real scale dimension of trimaran pontoon will be used in shallow water near the coastline.

(a)

(b)

Figure 2. Trimaran Pontoon (a) and Jukung Fisherman Boat (b).

In Figure 3 below, the pontoon arrangement. There are four anchors; each anchor has 1 kg of mass. The anchors made from rock stone, the properties from anchor will be ignored. All cables mooring located 135° apart on portside and starboard. Wave direction (μ) come 0° from the pontoon, known as *following waves* [12]. Look for Figure 4, Inertial Measurement Unit (IMU) sensor we used is MPU-9250. It has three features gyroscope, accelerometer, and magnetometer. It can capture 20 data for each second [13]. The authors average the data to make conclude every second. The sensor located over the pontoon, we used only gyroscope and accelerometer for finding heave, pitch, and roll motion.

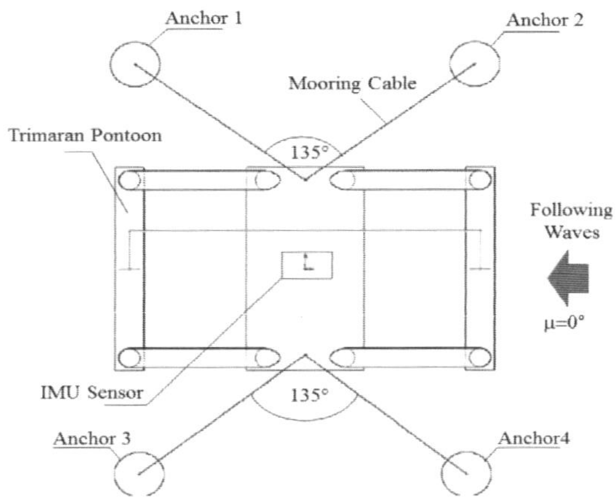

Figure 3. Pontoon arrangement from the top view.

Figure 4. IMU Sensor MPU-9250.

2.2 Scaled Up Equation

Strouhal equation used to finding real period and wave height in real sea condition. However, the equation derived from the Froude number as a comparison between Velocity (V) and Length (L) of the pontoon in real dimension and scaled. Froude Number Thus,

$$\frac{V_m}{\sqrt{gL_m}} = \frac{V_p}{\sqrt{gL_p}} \tag{1}$$

L_m is the length of a real model dimension, and L_p is Length of prototype scaled. Through gravitational constant its similar, so the velocity ratio:

$$\frac{V_m}{V_p} = \sqrt{\frac{L_m}{L_p}} = \sqrt{n} \tag{2}$$

Results from equation (2) provide the scale factor of velocity be equivalent to the square root of the length scale factor known as *n*. As McCormick mention, *strouhal number* need *fL/V*. Where *f* is wave frequency from offshore floating structure [14]. From the equation (1) and (2) on the design model we need,

$$\frac{f_m L_m}{V_m} = \frac{f_p L_p}{V_p} \tag{3}$$

As, $f = 1 / T$, where T is wave period

$$\frac{L_m}{T_m V_m} = \frac{L_p}{T_p V_p} \tag{4}$$

Equation (4) can be used to finding a correlation of time-scaling,

$$\frac{T_m}{T_p} = \frac{L_m}{L_p}\left(\frac{V_p}{V_m}\right) = n^{3/2} \tag{5}$$

3. Results and Discussion

The experiment tests six times; Table 1 shows the description of the test. Its conducted the regular waves condition in Laboratory with two variations of wave height. The experiment does not move the pontoon because its only tested in one heading wave (μ) only. The waves come from after the pontoon for making a motion to the pontoon. When experimenting, the mooring cable attached occasionally shifting. It needs a place like a fairlead on the vessel to make the cable attached tightly. IMU Sensor while retrieving data use wireless tool to connect a laptop. We used a Java program called telemetry viewer v0.4 as shown in Figure 5; IMU sensor will process in real time whenever change motion conducted. Figure 6 until 8 provides the experiment results.

Table 1. Variation test during experiment.

Test	Height (m)	Wave Period (s)
1	0.2	0.4
2	0.2	0.6
3	0.2	0.8
4	0.4	0.4
5	0.4	0.6
6	0.4	0.8

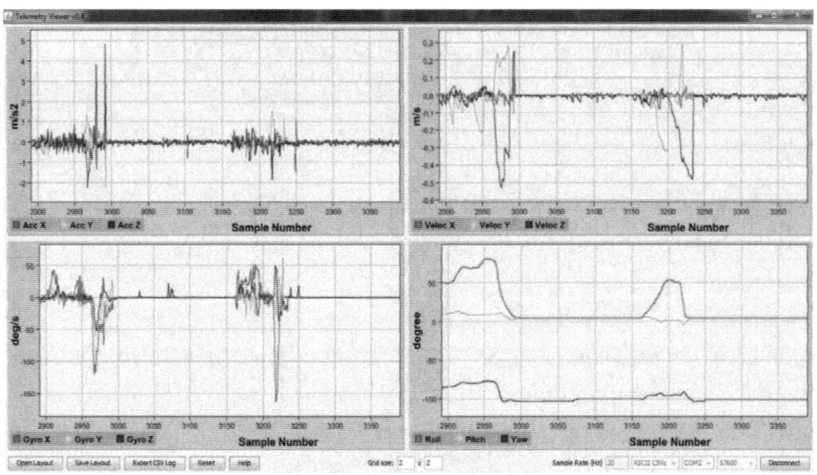

Figure 5. Telemetry viewer.

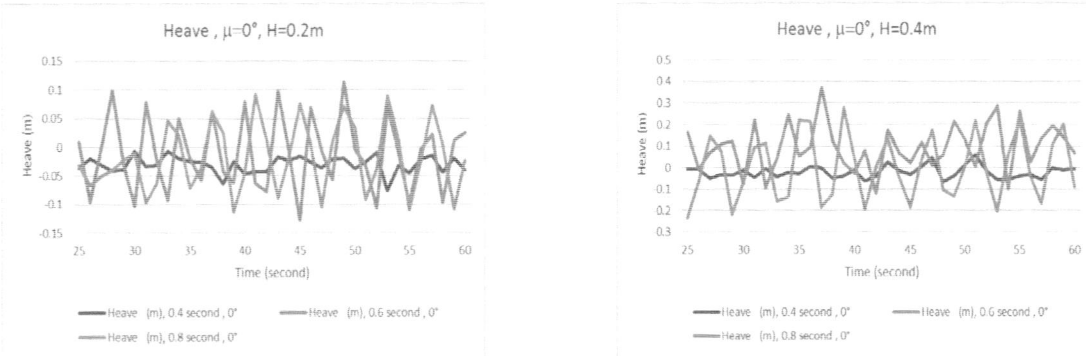

Figure 6. Time history of Heave Motion Response: H=0.2 m (left), H=0.4m (right)

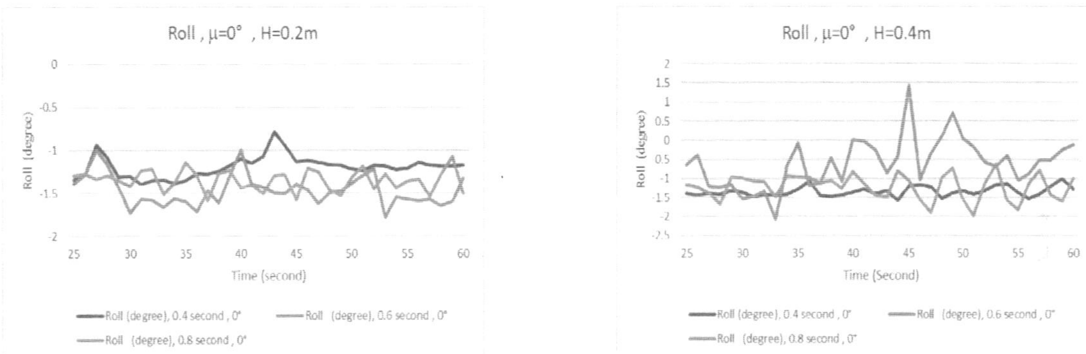

Figure 7. Time history of Roll Motion Response: H=0.2 m (left), H=0.4m (right)

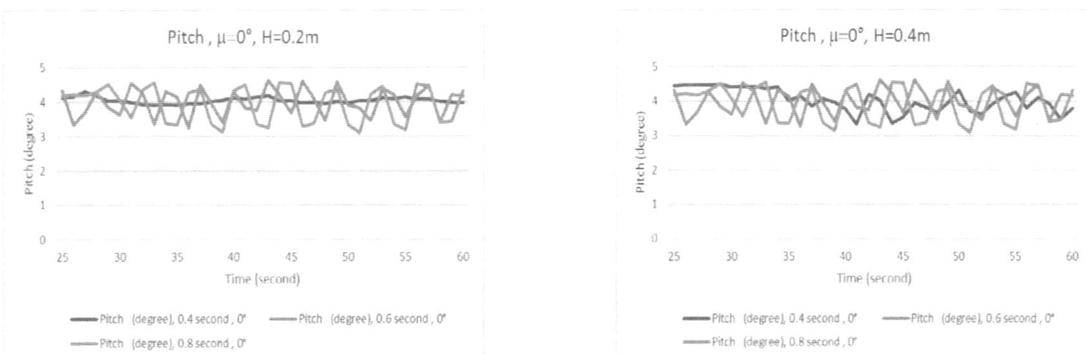

Figure 8. Time history of Pitch Motion Response: H=0.2 m (left), H=0.4m (right)

For all motion have a maximum value at 0.6 seconds period, except in roll motion. The roll motion at 0.2 meters has the highest value at 0.4 seconds wave period. Heave motion displacement affected depends on wave height and wave period change value. As we look on Figure 6 when H is 0.4 m, the highest displacement conducted at 0.4 m. When H is 0.2 m, the maximum heave displacement took only 0.1 m. Wave period at 0.6 s is the most optimum value for heave displacement. Significant value occurred at 37 seconds, the heave displacement up to 0.372 meters when the wave height is 0.4 meters. Based on Figure 7 roll motion when at H = 0.4 m have significant resonances from -2 to near 1.5 degrees. Big different occurred when compared at H = 0.2 m, the resonances significantly different. At 45 seconds, roll degree rose up to 1.425 degrees. It conducted because the attachment on the pontoon for cable mooring does not string tightly, so make the centre gravity of x-axis move slightly and affect the pontoon motion. Pitch motion graph investigation shown in Figure 8, the graphs show quite different value on pitch degree either wave height condition. The pontoon motion around 4 degrees on the y-axis. The biggest value of pitch degree is 4.6165 when at 43 seconds, the wave height is 0.4 m the period is 0.6 seconds.

Based on the calculation of scaled-up pontoon in section 2.2 the pontoon working in condition at 2–4 meters of ocean wave height. The equation compared between period test in the laboratory and comparison the length of the scaled pontoon to the real dimension. If the real rough sea condition occurred more than 4 meters, It could damage the trimaran pontoon.

4. Conclusion

The motion response results between heave, pitch and roll motion have maximum value while wave height is 0.4 m. Heave and roll motion of the pontoon has optimum when at wave period 0.6 s, where the pitch motion has the best period at 0.8 s. Highest value on heave displacement is 0.372 m, roll motion degree is 1.425 and Pitch motion is 4.6165 degrees. For further research, the pontoon needs to be tested in irregular wave's condition to validation in the real rough sea condition. The variation of wave heading also becomes an important factor for WECs while in operation. The fairlead on pontoon has to be designed well to avoid shifting the cable.

5. References

[1] BP 2017 *Statistical Review of World Energy* [Accessed 21.07.18]
[2] Mukhlis M and Permana A D 2003 *Projection of PLN Electricity Demands Year 2003 till 2020* BPPT p 28-29
[3] Indonesia Ministry of Energy and Mineral 2016 *Electricity Statistics for fiscal in 2016 no. 29* [Accessed 25.07.18]
[4] Government Regulation No. 79 of 2014 on National Energy Policy [Accessed 21.07.18]
[5] World Energy Council 2016 *world energy resources marine energy 2016* [Accessed 22.07.18]
[6] Mukhtasor 2012 *Ocean Energy in Indonesia* Ocean Energy Workshop Presentation by Indonesia Ministry of Energy and Mineral [Accessed 21.07.18]
[7] World Resources Institute *Coastal and Marine Ecosystem-Marine Jurisdictions: Coastline length* [Accessed 19.07.18]
[8] *Arif I S et al. 2017 proc. Int. Conf. on Energy Astechnova* vol 2 (Yogyakarta: E3S Web of Conferences) p 43
[9] Falnes J 2007 *A review of wave-energy extraction Norway. J. Marine Struc.*, **20**, p 185-201
[10] SI Ocean 2012 *Ocean Energy: State of the Art.* EU project Report p 79
[11] Paredes G M et al. Experimental investigation of mooring configurations for wave energy converters Portugal. Int. J. Marine Energy., 15, p 56-67
[12] Djatmiko E B 2012 *Response Operator of Offshore Structure on Irregular Waves* (Surabaya: ITS Press)
[13] InvenSense Inc 2014 *MPU-9250 Product Specification Revision* Doc. No. PS-MPU-9250A-011.0
[14] Mccormick M E. 1981 *Ocean Wave Energy Conversion* (New York: Wiley)

Acknowledgement

Authors' admirably thanks to Indonesia Ministry Research Technology and Higher Education through research grant for funding support in this research.

IOP Conf. Series: Materials Science and Engineering **462** (2019) 012011 doi:10.1088/1757-899X/462/1/012011

Static Photovoltaic Array Partially Shaded Condition With Boost Converter Using Perturb & Observe Algorithm

H A Sujono[1], **Ariadi**[1], **R Sulistyowati**[1] and **H Suryoatmojo**[2]

[1]Departement of Electrical Engineering, Institut Teknologi Adhi Tama Surabaya, Surabaya, Indonesia
[2]Departement of Electrical Engineering, Institut Teknologi Sepuluh Nopember, Surabaya, Indonesia
hari.agus17@itats.ac.id

Abstract. Photovoltaic is a semiconductor component that can produce electrical energy when exposed to energy from sunlight. Electricity generation using photovoltaics require a maximum power point tracker embedded in a microcontroller to control boost converter. Boost converter is a type of dc-dc converter which functions to increase dc voltage at a certain voltage level and is kept stable. A simulation for a power generation system using photovoltaics is needed to determine the level of system success that will be realized. In this article we discuss the system simulation using PSIM to find out the maximum output power on partially shaded static photovoltaics using Perturb and Observe algorithm at the maximum power point tracker. Testing on this simulation is done for several different resistance loads. Simulation results without using the Peturb and Observe algorithm show that the maximum output power of 12.26 watts with a load resistance of 36 Ohm, while the simulation results using this algorithm obtained a maximum output power of 55.55 watts. The comparison results for partially shaded 50% showed a maximum output power smaller than partially shaded 25%.

1. Introduction

Photovoltaic (PV) is a module with semiconductor material that can convert solar energy into electrical energy. In its implementation, PVs only produce low power and change periodically. This is caused by factors of light intensity and solar temperature and geographical location (longitude and lattitude). In addition, PV output power will be affected if the surface of this PV is covered by trees, buildings or passing clouds. When viewed from its characteristics, PV has nonlinear characteristics and changes to radiation and PV surface temperature [1]. So a control system is needed that can track the maximum power point of PV, the system is called the Maximum Power Point Tracker (MPPT). This is an electronic system that seeks maximum power by monitoring and controlling voltage and current. MPPT requires an algorithm to obtain and maintain a PV working point at maximum efficiency and produce the greatest output power called Maximum Power Point (MPP).

Many algorithms can be used in MPPT [2-3]. The Peturb and Observe (P&O) algorithm is one of the most common algorithms for MPPT used in research for MPP tracking techniques. This iterative-based method measures the output characteristics of a PV system and changes the work cycle of a DC - DC converter.

This paper shows the simulation results using PSIM to determine the maximum output power for partially shaded static PVs using the P&O algorithm. As a comparison, PV testing was carried out without and using the algorithm.

IOP Conf. Series: Materials Science and Engineering **462** (2019) 012011 doi:10.1088/1757-899X/462/1/012011

2. Literature Review

When PV gets input in the form of sunlight intensity and temperature can produce electric current. The amount of electric current produced by PV is directly proportional to the amount of intensity of sunlight entering the solar cell. The amount of sunlight intensity changes according to the shift in the position of the sun and weather. Weather factors such as cloudy weather and sunny weather will affect the intensity of the sun's light [4]. The mathematical models are developed to simulate the electrical characteristics of PV. Figure 1 shows the PV substitute circuit, where I and V are the current and voltage of the solar cell, then, I_L is the cell's photocurrent. R_{sh} and R_s is the shunt resistance and series resistance of solar cells [5]. The equation of the substitute circuit is shown in formula (1).

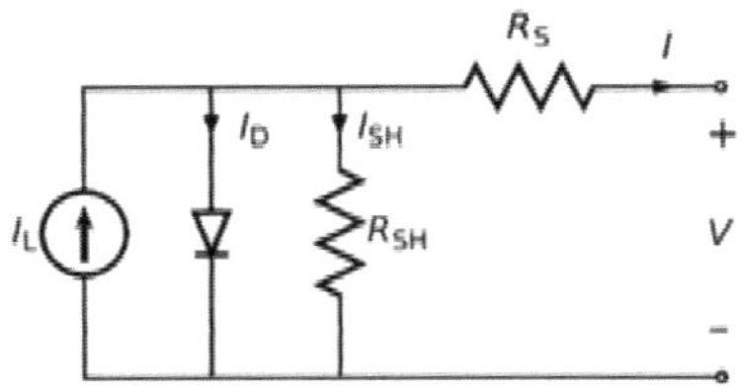

Figure 1. PV substitute circuit [5]

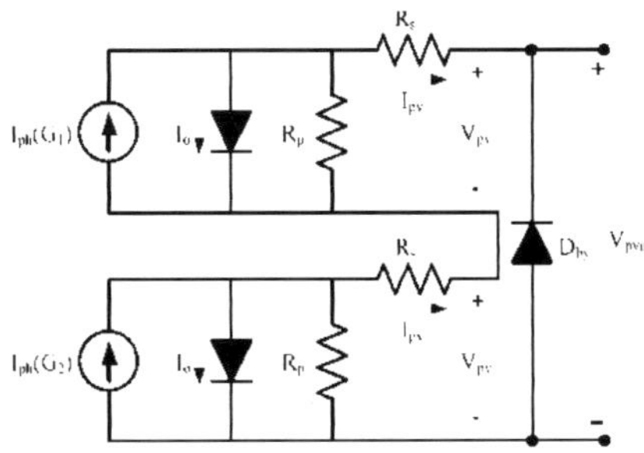

Figure 2. Partial shaded circuit modul [6]

$$I = I_L - I_D = I_L - I_0\left[e^{\frac{q(V+IR_S)}{nkT}} - 1\right] - \frac{V+R_S I}{R_{sh}} \quad (1)$$

PVs are often shaded by interference in the form of tree branches, through clouds, poles and buildings, etc., which produce partial shadows on PV systems. In PVs configured in series will make a constant moment in which the shaded cells will operate by reverse bias and generate an inverted polarity power leading to a decrease in power and thereby reducing the net power conversion efficiency [6]. A partially shaded module can be modeled two groups of PV connected in series in the module. Each group received a different level of radiation. It can be assumed that there are bypass diodes for the cells inside the module. Figure 2 shows the circuit model for the partially shaded module. This module consists of a connected cell series r in which s shaded cells receive G_1 radiation and - shaded cells (r - s) receive G_2 radiation. PV parameters can be represented in formula (2):

$$I_{ph1} = I_{ph}(G_1)\,,\ I_{ph2} = I_{ph}(G_2),\ N_{s1}\,, N_{s2} = (r-s)N_{s1} \quad (2)$$

where, subscripts 1 and 2 refer to the cells receiving radiation from G_1 and G_2 respectively.
In accordance with the existence of a single cut diode for a PV module not partially shaded, the output current and voltage of the PV is equal to following the following formula [6]:

$$I_{pvm} = Min\ (I_{pv1}\,, I_{pv2}) \quad (3)$$

$$V_{pvm} = \sum V_{pv} \quad (4)$$

Boost Converter is one type of DC - DC converter that works to increase the DC voltage [7-8]. Boost converter has an output voltage that is always greater than the input voltage. Boost converter generally consists of an inductor (L), a diode (D), a capacitor (C), and a switch. For the inductor and diode on the boost converter serves as a current source and voltage and to limit the ripple of the output current. Capacitors in the boost converter serve to limit the output voltage ripple. Figure 3 shows the boost converter circuit.

MPPT is an electronic system that controls the PV system so it can operate at maximum power. At the time of operation the PV will receive the intensity of the changing sunlight so that it affects the working point of the resulting current and others. Therefore the MPPT system allows the condition that the variable can be traced to its maximum power at a certain time and moment. Figure 4 shows the PV curve with different levels of light intensity.

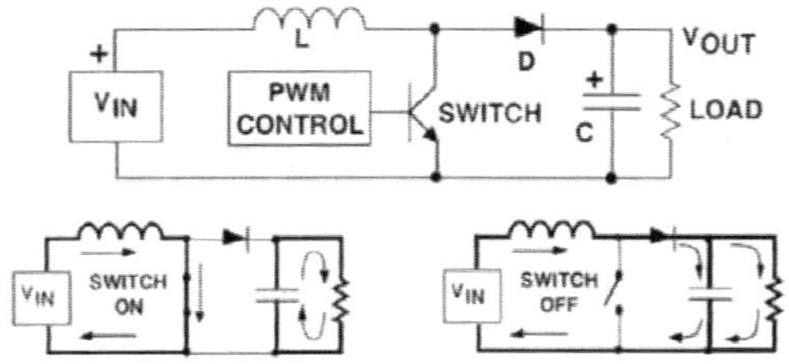

Figure 3. Boost converter circuit [9]

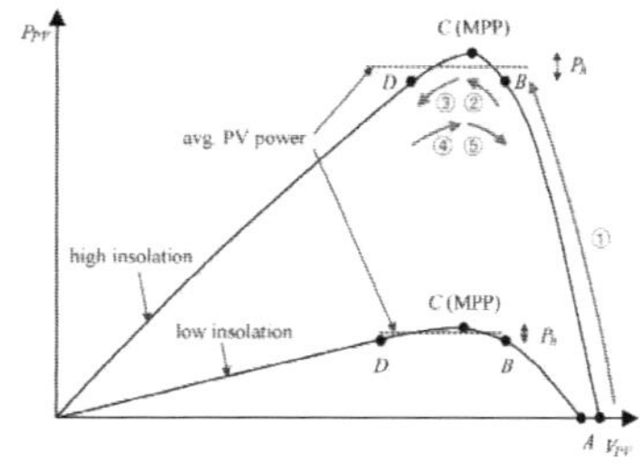

Figure 4. PV curve with different light intensity [9]

3. Method

3.1. *Partially Shaded Conditioning Planning*
In planning partially shaded conditions, PVs will be shaded by a quarter and a half using cardboard to compare the direction of power loss generated by PV with boost converter when partially shaded conditions.

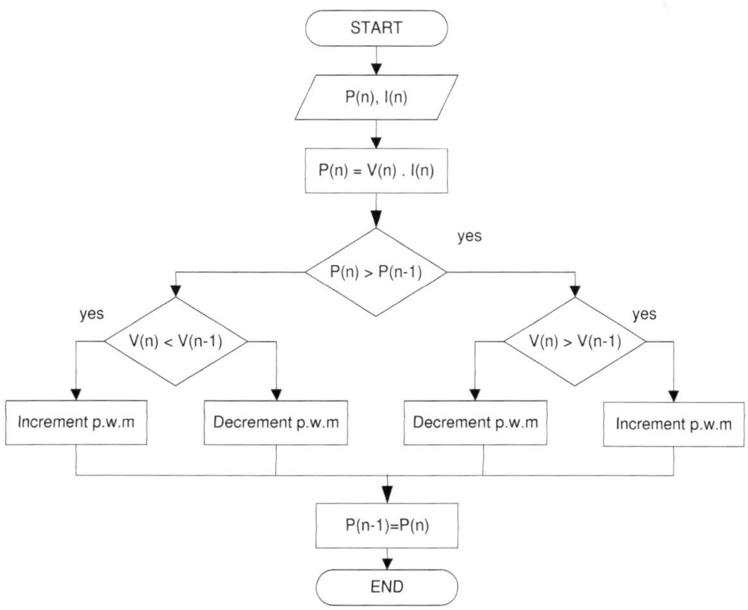

Figure 5. Flowchart of P&O algorithm

3.2. *Designing the Peturb and Observe Methode*
The design of the P&O algorithm in the MPPT system can be interpreted to look for the maximum power value generated by PV at a time. To find the power capability that can be provided by PV at time (t) is shown in Figure 5.

4. Results and Discussion

4.1. *Simulation without Perturb & Observe method*

The power output testing without the P&O algorithm is that power testing using PVs directly supplies to the load. In Figure 6 shows the test circuit without P&O algorithm in PSIM simulation. In this test using the PV parameters adjusted to the actual 100Wp PV parameters, using the light intensity variables are DC voltage source with value 600 and constant as temperature 25°C, and using variable resistor (rheostat) with the load resistance value of 36 Ω. The characteristic curves are shown in Figure 7.

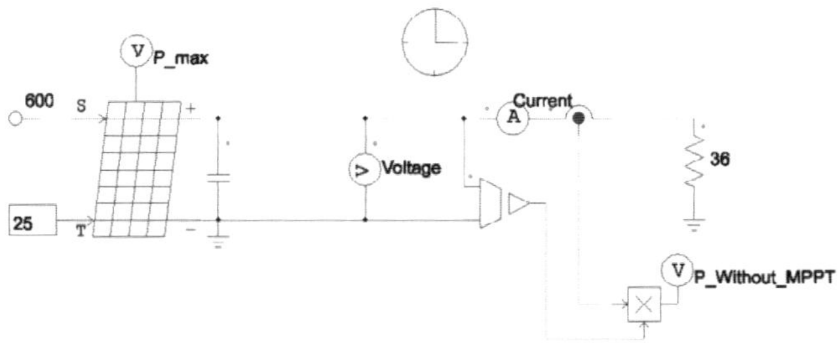

Figure 6. The simulation circuit without P&O algorithm

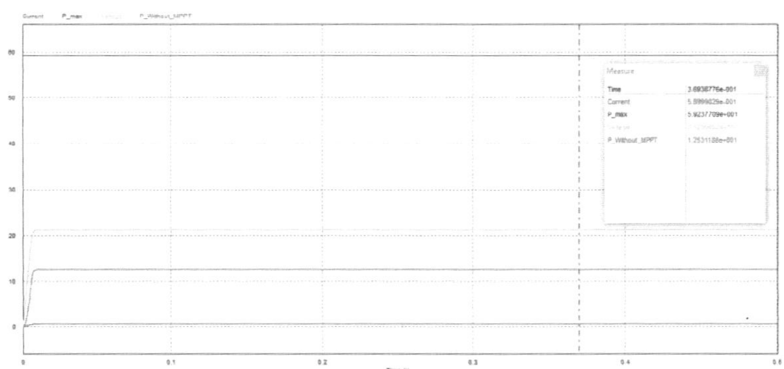

Figure 7. The simulation result without using P&O algorithm with load resistance 36 Ω

The power output value without P&O algorithm when compared to the maximum output power of PV look different. The maximum PV output power shown in Figure 7 is 59.23 watts, and the absorption power obtained is 12.26 watts at 36 Ω load resistance. For load resistance 28.8 Ω obtained the value of pmax 15.37 watt. While the load resistance of 11.3 Ω obtained a power of 36.63 watts.

4.2. *Simulation using P&O algorithm*

In this test using P&O algorithm, this method looks for the maximum power value by setting the modulation index value on the boost converter. This test is intended to prove that in theory P&O algorithm can indeed be used as peak power tracker on PV. Figure 8 is a simulation circuit using P&O method in PSIM simulation.

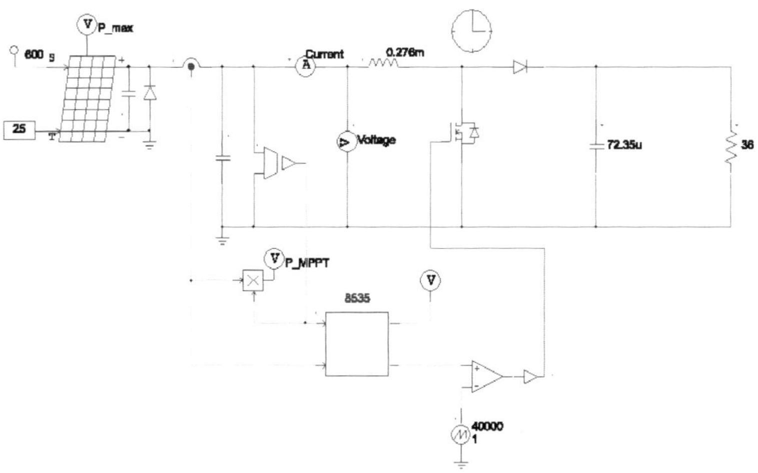

Figure 8. P&O algorithm simulation circuit

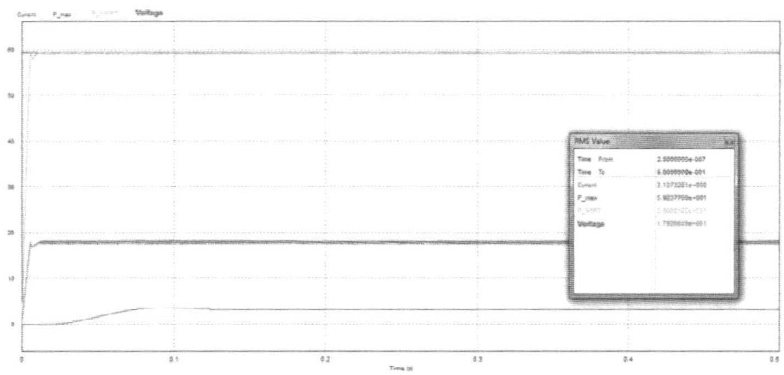

Figure 9. The simulation results using P&O algorithm with load resistance of 36 Ω

Figures 9 show the maximum PV output power of 59.23 watts and power optimized by the system at a 38 Ω weight resistance of 55.55 watts, at load resistance 28.8 Ω obtained power equal to 51.30 watt and at load resistance 11.3 Ω obtained the optimization of power of 24.92 watts.

4.3. *Simulation P&O algorithm condition Partially Shaded 50% and 25%*
In this simulation, a total of 12 PV units in the parallel series are shown in Figure 10. The DC source voltage of 6 is zero as a simulator that is partially shaded by 50% and the DC source voltage of 3 is zero as a simulator that is partially shaded by 25%.

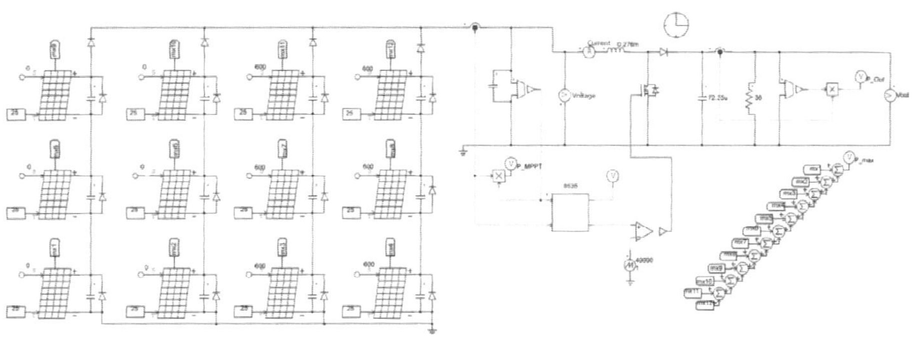

Figure 10. The P&O algorithm simulation circuit Partially Shaded condition 50%

IOP Conf. Series: Materials Science and Engineering **462** (2019) 012011 doi:10.1088/1757-899X/462/1/012011

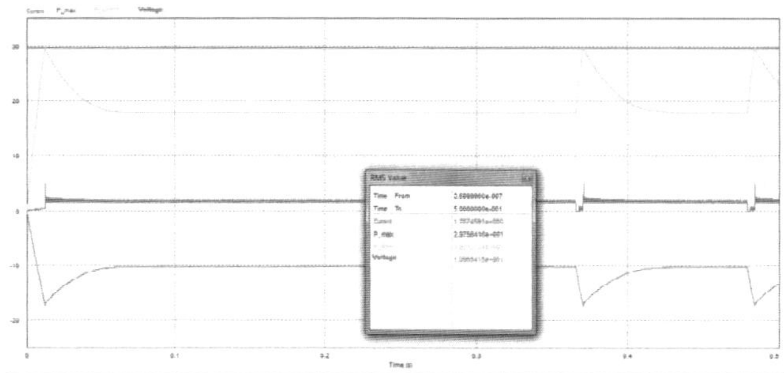

Figure 11. The simulation result using P&O algorithm of Partially Shaded 50% condition with load resistance 36 Ω

In Figure 11 the maximum PV power of 29.75 watts and the power optimized by the system at 38 beban load resistance is 19.33 watts. At a load resistance of 28.8 Ω obtained a power of 17.31 watts. While for load resistance of 11.3 Ω, the optimization of power is 10.38 watts.

Table 1. Resume Test Result of Simulation

Load (Ω)	Without P&O algorithm			With P&O algorithm		
	Voltage (V)	Current (I)	Power (P)	Voltage (V)	Current (I)	Power (P)
36	21.14	0.58	12.26	17.92	3.1	55.55
28.8	21.06	0.73	15.37	16.03	3.2	51.30
11.3	20.35	1.80	36.63	7.46	3.34	24.92

Load (Ω)	With P&O algorithm, Partially shaded 50%			With P&O algorithm, Partially shaded 25%		
	Voltage (V)	Current (I)	Power (P)	Voltage (V)	Current (I)	Power (P)
36	10.86	1.78	19.33	15.33	2.60	39.60
28.8	9.78	1.77	17.31	12.32	2.63	32.40
11.3	5.90	1.78	10.38	6.26	264	16.53

In Table 1, the power ratio measured by PSIM is shown without using the P&O algorithm until Partially Shaded. The value recorded in this simulation is the voltage and current in the input boost converter. For the sum of the power consumption values, the equation $P = V_{in}\ x\ I_{in}$ is used.

5. Conclusions

Testing has been done by simulating using PSIM between PV systems without and using the P&O algorithm. The results for the PV system using the P&O algorithm indicate a significant power optimization. While testing PV systems using partially shaded 50% P&O algorithm lower output power than 25% partially shaded. For load 36 Ω in PV system without P&O algorithm shows output power of 12.26 watts, while using P&O algorithm the output power is 55.55 watts. For load 36 Ω in PV systems using the P&O algorithm with partially shaded 50% output power of 19.33 watts, while the system is partially shaded 25% output power of 39.60 watts. These results are as expected so that partially shaded PV designs using P&O algorithm can be realized in a hardware framework for planning PV farms.

6. References

[1] S Amara, A Bouallegue and A Khedher, 2014, *Theoretical and Practical Study of a Photovoltaic MPPT Algorithm Applied to Voltage Battery Regulation*, International Journal of Renewable Energy Research Vol. 4, No 1.

[2] Bouselham L, Hajji B and Hajji H, 2015, *Comparative Study of Different MPPT Methods for Photovoltaic System*, Renewable and Sustainable Energy Conference (IRSEC) 3rd International (pp. 1-5). IEEE.

[3] Mohammadmehdi Seyedmahmoudian, Arash Mohamadi, Swarna Kumary, Aman Maung Than Oo, Alex Stojcevski, 2014, *A Comparative Study on Procedure and State of the Art of Conventional Maximum Power Point Tracking Techniques for Photovoltaic System*, International Journal of Computer and Electrical Engineering, Vol. 6, No. 5, pp. 402-414.

[4] Danie Novin Sugiharto, 2010, *Implementasi Algoritma Maximum Power Point Tracking Pada Panel Photovoltaic Menggunakan Metode Perturb and Observe*, Universitas Indonesia

[5] Rusminto Tjatur Widodo, Rugianto, Asmuniv, Purnomo Sejati, 2011, *Maximum Power Point Tracker sel Surya Menggunakan Algoritma Perturb and Observe* ITS Surabaya.

[6] Mohammadmehdi Seyedmahmoudian, Saad Mekhilef, Rasoul Rahmani, Rubiyah Yusof and Ehsan Taslimi Renani, 2013, *Analytical Modelling of Partially Shaded Photovoltaic Systems*, Journal of Energies.

[7] Daniel W Hart, 2009, *Power Electronics,* McGraw-Hill US.

[8] Articles, 2004, *Cascade DC-DC converter Connection of Photovoltaic Modules*, IEEE Transaction on Power Electronic.

[9] Jung Min Kwon, Bong Hwan Kwon and Kwang Hee Nam, 2008, *Three-Phase Photovoltaic System With Three-Level Boosting MPPT Control*, IEEE Transaction on Power Electronics.

IOP Conf. Series: Materials Science and Engineering **462** (2019) 012012 doi:10.1088/1757-899X/462/1/012012

Normalising of 316L Stainless Steel using Temperature and Holding Time Variations

V A Setyowati[1], Suheni[1], E W R Widodo[1], S A Hermanto[1]

[1]Mechanical Engineering Department, Institut Teknologi Adhi Tama Surabaya, Indonesia

vuri@itats.ac.id

Abstract. Materials for some component must be through machining processes, for example, is a lathe. Lathe causes internal stress. One of many ways to avoid the internal stress is normalising. Normalising of the heat treatment is heating materials above austenite temperature, holding for a period for transformation and air cooling. This research was conducted to analyse the effect of holding time and temperature normalising of 316L stainless steel toward hardness and microstructure. Hardness testing was measured by microhardness Vickers tester, and the characterisation was used optical microscope and XRD. The highest hardness was obtained 193.2 HVN from normalising with a temperature of 800°C and holding time of 30 minutes, while the lowest was 166.3 HV using 950°C and 90 minutes. The microstructure was obtained austenite and ferrite phase. The normalised specimen had a wider peak profile than the unnormalised specimen, can be observed at 50.9° of the 2θ angle which had bigger FWHM.

1. Introduction

The use of metal as the main material of production has an increasing value for industrial. Steel of raw materials was used as tool steel and automotive components. Kinds of steel can be applied as industrial components such as stainless steel. Stainless steel is widely used as engine components such as gears, tableware, shafts and other components [1]. Stainless steels certainly cannot be separated from the machining process, especially the lathe process. The lathe process is done by slicing the specimen by the chisel edge to form a product using a rotating process. In the process of the lathe, the chisel edge often damaged because abrasive resistance decreased. The shifts between chisel edge and workpiece cause the decreasing of abrasive resistance. As a result of the continuous shift, the temperature of the material is highly increased. The temperature differences on the material surface and the depth of the material causes residual stress. Residual stress effect to mechanical properties such as hardness properties [2].

One way to improve the mechanical properties is heat treatment such as normalising. Normalising is a process that is done by heating the steel until it reaches the austenite temperature. The temperature is held for a while and then cooled slowly by using air cooling medium. The purpose of normalising treatment is to increase uniformity and to eliminate residual stress. The heating temperature of normalising is about 50° C above the critical temperature Ac3 for hypereutectoid steels to obtain homogeneous Austenite [3]. Some factors during heat treatments contribute to the mechanical properties of the treated material. Carbide phases were produced after heat treatment. Some aspects of carbides including size, type, volume percentage, microstructures have important roles in mechanical properties [4]. Therefore, it should be controlled to avoid decreasing of

IOP Conf. Series: Materials Science and Engineering **462** (2019) 012012 doi:10.1088/1757-899X/462/1/012012

mechanical properties. This study aims to investigate the temperature and to hold time effect to hardness and microstructure for 316L SS using normalising treatment.

2. Materials and Method

2.1. Materials

The materials used in this research are a rounded bar of Stainless steel type 316L. The compositions of 316L stainless steel material can be seen in Table 1. Sample preparation was done by cutting stainless steel that had been in the lathe using a hacksaw with a length of 10 mm, and diameter 25 mm for nine specimens. The dimension for the sample without normalising is 10 mm long, and 10 mm diameter of 1 specimen. The schematic of the sample dimension was shown by figure 1.

Table 1. The compositions of 316L stainless steel material.

Elements percentage for 316L SS (%)								
C	Si	Mn	P	S	Cr	Mo	Ni	N
0.020	0.49	1.79	0.040	0.010	16.88	2.06	10.6	0.0655

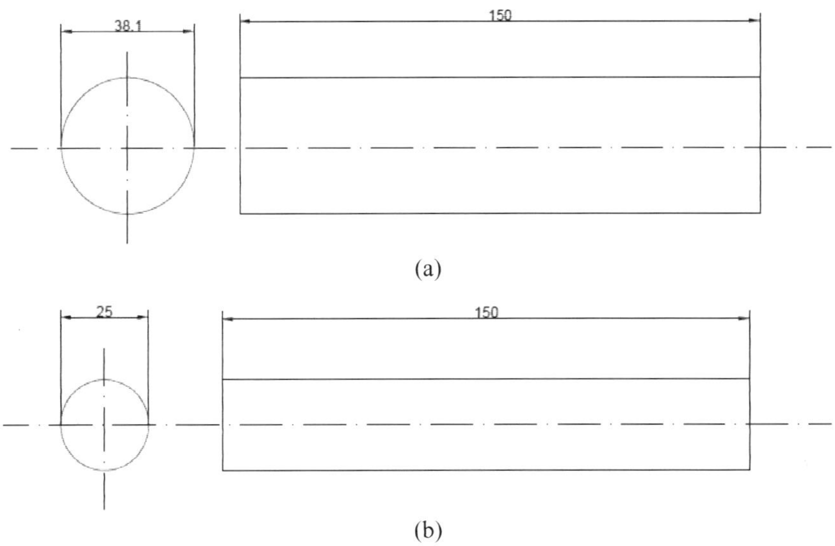

Figure 1. The dimension (mm scale) (a) before and (b) after shaped by lathe machine.

2.2. Method

The sample was prepared by the desired dimension. The following step is heat treatment. The sample was heated by muffle furnace at different temperatures which were 800°C, 950°C, and 1100°C. Those temperatures formed single phase which improved the mechanical properties. Holding time of heat treatment was also used as a variation for 30, 60, and 90 minutes. Then, the sample was cooled by air as normalisation media. Prepared samples were characterized by hardness test using microhardness test, microstructure analysis, and X-Ray Diffraction (XRD) analysis. For hardness test, samples should be smoothed by grinding and polishing step. The mechanism of testing materials based on ASTM E 384. Vickers's microhardness can be calculated from the diagonal of the indentation area. Polished samples were etched to appear grain boundaries. The microstructure was observed by a microscope optic, and XRD identified the sample component.

3. Results and Discussion

3.1. Hardness Properties

Figure 2 shows the hardness number of each specimen treated by temperature variations. From the figure was known that high temperatures treatment decrease the hardness. Phase transformation at higher temperatures is faster than the phase transfers at low temperatures, so the phase formed on the specimen effects the hardness value. The austenite phase dominates the phase, and the characteristic of the austenite phase is soft. The hardness of Normalised sample at 1100°C and held for 30 minutes achieved 181.4 HV. The raw material of 316L SS was tested by a hardness test to compare the effect of heat treatment, and it has 189.5 HV. Temperature treatment caused decreasing of hardness at 950 °C but increased at 1100°C [5]. Increasing temperatures promote the formation of carbide precipitate and lead to decrease the hardness. Otherwise, decreasing of hardness may due to grain size. Fine grain size has more grain boundaries and has a large number of carbides. Coarsening grain above 950°C of temperature treatment promote lower precipitation carbides to increase the hardness of the normalised material. For further reasons, increasing lath size, precipitate size, subgrain in dislocation have an important role to decrease in hardness while increasing normalising temperature [6]. It has mentioned that increasing dislocation density at higher normalising temperature can decrease in hardness.

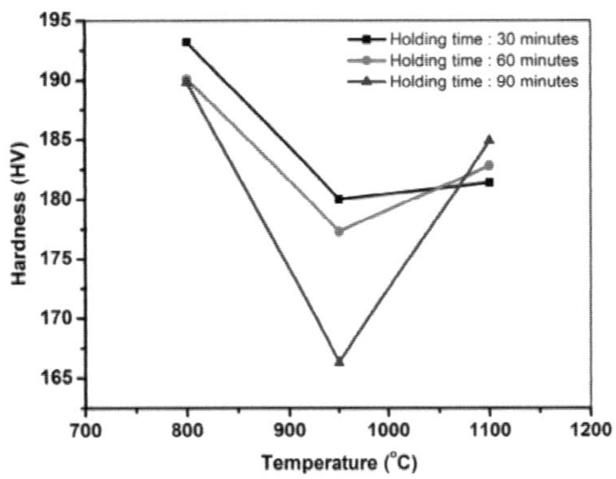

Figure 2. The hardness number with normalising temperature.

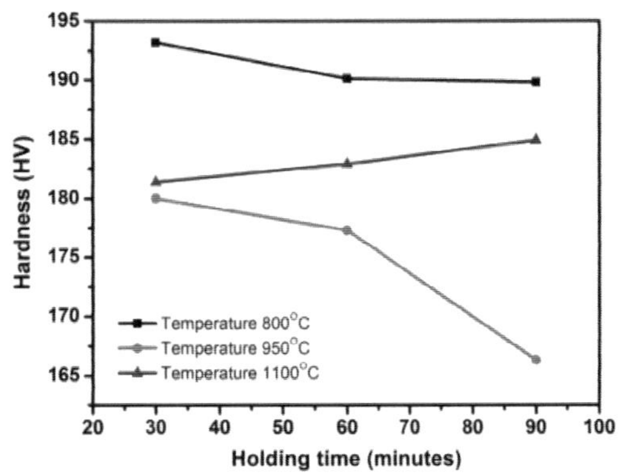

Figure 3. The hardness number with holding time variations of normalising.

Figure 3 shows that longer time to hold the samples during normalising promotes gradually decrease of hardness. The normalised sample reaches the highest hardness at 800 °C held for 30 minutes which is 193.2 HV. The temperature of 950 °C achieves the maximum hardness value of 180 HV with holding time 30 minutes and decreased while increasing the holding time. Regarding from hardness testing, holding time of normalising decreases in hardness but significantly increase during 1100°C of normalising temperature. Accumulations of nitrogen content increase on the surface area. High amounts of nitrogen cause differences of concentration inside and outside materials. It promotes diffusion from the surface to inside materials. Increasing of nitrogen content enhance the hardness. Detail information about the hardness of normalised samples is presented by Table 2.

Table 2. Data from microhardness measurement.

Temperature	Holding time	Repetition	d1	d2	HV	Average ± *Dev*
		1	54.29	54.31	188.6	
Raw Material		2	53.83	53.83	191.9	189 ± 2.14
		3	54.4	54.4	187.9	
	30 minutes	1	53.48	53.46	194.5	
		2	54.09	54.09	190.1	193.2 ± 2.73
		3	53.38	53.4	195.1	
	60 minutes	1	54.56	54.56	186.8	
800°C		2	53.56	53.6	193.7	190.1 ± 3.46
		3	54.13	54.15	189.7	
	90 Minutes	1	53.77	53.79	192.3	
		2	54.4	54.4	187.9	189.8 ± 2.25
		3	54.2	54.2	189.3	

Temperature	Holding time	Repetition	d1	d2	HV	Average ± *Dev*
950°C	30 minutes	1	56.1	56.1	176.7	180 ± 2.88
		2	55.37	55.39	181.3	
		3	55.26	55.3	182	
	60 minutes	1	56.18	56.18	176.2	177.3 ± 2.48
		2	56.26	56.32	175.5	
		3	55.57	55.57	180.1	
	90 minutes	1	57.66	57.64	167.3	166.3 ± 0.91
		2	57.85	57.85	166.2	
		3	57.97	57.97	165.5	
1100°C	30 minutes	1	55.25	55.25	182.2	181.4 ± 1.74
		2	55,67	55.69	179.4	
		3	55.19	55.19	182.6	
	60 minutes	1	55.86	55.86	178.2	182.9 ± 4,32
		2	55.01	55.01	183.3	
		3	54.55	54.57	186.8	
	90 minutes	1	54.4	54.4	187.9	184.9 ± 3.33
		2	55.37	55.39	181.3	
		3	54.75	54.79	185.4	

3.2. Morphological Analysis

From the metallography test results obtained that microstructure of specimens contains austenite (white colour), ferrite (dark) and chrome carbide (black spots). In the non-normalising specimen shown in figure 4 has a microstructure of austenite and ferrite phases but is dominated by the ferrite phase.

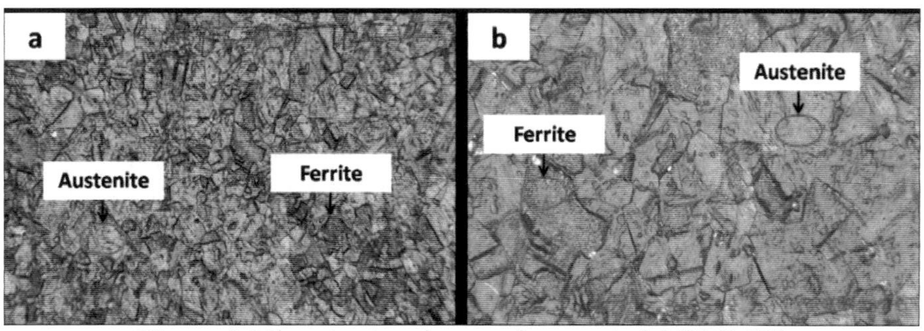

Figure 4. Microstructure image of raw material (316L SS)

The 316 L stainless steel microstructure with normalising treatment at 800° C is shown by Figure 5. There are austenite, ferrite and chromium carbide phases. Austenite phase dominates microstructure at normalised 800°C. The appearance of chromium carbide is due to the normalising process of slow cooling. The phases are still dominated by austenite and low amount of Chromium carbides. This sample has the highest hardness because it has low Cr carbides. They appear at grain boundaries as fine carbides. It is considered that the presence of Cr contributes to hardness decreasing [7].

The increasing normalising temperature at 950°C promotes more carbides and ferrite compared to normalised at 800°C. It is presented by figure 6. Austenite grain size is larger than ferrite and shows the obvious grain boundaries. Cr carbides appear as a precipitate. Ferrite and Cr carbides uniformly distributed in the austenite matrix because the diffusion rate of a solute atom is faster at the higher temperature [8]. Precipitation occurred near 950°C. Therefore, the hardness gradually decreased after 950°C. It is proven by Figure 7 that show more Cr carbides for normalised at 1100°C. Coarsening grain was observed for normalised temperature at 1100°C. It composed carbides and austenite matrix. The grain size is increased with increasing normalising temperature and Cr carbides distributed on the austenite matrix. The main carbides distributed inside the grains or at grain boundaries. Uniformly distributions of Cr carbides are dissolved into the matrix after treated at higher temperature [9]. More Cr carbides at high temperature caused by the good stability of Cr at high temperatures [10]

Figure 5. Microstructure image of normalised 316L SS at 800°C for 30 minutes

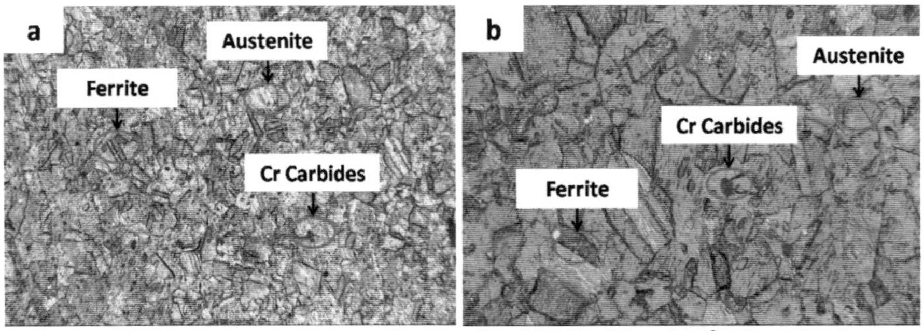

Figure 6. Microstructure image of normalised 316L SS at 950°C for 30 minutes

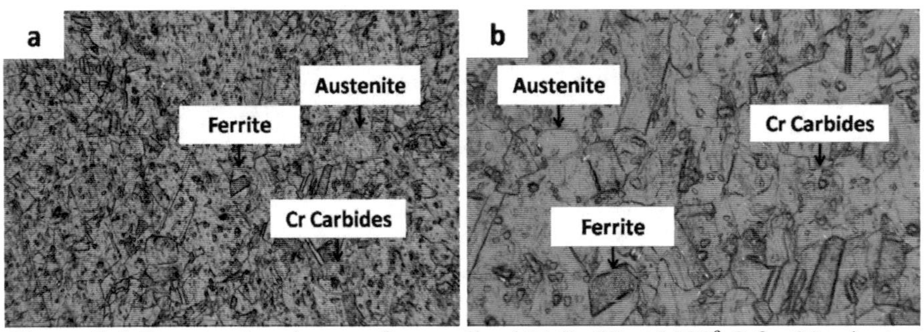

Figure 7. Microstructure image of normalised 316L SS at 1100° C for 30 minutes

3.3. XRD Analysis

Figure 8 shows the diffraction patterns of normalised 316L SS using temperature variations for 30 minutes. The strongest peaks of XRD indicate as austenite (ɣ) peaks. The position of ɣ peaks at 2θ=43.5°, 50.9°, 74.7°, 90.5° [10]. XRD pattern confirmed the presence of austenite phase from microstructure as mentioned above but XRD can not identify Cr carbides. Low amount of carbides may cause it. The diffraction peak of 2θ= 95° showed the presence of δ-ferrite phase. The lowest intensity explained low amount δ-ferrite phase. XRD cannot detect the phase under 5wt%. During the cooling at air media, phase transformation occurs from δ → ɣ and it was controlled by diffusion. The existence of δ-ferrite phase shows that transformation to the austenite phase is not completed [11].

The diffraction peak of normalised sample at 2θ = 50.9° has a wider peak shift than a 316 L stainless steel specimen without normalising treatment. The shift and widening peak are indicated by the magnitude of the widening number (FWHM) as shown in table 3. The slight peaks of XRD pattern mean crystalline structure.

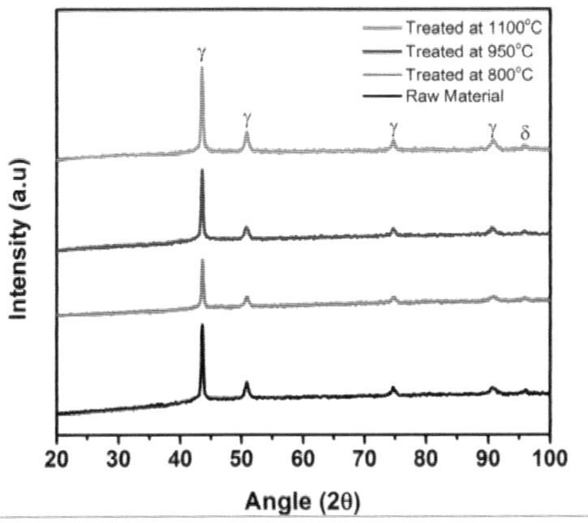

Figure 8. XRD Pattern of normalised 316L SS using temperature variations for 30 minutes.

Table 3. XRD data for FWHM left

Raw Material		Normalised at 800^{0}C		Normalised at 950^{0}C		Normalised at 1100^{0}C	
Pos. [°2Th.]	FWHM Left [°2Th.]	Pos. [°2Th.]	FWHM Left [°2Th.]	Pos. [°2Th.]	FWHM Left [°2Th.]	Pos. [°2Th.]	FWHM Left [°2Th.]
43.5792	0.1004	43.5789	0.1004	43.5743	0.1338	43.5756	0.0836
50.9097	0.3346	50.9154	0.4015	50.7653	0.4015	50.8404	0.4684
74.7015	0.4015	72.5446	0.4015	74.6664	0.3346	74.7207	0.4684
90.5995	0.5353	74.7500	0.5353	90.6769	0.8029	90.7551	0.6691
95.9431	0.8029	90.8488	0.9368	95.9333	0.6691	95.9977	0.8029
		96.0434	0.8029				

4. Conclusions

Based on the investigation of normalising treatment for hardness and microstructure analysis, can be concluded that the high normalisation temperature affects the decreasing of hardness. For normalised 316L stainless steel at 1100°C has a higher hardness than normalised at 950°C because of more Cr carbides and ferrite phases. Moreover, holding time also promotes the decreasing of hardness. Longer time for normalised 316L stainless decreases the hardness but increases at a temperature higher than 950°C. Moreover, macrostructures of normalised 316L stainless steel using temperature and holding time variations present austenite, ferrite, and Cr carbides. High Cr Carbides amount occurred while increasing normalisation temperature.

5. References

[1] X. Yuan, L. Chen, Y. Zhao, H. Di, F. Zhu, Journal of Materials Processing Technology **217** (2015) 278.
[2] A. Sorsa, K. Leiviskä, S. Santa-aho, T. Lepistö, Ndt & E International **46** (2012) 100.
[3] G.E. Totten, Steel heat treatment: metallurgy and technologies, CRC press, 2006.
[4] C. Rodenburg, W. Rainforth, Acta Materialia **55** (2007) 2443.
[5] N. Saini, C. Pandey, M.M. Mahapatra, Materials Science and Engineering: A **688** (2017) 250.
[6] D.R. Barbadikar, G. Deshmukh, L. Maddi, K. Laha, P. Parameswaran, A. Ballal, D. Peshwe, R. Paretkar, M. Nandagopal, M. Mathew, International Journal of Pressure Vessels and Piping **132** (2015) 97.
[7] C. Köse, R. Kaçar, Materials & Design **64** (2014) 221.
[8] H. Sun, Y. Sun, R. Zhang, M. Wang, R. Tang, Z. Zhou, Materials & Design **67** (2015) 165.
[9] H. Peng, L. Hu, L. Li, L. Zhang, X. Zhang, Journal of Alloys and Compounds **740** (2018) 766.
[10] S. Jung, C. Jeon, Y.H. Jo, W.-M. Choi, B.-J. Lee, Y.-J. Oh, S. Jang, S. Lee, Materials Science and Engineering: A **656** (2016) 190.
[11] V.A. Setyowati, E.W.R. Widodo, Jurnal Teknik Mesin **7** (2017) 74.

Design Steering System with Independent Front Wheel Drive of The Hybrid Vehicle-Air Pressure and Electrical

B Setyono[1], D A Patriawan[1], A W Putra[1], H Irawan[1], and E A Zuliari[2]

[1]Department of Mechanical Engineering, Institut Teknologi Adhi Tama Surabaya, Surabaya, Indonesia
[2]Department of Electrical Engineering, Institut Teknologi Adhi Tama Surabaya, Surabaya, Indonesia

desmas@itats.ac.id

Abstract. The development of alternative vehicle were continued by designing hybrid vehicle that using compressed air and electrical propulsion. This compressed air propulsion on the rear wheels and two electric motor as a front wheel drive (FWD) were installed independently, where left and right can rotate with different speeds and direction. Independent FWD could be a problem when two wheels has a different rotation. The first experiment was tested by measuring the rotation per minute (RPM) on each wheel. Measurement were performed by varying the position of the throttle pedal. The result of measurements on both wheels rotating speed showed a difference of 1 %. The second test was to compare the space requirement, which derived from the function of cornering radius. Space requirement for inactive vehicle cornering system is 4.52 meters and 3.39 meters when system is active.

1. Introduction

Environmental issue is an important factor in the development of automotive technology. In decades, specific attention were given to vehicles that utilize fossil fuels, because these vehicles produce carbon dioxide (CO_2). CO_2 is also considered as major greenhouse gas which has contributing to global warming and climate change [1].

There are about six types of gases that cause the greenhouse effect. Carbon dioxide (CO_2) and methane (CH_4) most cause the greenhouse effect [2] and both are produced from the combustion of fossil fuel vehicles.

Environmentally friendly vehicles began developed by automotive companies and academia. Hybrid, electric and hydrogen propulsions are the type of vehicle propulsion which is commonly produced and developed. Hybrid car driven by wood-gas and electric were first manufactured in 1916 [3].

This paper discusses the steering system on hybrid vehicle which is driven by pressurized air and electrical propulsion. Compressed air is used as the rear wheel drive (RWD) whilst electric motor is used as front wheel drive (FWD). This vehicle can be seen in Figure 1.

Steering system made to solve two (2) major issues. The first issue is synchronization on both front wheels when vehicle is running straight. The second issue is minimizing space requirements when the vehicle is turning. The rotation speed on the wheels can be adjusted by Traction Control System

(TCS), so as to minimize slip on the vehicle [4]. TCS developed for steering system used on this vehicle, so the vehicle is able to regulate the rotation of the front wheel as desired.

Front wheel drive (FWD) and rear wheel drive (RWD) has advantages and disadvantages. Test was conducted to determine the advantages and disadvantages of FWD and RWD. There are three variables of test performed: horizontal, sloping and low friction road. On horizontal road, both layout showed understeering behavior's up to the limit lateral accelerations. On sloping roads, RWD vehicle tended oversteer and FWD still understeer [5]. From this study, both RWD and FWD was not affecting the horizontal road conditions.

Figure 1. Hybrid vehicle-pressure air and electric

Safety is an important issue in the vehicle, so it needs to be developed, even for prototype. In electric vehicle when an unexpected sudden stop occurs due to failure, there is a possibility a serious traffic accident may be caused [6]. This vehicle two DC motors mounted on the wheel, so it needs to be taken into account when the wheels to propulsion systems fail.

2. Design of Steering System

This vehicle has three wheels where there is a rear wheel drive using pressurized air. At the front using two wheels with each mounted brushless DC motor. If the compressed air does not generate thrust, the propulsion function is replaced by electric drive.

Brushless DC Motor mounted in wheel on the front wheels of the vehicle. Motor mounted on both front wheels and each motor is capable of rotating independently. Independent designs DC motor system has the advantage of distribution of torque evenly so that the vehicle is able to accelerate faster and more efficiently if you use 1 DC motor with the same power. However independent DC motor system also has the disadvantage that synchronization of the rotation of the wheel, because the left and right wheels has an independent system so that both wheels have a tendency different rotational speed.

The steering system is an issue on this vehicle. Results obtained manually steering turning radius a fairly wide, therefore it is necessary to design a new steering system with TCS. Overview of the steering system can use Ackerman Geometry.

The necessary steering angle of front wheels can be constructed via given momentary pivot pole O, can be seen in Figure 2. Ackermann geometry can only be analysed by vehicle moving in low lateral acceleration range. This hybrid vehicles moving in the low lateral acceleration. Then, the following relations apply:

$$tan\delta_i = \frac{l}{R_1} \text{ and } tan\delta_o = \frac{l}{R+S} \quad (1)$$

where w label the track width and l denotes the wheel base. Eliminating the curve radius R_1 we get:

$$tan\delta_o = \frac{l}{\frac{l}{tan\delta_i}+w} \text{ or } tan\delta_o = \frac{l\,tan\,\delta_i}{l+S\,tan\delta_i} \quad (2)$$

The deviation $\Delta\delta_o = \delta_o^a - \delta_o^A$ obtained from δ_o^a as the actual steering angle and δ_o^A is Ackermann steering angle. Result δ_o^a can be obtained from equation (2), Differences $\Delta\delta_o$ is used to judge the quality of the steering system.

Ackermann approaches can also be used to calculate the space requirement of a vehicle during cornering. The hybrid vehicle uses two front wheels in accordance with Ackermann geometry, outer point on the front of the vehicle used as a maximum radiusR_{max}. Inner side of the vehicle at location of the rear axle will run on the minimum radiusR_{min}, the equation can we get:

$$R_{max}^2 = (R_{min} + w)^2 + (l)^2 \quad (3)$$

where l, w adalah wheel base dan width of the vehicle. Then, space requirement $\Delta R = R_{max} - R_{min}$. More specifically can be described as follows:

$$\Delta R = R_{max} - R_{min}$$
$$\Delta R = \sqrt{(R_{min} + W)^2 + (l)^2} - R_{min} \quad (4)$$

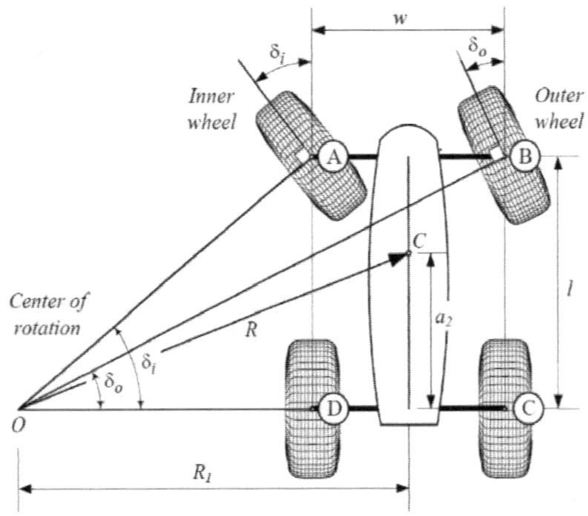

Figure 2. Calculate space requirement [11]

Turning radius calculation with the Ackermann cannot be done when the vehicle is on the move at high speeds, so the testing is done at low speed. Problems can also occur because the two front wheels are fitted independent motors. The problem is the left and right wheels can produce different rotational speeds. However, when turning this could be an advantage.

Rotation on FWD can be set by placing a controller in each motor, so that rotation of the left and right wheels can be set independently. Figure 3 can be seen wiring diagram FWD.

From Figure-3 it can be seen that each motor has its own controller. The direction and speed of rotation of the wheel can be made different by adding a relay on steering. This steering system work when the vehicle is turning right, the controller gives a signal for the motor right wheel had no electricity, so that the left wheel is still working and the right wheel only follow. It also happens conversely.

The steering system is expected to shorten the cornering radius in this hybrid vehicle. Initial testing of synchronization between the left and right wheels to get the same RPM.

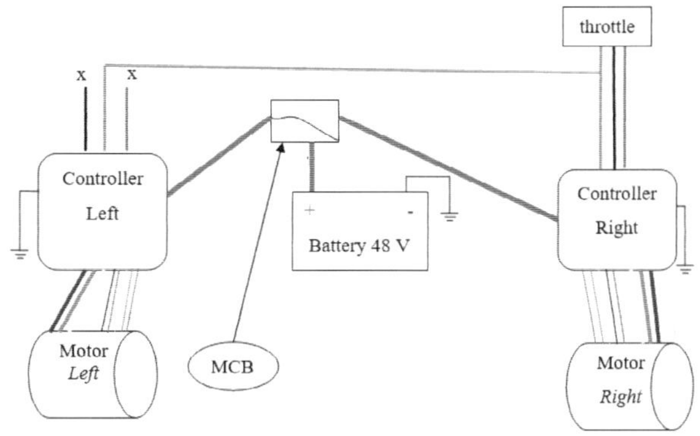

Figure 3. Wiring diagram FWD

From Figure-3 it can be seen that each motor has its own controller. The direction and speed of rotation of the wheel can be made different by adding a relay on steering. This steering system work when the vehicle is turning right, the controller gives a signal for the motor right wheel had no electricity, so that the left wheel is still working and the right wheel only follow. It also happens conversely.

The steering system is expected to shorten the cornering radius in this hybrid vehicle. Initial testing of synchronization between the left and right wheels to get the same RPM.

3. Experiment Result

The first test is done by observing the rotation of the two front wheels. This test was conducted to determine the RPM of the two wheels, in case of the rotation speed difference between left and right wheels may lead to accidents. Testing is done by pressing the throttle pedal in position 1/4, 1/2, 3/4 and full throttle. RPM measurement results on the left and right wheels can be seen at table 1.

Tabel 1. RPM left and right wheel.

Throttle Position	RPM on Left Wheel	RPM on Right Wheel
¼	213,4	215,7
½	298.8	300.2
¾	455.5	457.5
Full	588.1	589.5

From Table 1 it can be seen there is little difference in RPM on the left and right wheels of vehicles. However, the overall results showed there only 1% difference between left wheels with wheel RPM right, it can still be tolerated.

Results from Table-1 showed no significant difference, so it does not need special control or using traction control to synchronize the RPM on the front wheels. The next test is the effect of adding mass to the space requirement. Mass of vehicle is 250 kg. Figure 4 shows the effect of adding mass to the space requirement. The effect of adding mass to the space requirement is because there is a displacement of the centre of gravity of the vehicle. Data taken at a speed of 20 km / h.

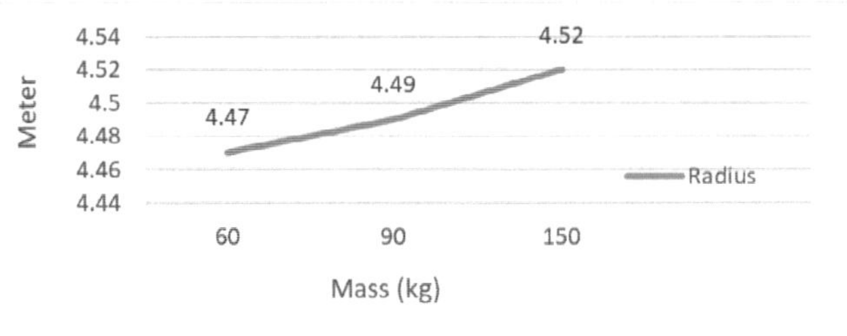

Figure 4. The effect of adding mass to the space requirement when steering system off

The next test was to determine the effect of speed on the space requirement. In this test done when the steering system is powered off, this is done to determine how much space requirement the default condition. The radius of the wheel is 12 inch, the addition of a mass of 90 kg and with RPM obtained in Table-1. The effect of speed on the large turning radius can be seen in Figure 5.

Figure 5 shows that the large space requirement when the vehicle turn is influenced by the speed of the vehicle. The greater space requirement is influenced because of the centrifugal force.

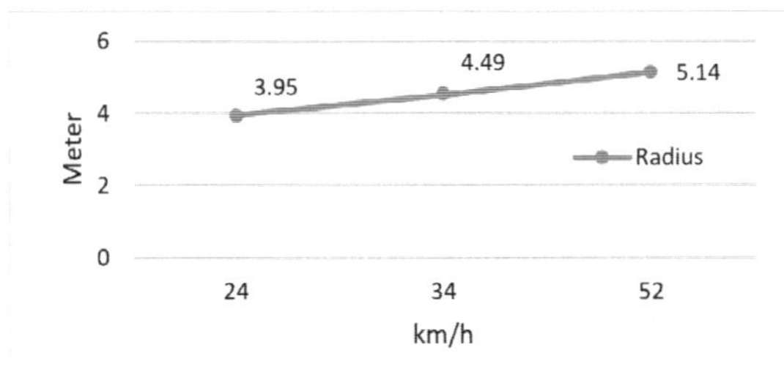

Figure 5. Effect of speed on corning radius space requirement when system is off

The workings of the steering system can be seen in Figure 6, wherein when the vehicle is moving straight, the both wheel will move with the same rotation. However, when turning the front wheels will have different speeds. When the vehicle is turning right, then the left wheel will spin faster than the right wheel. This also happens conversely. Comparison large space requirement between the systems steering off and on with variable mass can be seen in Figure 7.

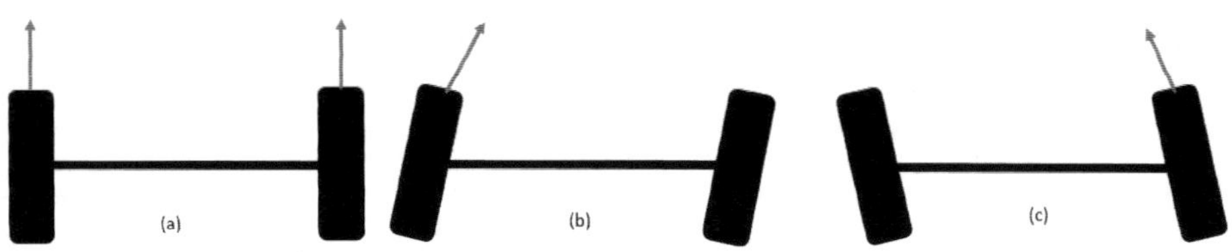

Figure 6. Steering system working (a) when vehicle move straight (b) turn right and (c) turn left

The results in Figure 7 does not significantly affect space requirement. Active steering system is able to reduce the space requirement of about 1 meter compared to when the system is off. Mass variation is not very influential, both when the system off or on.

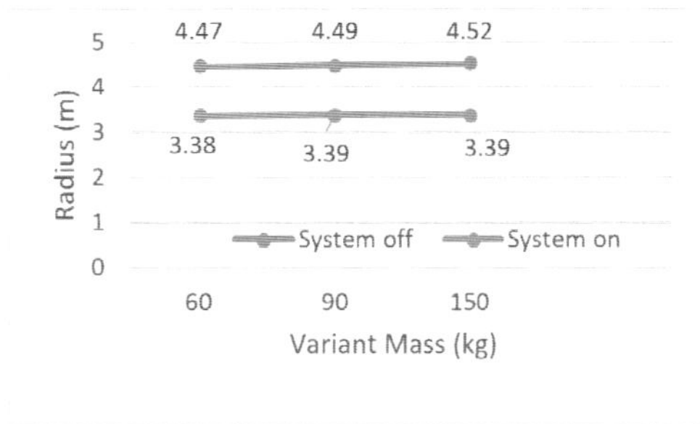

Figure 7. Comparison of space requirement when the system off and on with the mass variations

The next test is to find the effect of variable speed and steering system on the space requirement. Figure-5 shows the change in space requirements by speed variation. Figure 8 shows the comparison of space requirement to variations in speed with the steering system on and off. Steering system in an active condition has a smaller space requirement compared with the steering system on the condition off. Active steering system is able to reduce the space requirement of 1.46 meters when speed of 26 km / h.

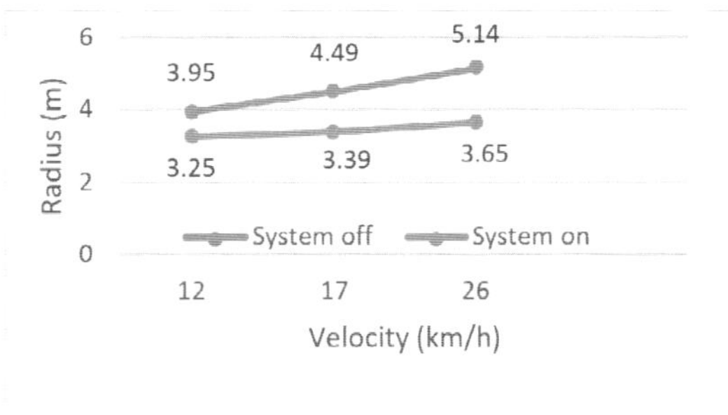

Figure 8. Velocity vs turning radius

4. Conclusion

This study shows that the front wheel drive (FWD) with independent propulsion has some advantages. One of them is in reduce space requirement when the vehicle turns, so the vehicle can turn with more narrow range. FWD with independent motors facilitate vehicle to control the speed and direction of rotation are different on each motor. Different rotation on the left and right wheels can influence space requirement cornering radius. Differences space requirement when the system is activated reaches 1.49 meters shorter than the system off. Space requirement could still be further reduced if the direction of rotation between left and right wheels made differently. However independent propulsion FWD Also there are still weaknesses that synchronization between both wheels. The rotational difference between left and right wheels is only 1%, but this could be a problem when high RPM.

5. References

[1] Azad A.K ., Nashreen S.W. and Sultana J. 2006. State of Energy Consumption and CO_2 Emission in Bangladesh. Royal Swedish Academy of Science. 35(2):86-88.

[2] National Pharmacopoeia Committee. 2010**.** Pharmacopoeia of People's Republic of China. Beijing: Chinese Medical Science and Technology Press. 224.

[3] Fush A. 2009. Hybrid Vehicle and the Future of Personal Transportation. Boca Raton, CRC Press.

[4] National Energy Council. 2016. Indonesia Energy Outlook 2016. ISSN: 2503-15.

[5] Kaloko, B. S., Soebagio and M. H. Purnomo. 2011. Design and Development of small Electric Vehicle using MATLAB/Simulink. International Journal of Computer Applications. 24(6): 19-23.

[6] Rijanto, E., A. Hartanto., N. Huda and M. I. Devi. 2012. Perencanaan Pengembangan Prototipe Produksi Bus Listrik 15 Penumpang. Berita Iptek. 5(1): 51-66.
[7] Sudaryono., Soebagio and M. Ashari. 2013. Modeling and Simulation of Electric Vehicle fed by PEM Fuel Cell. International Journal of Science Invention. 2(6):121-126.
[8] Song J. and Boo K. 2004. Perfomance evaluation of traction control system using a vehicle dynamic model. Proceedings of the Institution of Mechanical Engineers. 218(7):685-696.
[9] Mashadi B., Ebrahimi N. and Marzbanrad. 2007. Effect of front-wheel drive or rear-wheel drive on the limit handling behavior of passenger vehicle. Proceedings of the Institution of Mechanical Engineers. 221(D4):393-403.
[10] Mutoh N. 2009. Front-and-Rear-Wheel-Independent –Drive –Type Electric Vehicle (*FRID EV*) with Compatible Driving Performance and Safety. World Electric Vehicle Journal. 3:0017-0026.
[11] Khristamto M., Praptijanto A. and Kaleg S. 2015. Measuring geometric and kinematic properties to design steering axis to angle turn of the electric golf car. Energy Procedia. 68:463-470.

Acknowledgments
The Authors acknowledge the support from Ristekdikti (Ministries of Research, Technology and Higher Education) grant.

IOP Conf. Series: Materials Science and Engineering **462** (2019) 012014 doi:10.1088/1757-899X/462/1/012014

Slope Stability Analysis Under a Complex Geotechnical Condition - A Case Study

G S Utami[1], B A M Bali[2]

[1]Civil Engineering Department, Institut Teknologi Adhi Tama Surabaya, Surabaya, East Java, Indonesia
[2]Mining Engineering Officer, Mineral Directorate, Autoridade Nacional do Petroleo e Minerais, Dili, Timor Leste

fita.fina3@gmail.com

Abstract. Slope failures in a mining project will contribute to significant safety impacts and economic losses. Therefore, it is imperative for mining engineers to analyse and design an open pit mine considering safety purposes. In this study, a limit equilibrium method using GeoSlope-Slope/W program was used to investigate the stability of an open pit mining slope for a proposed mining activity based on a case study in Tanjung Enim, District of Muara Enim, Province of South Sumatera - Indonesia. The results of this study reveal that the slope design for the proposed mining activity is technically not in stable condition due to the shear force of the slope is higher than its shear strength. Such mechanism occurs due to the existence of a complex geotechnical condition beneath the surface, such as faults and discontinuities, which then lead to the reduction of slope strength. Having an inapplicable proposed slope design, this study proposes a re-design of slope for safety purposes in order to support the future mining operation. The analysis approach proposed in this study can be a viable alternative to assess the stability of slopes in any mining operation when a complex geotechnical condition exists.

1. Introduction

Slope stability analysis forms a critical part of any open pit mining operations during the life cycle of a mining project. Earlier studies present that the analysis and assessment of slope stability should be considered in an open pit mine operation [1]. Also, the previous extensive study reveals that slope failures in a mining project will contribute to significant safety impacts and economic losses [2–4]. Therefore, the slope stability analysis is imperative to be investigated before implementing an open pit project in a permitted mining area.

The case study of this project is located at the East Site of Tambang Air Laya (TAL), a mining operation site run by PT Bukit Asam, Persero, Tbk. The proposed mining operation at the project site was to excavate various lithology from the depth of 57-meters above sea level down to the lowest depth of -128 meters, which will form various multiple open pit slopes. From geological aspects, the geological structures at the mining site consist of various joints and normal faulting having a radial pattern due to intrusion in the subsurface. The effect of such subsurface intrusion resulted in the existence of various continuous faults at the bottom side of existing coal and rock layers. Having such complex of geotechnical and geological condition, the slope stability analysis for the proposed mining activity is a very critical aspect to be assessed in order to ensure that the proposed mining operation will be carried out in a proper manner considering safety purposes. This is according to the earlier studies' results stating that the complexity of a geological and geotechnical condition may control the

IOP Conf. Series: Materials Science and Engineering **462** (2019) 012014 doi:10.1088/1757-899X/462/1/012014

behaviour of rock and soil that can lead to a slope failure in a mining project [5–6]. Based on the problem statement, the current study aims to investigate and optimise the stability of the proposed slope design in order to propose an optimal final slopes design for the proposed mining activity.

Earlier studies mainly focus on analysing the stability of an open pit mining slope without giving any alternative should the slope stability analysis results present that the slope is not safe [1–3]. The current study will not only focus on slope stability analysis aspect but will also propose an optimal design of slope for safety purposes in order to support the proposed mining operation given by the company. A cross-section representing the field condition was selected for analysis purpose. The Geo Slope-Slope/W program was used as a computational analysis tool due to its simplicity, and this program has been widely accepted in various slope stability analyses for the mining operation.

2. Case Study Description

2.1. Location and Accessibility

The current study is focussed on the East site of Tambang Air Laya (TAL), one of the mining operation sites operated by PT Bukit Asam, Persero, Tbk. It is geographically located in Tanjung Enim, District of Muara Enim, Province of South Sumatera, Indonesia. As shown in Figure 1, the distance from the Capital city of South Sumatera Province, Palembang to the project site is around 150 km and can be reached by terrain transportation in approximately four to five hours.

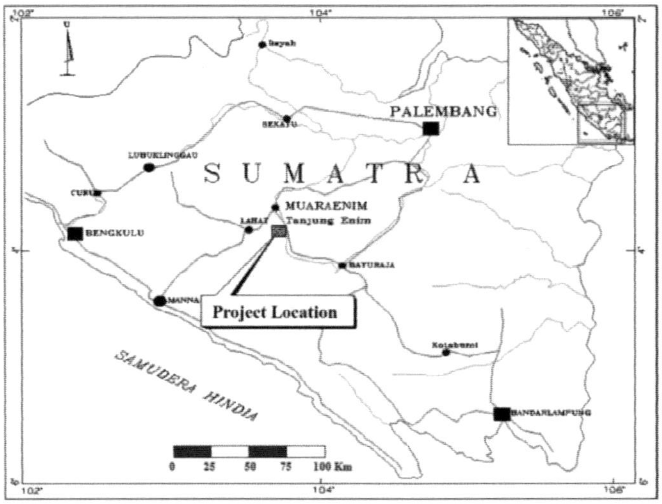

Figure 1. Location of the project site.

2.2. Local Geology

The geological structures at the mining site include various joints and normal faulting having a radial pattern due to intrusion in the subsurface. The effect of such intrusion also resulted in the existence of various continuous faults at the bottom side of the existing rock layer (Figure 2).

The formation distribution at the project site follows the subsurface rock formation with a general direction from north to south at an angle of 15 to 20 degrees. The previous study identifies the three main intrusive bodies near the Air Laya and Suban mining operation, namely: Bukit Asam dyke, Suban sill and a vertical parasitic cone to the west of Air Laya Dome [8]. The Bukit Asam dyke was found to be the largest intrusive body, and its outcrop forms a hill. The presence of the andesite intrusion in the Bukit Asam area has resulted in locally change the rank of the coals [9]. These coal seams can be classified into medium-volatile bituminous to anthracite coals [10].

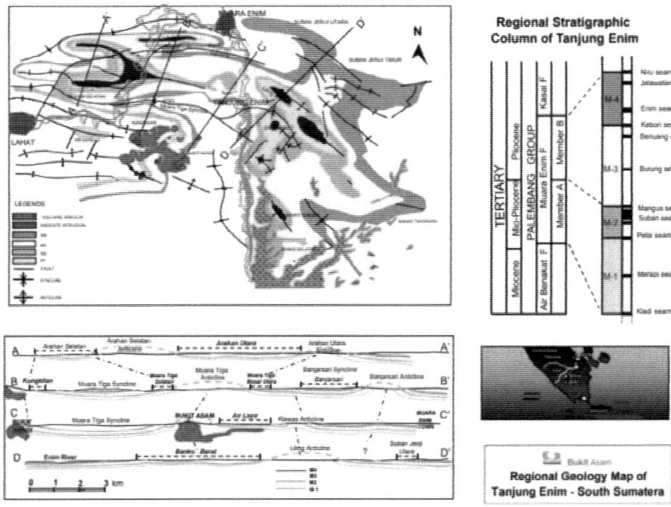

Figure 2. Regional geology map of Tanjung Enim, South Sumatera.

3. Methodology

The methodology used in this study involves three main works, including field investigation, data collection and data analysis. The three main works and their details are described in the subsequent sections.

1) A field investigation was conducted to assess the slope geometry and identify the lithology of the project site. The individual slope has a height and angle of 9 meters and 45 degrees, respectively, with a total overall slope height of 184 meters. The lithology at the project site was identified by assessing borehole information from fifteen exploration borehole data.
2) Data collection was conducted to obtain the required input data parameter for the analysis. The required data for the current study is the geometry of the proposed open pit slope (height and angle of slopes and berm width), physical and mechanical properties of soil and rock, water level condition, and normal load caused by the weight of mining equipment. The physical and mechanical properties data were obtained from the laboratory analysis results for various soil and rock samples from fifteen exploration boreholes.
3) The slope was assumed to be at the fully saturated condition in order to obtain the most conservative analysis results, while the normal load for the weight of equipment was identified to be 600 kN/m^2. The representative cross-section for the proposed open pit slope design and required input data parameters be shown in Figure 3 and Table 1, respectively.

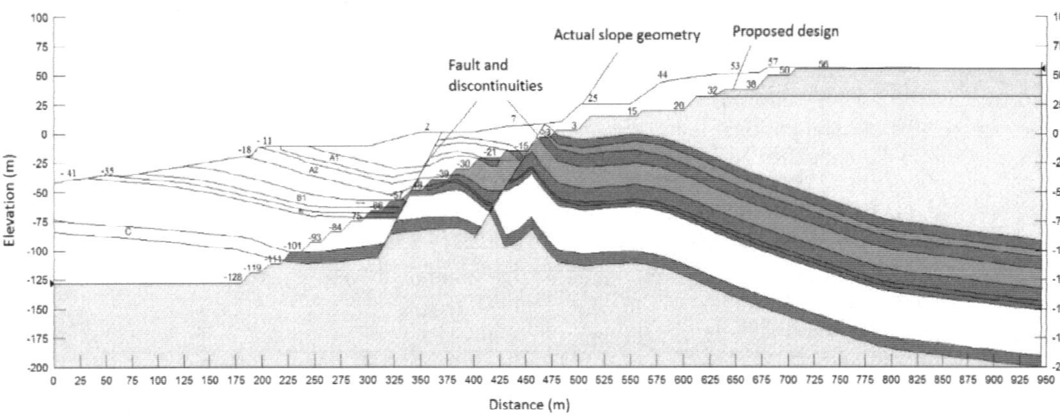

Figure 3. A typical representative cross-section for the proposed mining operation.

4) After all required input data parameter for analysis are obtained, slope stability analysis was performed using the Limit Equilibrium Method in GeoSlope/Slope-W program. This program was selected for this study due to its simplicity, and it has been globally accepted for analysing the stability of various open pit mining slopes. The output of the analysis is a factor of safety, which defined as the ratio of the shear strength to the shear stress required for equilibrium. For this study, the minimum required mining safety factor is 1.25, standardised by the Geotechnical Department of PT Bukit Asam (Persero) Tbk.

Table 1. Physical and Mechanical properties of rock strata, Bali [7].

Lithology	Density (kN/m^3)	Cohesion (MPa)	Friction angle (degree)
Top soil	19.5	37.2	12
Overburden	20.2	79	23
Coal layer (A1)	11.3	64.8	25
Interburden A1 - A2	20.7	132.2	30
Coal layer (A2)	11.9	64.8	25
Interburden A2 - B1	21.7	114.1	26
Coal layer (B1)	12.1	64.8	25
Interburden B1 - B2	22.2	62.2	25
Coal layer (B2)	11.9	64.8	25
Interburden B2 - C	22.2	144.5	27
Coal layer (C)	13.7	64.8	25
Underburden C layer	22.4	86.1	22

Note: Interburden layers is the lithology located in between two coal seam layers.

4. Results Analysis and Discussion

4.1. Results Analysis

Scholarly literature presents that the predicted safety factor of 1.5 indicates a high safety condition of a slope [11]. The possibility for slope to fail is high when the predicted safety factor is close to 1.0 [12]. For this study, the minimum safety factor for slope to be considered in technically safe condition is 1.25, based on the standard operation guideline given by the geotechnical department team of PT Bukit Asam (Persero) Tbk.

The slope stability analysis results for this study are shown in Table 2. From the analysis results showing on Figure 3, it can be noted that the factor of safety from the depth of 56-meters down to the depth of -15-meters sea level is 1.436, which account for 71-meters height of multiple slopes, while the safety factor from the depth of -15 meters to -128 meters sea level is 1.0. The predicted overall slope safety factor is 1.018. In other words, the top portion of the proposed open pit slopes from the depth of 56-meters down to the depth of -15-meters sea level is predicted to be in stable condition. However, the analysis results reveal that the bottom portion of the proposed open pit slopes from the depth of -15 to -128 meters sea level is not in stable condition due to the shear strength of the slope is lower than its shear force, which than lead to a low safety factor.

Table 2. Slope stability analysis results for the proposed mining activity.

Slope Geometry				
Height (meter)	Slope angle (degree)	Maximum altitude (meter)	Minimum altitude (meter)	Predicted safety factor (SF)
71	15	56	-15	1.436
113	24	-15	-128	1.000
184	19	56	-128	1.018

This study found out that the mechanism leading to a low safety factor for the bottom portion of the open pit is due to the presence of faults and discontinuities. Faults and discontinuities persistence reduce the shear strength of the slope formation, which then leads to the reduction of slope shear strength. This study concludes that the critical horizon where a slope could fail is concentrated at the slope horizon where a complex geologic structure, such as faults and discontinuities, exist.

Since the analysis results reveal that the proposed mining activity is in the high possibility to fail especially at the bottom part of the proposed slope design, a redesign of slopes for safety purposes is required. This step should be considered in order to ensure that the future mining operation will be carried out in a proper manner considering safety purposes.

4.2. Re-design of Slopes for Safety Purposes

The general pit slope design rules presented by Hoek and Bray [13] is that the ultimate objective of an open pit slope design is to achieve an optimum design, which should compromise between a slope that is steep enough to be economically acceptable and one which is flat enough to be technically safe. Both design aspects should be considered when designing a slope for a mining operation. The approach for re-design of slopes for this study is to design the slope to be flat enough for safety purposes as proposed by Hoek and Bray [13].

Based on the results of slope stability analysis, the slope design, especially in the bottom part of the overall slope height, is technically not stable due to the existence of fault and discontinuities, which then lead to a low predicted safety factor. Therefore, it is imperative to re-design the current proposed slope design in order to support the future mining operation. The re-design of for safety purpose is presented in Figure 4.

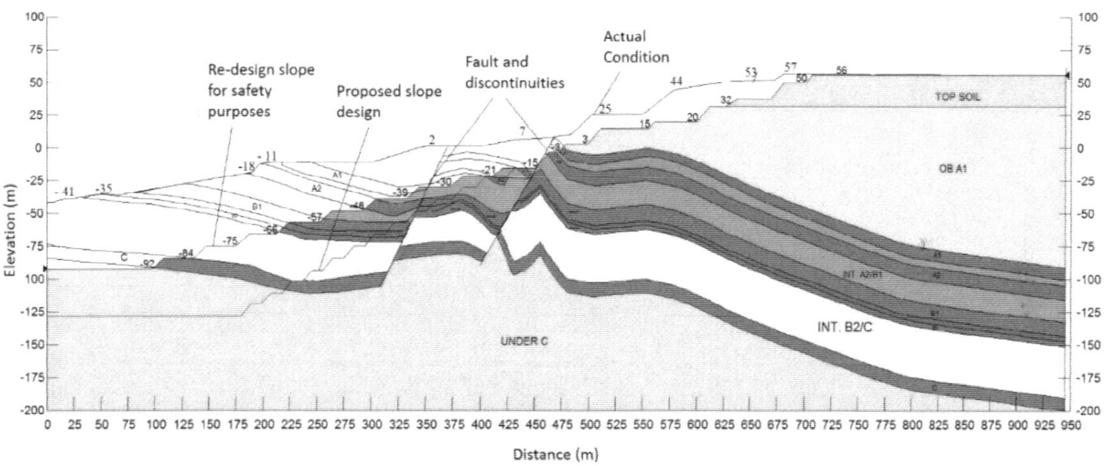

Figure 4. A typical cross-section for the re-design of overall slope geometry.

The stability analysis result for the re-design of the slope is shown in Table 3. From the analysis results shown in Figure 4, it shows that after re-design slope is made, the predicted factor of safety for the overall slope is increased to 1.388, while the predicted factor of safety for the top and bottom side of the overall slope height is 1.436 and 1.360, respectively. Therefore, it can be concluded that the proposed re-design slope is technically stable for the proposed mining operation. This study suggests that even though the re-design of slope approach proposed in this study will reduce the target of production of the company, but this approach still can be implemented for the safety of mining workers and equipment purposes.

Table 3. Slope stability analysis results for the proposed re-design of overall slope height.

Slope Geometry				Predicted safety factor (SF)
Height (meter)	Slope angle (degree)	Maximum altitude (meter)	Minimum altitude (meter)	
71	15	56	-15	1.436
77	13	-15	-92	1.360
148	14	56	-92	1.388

5. Conclusion

In the current study, a Limit Equilibrium Method in GeoSlope-Slope/W program was used to investigate and optimise the proposed open pit slopes design, based on a case study in Tambang Air Laya mining Site, South Sumatera, Indonesia.

This study reveals that the critical horizon where a slope could fail is concentrated at the slope horizon where a complex geologic structures such as faults and discontinuities exist. In addition, it was assessed that the slope design for the proposed mining operation cannot be applied for future mining operation due to the proposed open pit slopes from the depth of -15 to -128-meters sea level is technically not in stable condition. This happen due to the shear strength of the slope is lower than its shear force, which then lead to a low safety factor. Such mechanism occurs due to the existence of faults and discontinuities in the subsurface at the mining site which than lead to the reduction of the shear strength of the slope. Due to having an inapplicable slope design for the proposed mining operation, a re-design of slope approach was proposed and its analysis was performed accordingly in order to ensure that the future mining operation will be carried out in a proper manner considering safety purposes.

The main significance of this study is that it not only proposes an analysis approach for assessing the stability of slopes for the case study, but it can also be used for other similar projects where a complex geotechnical condition such as faults, discontinuities and folds exist and require a geotechnical investigation and assessment for mine safety purposes.

6. Reference

[1] Satyanarayana. I., Budi. G., Sen. P., & Sinha, A.K. (2018). Stability Evaluation of Highwall Slope in an Opencast Coal Mine - A Case Study. *AMSE JOURNALS - AMSE IIETA publication. Vol.* **78**, 253-273. Retrieved from http://iieta.org/sites/default/files/Journals/MMC/MMC_C/78.03_01.pdf.

[2] Kumar. V, and Parkash. V. (2015). A Model Study of Slope Stability in Mines Situated in South India. *Pelagia Research Library - Advances in Applied Science Research*, **6** (8), 82-90. Retrieved from http://www.imedpub.com/articles/a-model-study-of-slope-stability-in-mines-situated-in-south-india.pdf.

[3] Bye. A. R., & Bel. F. G (2001). Stability assessment and slope design at Sandsloot open pit, South Africa. *International Journal of Rock Mechanics and Mining Science*, *Vol.* **38**, 449-466. Retrieved from www.elsevier.com/locate/ijrmms.

[4] He. M. C. Feng. J. L., & Sun. X. M. (2008). Stability evaluation and optimal excavated design of rock slope at Antaibao open pit coal mine, China. *International Journal of Rock Mechanics and Mining Science*, *Vol.* **38**, 289-302. doi:10.1016/j.ijrmms.2007.05.007.

[5] Al-saadi. S. N., Hijab. R. B. and Ajjotheri. J. H. (2007). The Relationship between Geological Structure and Failure Type in the Rock Slopes of Parts of Anah Monocline, West of Iraq. *Journal of Kirkuk Universit-Scientific Studies*, *Vol.* **2**, 1-16. Retrieved from https://www.iasj.net/iasj?func=fulltext&aId=43432

[6] Fleurisson. J. (2012). Slope Design and Implementation in Open Pit Mines: Geological and Geomechanical Approach. *Procedia Engineering*, *Vol.* **46**, 27-38. doi: 10.1016/j.proeng.2012.09.442

[7] Bali. B. A. M. (2013). Analisa Kestabilan Lereng Rencana Penambangan Pada Desain Geometri Lereng Rencana Penambangan di PT. Bukit Asam Unit Pertambangan Tanjung Enim Site Tambang Air Laya (TAL) Ekstension Sisi Timur Sumatera Selatan. *Bachelor Degree Thesis submitted to Mining Engineering Department at Insitut Teknologi Adhi Tama Surabaya.*

[8] Pujobroto, A. and Hutton, A.C. (2000). Influence of andesitic intrusions on Bukit Asam coal, South Sumatra Basin Indonesia. *Proceedings of Southeast Coal Geology Conference, Directorate General of Geology and Mineral Resources of Indonesia, Bandung*, 81-84.

[9] Santoso, B. and Daulay, B. (2005). Coalification trend in South Sumatra Basin, Proceedings Joint Convention Surabaya, *The 30th HAGI, The 34th IAGI and The 14th PERHAPI, Annual Conference and Exhibition, JCS2005-M004*.

[10] Amijaya, H. and Littke, R. (2006). Properties of thermally metamorphosed coal from Tanjung Enim area, South Sumatra basin, Indonesia with special reference to the coalification path of macerals. *International Journal of Coal Geology*, Vol. **66**, 271-295.

[11] Islam, M. R., & Faruque, M. O. (2013). Optimization of slope angle and its seismic stability: A case study for the proposed open pit coal mine in Phulbari, NW Bangladesh. *Journal of Mountain Science*, *Vol. 10*(6), 976-986. doi:10.1007/s11629-012-2483-6.

[12] Wyllie, D. C., & Mah, C. W (2004). Rock slope engineering: Civil and mining 4th edition. *London: Spon Press.* Retrieved from https://www.google.com/search?q=ISBN+0-203-57083-9&ie=utf-8&oe=utf-8#

[13] Hoek, E. and Bray J.W. (1981). Rock Slope Engineering, The Institute of Mining and Metallurgy, London, 3rd edition.

IOP Conf. Series: Materials Science and Engineering **462** (2019) 012015 doi:10.1088/1757-899X/462/1/012015

Asbestos-free Brake Pad Using Composite Polymer Strengthened with Rice Husk Powder

W E Primaningtyas [1], R R Sakura[2], Suheni [1], I Syafi'i[1], and A A G A D Adhyaksa[1]

[1] Department of Mechanical Engineering, Faculty of Industrial Engineering, Institut Teknologi Adhi Tama, 60111 Surabaya, Indonesia
[2] Department of Mechanical Engineering, Faculty of Engineering, Universitas Negeri Jember, 68121 Jember, Indonesia

Abstract – Indonesia is one of biggest country with agricultural product and its very abundant. Agricultural product would give effect on an agriculture waste. For the example of agricultural waste is rice husk (RH). RH is the outer layer of the paddy. RH as an organic waste is obtained from the outer covering of rice grains during the milling process. The mechanical properties of RH is low combustion value and potential damage to the environment. The recent uses of RH are as a fertilizer additive, animal food, stock breeding rugs, cooking fuel, paving applications, filling materials, and energy generation. RH is composed of lignin, cellulose, and SiO_2, therefore it can be considered as a natural organic-inorganic composite material. In the other hand, as the enormous spare part development nowadays, the use of asbestos-based materials in the manufacture of vehicle brake pad is considered to endanger health, especially brake pad. Asbestos-based materials are used because of their temperature resistance. However, the use of asbestos has been banned due to carcinogenic substances and substances that are harmful to the environment. Asbestos-based materials can also endanger respiratory health because they can endanger the lungs if inhaled. The objective of this study is to add value to rice husk by developing a sustainable material composite. The rice husk was combine with biodegradable polymer as matrix. This study aims to determine the effect of particle size of rice husk powders, and structural mechanisms on non-asbestos brake composites. The ratio of reinforcement and matrix as follows 20:80; 30:70; 40:60 with carbonized and uncarbonized condition. The expected result is the best wear properties conducted from rice husk powder composites brake pad are in close with commercial asbestos based brake pad.

1. Introduction

Indonesia is one of the biggest country for agricultural production, such as rice, corn, sugar cane, etc. Annual rice production worldwide is over 700 million tons per year [1]. It means that agricultural waste produced very abundant. Rice Husk (RH) is one of the agro-industrial waste products that is most abundantly accumulated. The mechanical properties of RH is low combustion value and potential damage to the environment [2-5]. RH as an organic waste is obtained from the outer covering of rice grains during the milling process [6]. The recent uses of RH are as a fertilizer additive, animal food, stock breeding rugs, cooking fuel, paving applications, filling materials, and energy generation [7]. The uses of is waiting to be elevate. In the other hand the development mechanical spare part production has occurred enormous nowadays, The use of asbestos-based materials in the manufacture of vehicle brake pad is considered to endanger health [24]. Brake pads that are commonly used are composites of asbestos-based materials bound by a polymer-based and some other materials [8]. Asbestos-based materials are used because of their temperature resistance [9]. However, the use of asbestos has been banned due to carcinogenic substances and substances that are harmful to the environment [10], since January 1, 1997 in France [11]. Asbestos-based materials can also endanger respiratory health because they can endanger the lungs if inhaled [9].

Physical, mechanical, and thermal properties of low strength concrete containing different amount of original rice husk [12]. RH is composed of lignin, cellulose, and SiO2, therefore it can be considered as a natural organic-inorganic composite material [13]. The composition of RH as follows cellulose (50%), lignin (25-

30%), silica (15-20%), and moisture (10-15%). Carbonization of RH in inert atmosphere yields highly porous carbons with large specific surface area [14-15]. RH was obtain from burning of rice husk at around 600 °C, leading to CO2 release and energy consumption [16]. Recently, porous carbons prepared from rice husk attracted attention in the field of energy storage [17], solar technology [18], and environment mental protection [19]. There are several application of RH such as activated carbon. In physical activation, carbonization and activation occur separately but in chemical activation both the process of carbonization and activation occur simultaneously by using a chemical agent [20-21]. Process of physical activation is very useful in the generation of activated carbon, which have low specific surface area [21]. The objective of this study is to add value to rice husk by developing a sustainable material composite with improve combustion conditions and the ratio of fabrication composite. The rice husk was combine with biodegradable polymer as matrix. The ratio of reinforcement and matrix as follows 20:80; 30:70; 40:60.

2. Materials and Methods

2.1 Materials and Equipment

The materials used are rice husk, epoxy resin, and rubber silicon along with their catalyst. The equipment used in this study are electric furnace, electric blender , a set of sieves of 297, 177, and 149 μm (50,80, and 100 mesh respectively) , digital weighing balance, 1 inch diameter PVC pipe, polishing machine along with abrasive paper grit size 600, ruler, vernier caliper, and vacuum cleaner.

2.2 Methods

Rice Husk were collected from some rice field area in Bali after sun dried within 3 days. To ensure the moisture was decrease towards zero percent, rice husk dried in the electric furnace using temperature 70°C within 3 hours. Dried rice husk was charged into electric blender to reduce the particle size to form rice husk powder. Rice husk powder sieve manually to get rice husk powder of 250, 177, 149, and 125 μm. The oversize rice husk powder return to electric blender for regrinding process to get the finer particle. To form carbonized rice husk powder (Figure 1b), the uncarbonized rice husk powder (Figure 1a) packed using aluminium foil and charged into electric furnace within 30 minutes in 500°C. Each variant size of particle mix with epoxy resin and its catalyst using variant volume percentage composition. Samples were produced by varying rice husk powder added from 20–40 Vt% with interval of 10 Vt% in. Two set of samples were produced using the uncarbonized and carbonized rice husk powder.

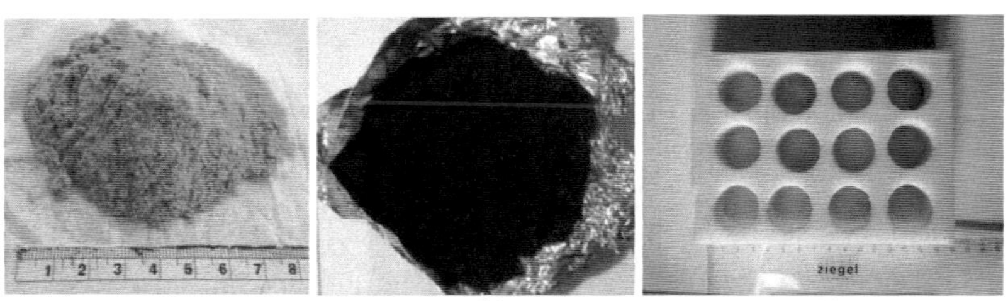

(a) (b) (c)

Figure 1. (a) uncarburized rice husk powder (b) carburized rice husk powder
(c) Mold of brake pad polymer composite model

Before casting the mixture, mold made using rubber silicon and 1 inch diameter PVC pipe as a pattern (Figure 1c). Mixture for each composition was blended manually for a period of 2 minutes and assumed homogeneous then cast inside the dies and let them dry in room temperature through agitation and vacuum process before. Model of brake pad polymer composite asbestos free shown in Figure 2a. Besides the model, the prototype of brake pad polymer composite asbestos free was also produced shown in Figure 2b.

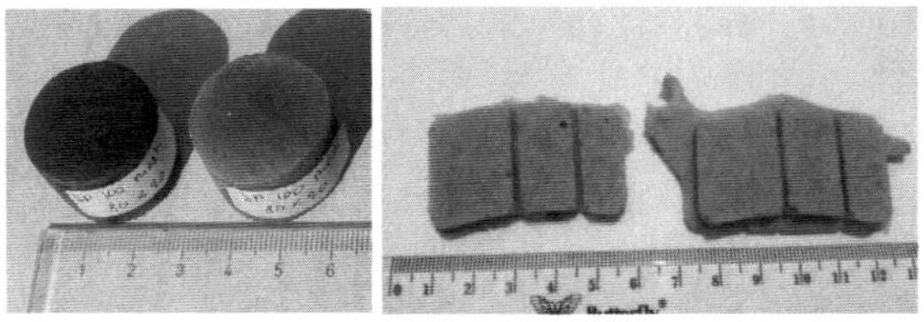

Figure 2. Brake pad polymer composite asbestos free specimen (a), prototype (b)

Polishing machine used for wear resistance characterization and the test sample was placed in the test setup, with following parameter, measuring period occurs for 10 minutes, velocity used equal 300 RPM, clockwise rotation, and 600 grit size abrasive paper drop of water needed along the wear resistance characterization procedure. To give the composite specimen well distributed load, 200 gr specimen dummy placed above the specimen while the procedure occurs. The principle idea doing wear resistance characterization using polishing machine was adopted from wear resistance characterization using Ogoshi method [22], whom use abrasive media to mention the wear resistance properties. Samples were weighed and height measured before and after the experiment to get weight and height loss. Figure 3 shown the process of wear resistance characterization. Wear resistance value provided with calculation using equation below:

$$W_s = \frac{\Delta V}{F_n . L} = \frac{\Delta m}{\rho . F_n . L} \text{ [23],} \qquad (1)$$

Where:

W_s : Wear resistance specific properties of samples
ΔV : Volumetric material loss (m^3)
Δm : Weight loss (kg)
F_n : Normal force applied ($kg.m/s^2$)
L : Total distance travelled (m)
ρ : Density (kg/m^3)

Wherefore the samples are tube then the volumetric material loss (ΔV) equal with volume of tube with height of difference of initial and final height of samples. Total distance travelled (L) got from amount of abrasive disc rotation on the procedure multiply with it circumstance.

Figure 3. Wear resistance characterization using polishing machine

3. Results and Discussion

The results obtained from the wear resistance tests conducted in this study are given below for all samples with 3 times repetition for carbonized and uncarbonized particle described in Table 1 and Table 2 respectively. Wear properties expressed as velocity of consuming composite brake pad per kilograms. As higher the value stated lower wear resistance express.

IOP Conf. Series: Materials Science and Engineering **462** (2019) 012015 doi:10.1088/1757-899X/462/1/012015

Table 1. Wear resistance properties for carbonized particle

Particle Size (mesh)		50				80				100			
Carbonized Particle	Repetition Number	ΔW (gr)	ΔH (mm)	Wear Resistance Value (Ws)	Wear Resistance Value Average (Ws)	ΔW (gram)	ΔH (mm)	Wear Resistance Value (Ws)	Wear Resistance Value Average (Ws)	ΔW (gram)	ΔH (mm)	Wear Resistance Value (Ws)	Wear Resistance Value Average (Ws)
Composition (Matrix : Filler) % Vt 80:20	1	0,67	0,7	0,00074	0,00085	1,12	1,05	0,0011	0,0015	0,25	0,61	0,00065	0,00058
80:20	2	0,4	0,9	0,00096	±	1	1,8	0,0019	±	0,24	0,55	0,00058	±
80:20	3	0,67	0,8	0,00085	0,00011	1,02	1,4	0,0015	0,0004	0,31	0,48	0,00051	0,00007
70:30	1	1	1,2	0,00128	0,00090	0,42	2	0,0021	0,0021	1,04	2,7	0,00287	0,00283
70:30	2	0,47	0,65	0,00069	±	0,3	2	0,0021	±	1,11	2,75	0,00292	±
70:30	3	0,67	0,7	0,00074	0,00032	0,36	2,05	0,0022	0,0000	1,92	2,55	0,00271	0,00011
60:40	1	1,33	1,2	0,00128	0,00094	1,19	1,8	0,0019	0,0017	0,12	1,25	0,00133	0,00128
60:40	2	0,67	0,75	0,00080	±	1,17	1,3	0,0014	±	0,13	1,2	0,00128	±
60:40	3	0,67	0,7	0,00074	0,00029	1,06	1,6	0,0017	0,0003	0,2	1,15	0,00122	0,00005

Table 2. Wear resistance properties for uncarbonized particle

Particle Size (mesh)		50				80				100			
Uncarbonized Particle	Repetition Number	ΔW (gr)	ΔH (mm)	Wear Resistance Value (Ws)	Wear Resistance Value Average (Ws)	ΔW (gram)	ΔH (mm)	Wear Resistance Value (Ws)	Wear Resistance Value Average (Ws)	ΔW (gram)	ΔH (mm)	Wear Resistance Value (Ws)	Wear Resistance Value Average (Ws)
Composition (Matrix : Filler) % Vt 80:20	1	0,33	1,2	0,00128	0,0011	0,95	2,7	0,0029	0,0028	1,72	2,7	0,0029	0,0031
80:20	2	0,62	1	0,00106	±	0,96	2,45	0,0026	±	2,11	2,2	0,0023	±
80:20	3	0,33	0,8	0,00085	0,0002	0,61	2,85	0,0030	0,0002	2,42	3,9	0,0041	0,0009
70:30	1	1,33	0,75	0,00080	0,0007	2,54	3,35	0,0036	0,0029	1,71	2,4	0,0026	0,0022
70:30	2	0,67	0,67	0,00071	±	2,3	2,5	0,0027	±	1,82	1,75	0,0019	±
70:30	3	0,34	0,5	0,00053	0,0001	2,28	2,35	0,0025	0,0006	1,69	2,1	0,0022	0,0003
60:40	1	0,67	0,6	0,00064	0,0009	0,32	1,7	0,0018	0,0014	1,96	1,51	0,0016	0,0016
60:40	2	0,34	0,8	0,00085	±	0,33	1	0,0011	±	1,83	1,49	0,0016	±
60:40	3	0,67	1,1	0,00117	0,0003	0,49	1,35	0,0014	0,0004	1,9	1,45	0,0015	0,0000

Unseen the big difference in wear properties of carbonized and uncarbonized filler in Figure 4, then the carbonizing processs in rice husk filler not give a significantly increasment of wear resistance. Finer size particle not give a better wear resistance. Unlike composite with corn cob powder filler [24] or other natural fiber, mostly 297 μm particle of rice husk powder (50 mesh) filler gift the best wear resistance in composite brake pad either in carbonized or uncarbonized condition. The best wear resistance provides from adding 20% filler carbonized in 100 mesh size of particle.

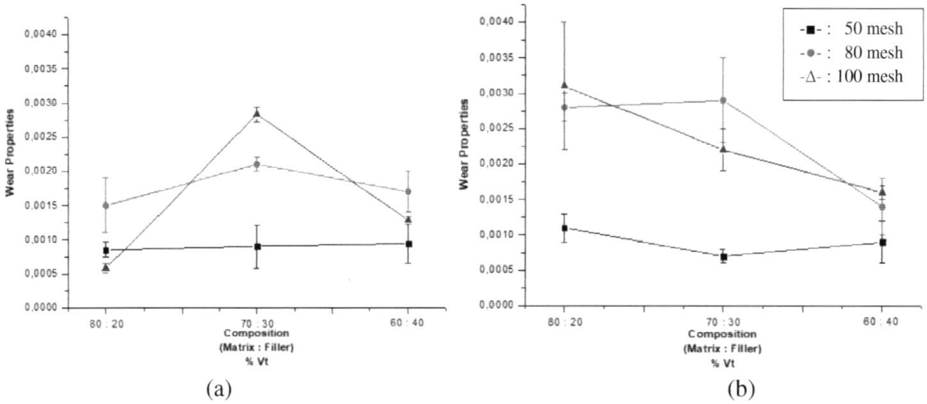

(a) (b)

Figure 4. Wear properties' graph of samples (a) carbonized and (b) uncarbonized filler.

Carbonized filler gift better bind ability in the mixing protocol, while the uncarbonized particle have tendency to coagulate as the increasing of particle size number. Effect of carbonize filler in wear properties value clearly shown in 80 mesh samples data. Found a similar trend among 2 set of data with better wear resistance for carbonized filler, also be in effect for other particle size in different filler condition. An addition of filler amount in carbonized filler decrease the wear resistance of the composite brake pad, in the opposite, in uncarbonized filler wear resistance indicates increase. An optimum result for carbonized filler conducted from composites with composition with the lower contain of filler. In the other hand, in uncarbonized filler optimum result shown from composite with higher contain of filler. The presence of high amount of uncarbonized rice husk particle in composite gift a harsh surface roughness and it gift they better wear resistance. Hence asbestos free brake pad can be produced using rice husk powder as the filler, the best wear properties conducted from rice husk powder composites brake pad are in close with commercial asbestos based brake pad.

4. Conclusion

From result and discussion in this work the following conclusions can be made: Rice Husk composite's wear resistance value shows where rice husk powder can be used for filler element in polymer epoxy composites. It provide better wear resistance value than asbestos brake pad in 20% filler carbonized in 100 mesh size particle, means it can replace the asbestos contained brake pad. Carbonized process gives slight increasing of wear resistance value for the composite.

REFERENCES

[1] FAO data, http://faostat3.fao.org, 2017 (accessed 8.08.2017).

[2] Liu, Y., et al., 2011. *A sustainable route for the preparation of activated carbon and silica from rice husk ash*. J. Hazard. Mater. 186 (2), 1314–1319.

[3] Shen, Y., Zhao, P., Shao, Q., 2014. *Porous silica and carbon derived materials fromrice husk pyrolysis char. Microporous Mesoporous Mater*. 188, 46–76.

[4] Alvarez, J., et al., 2014. *Upgrading the rice husk char obtained by flash pyrolysis for the production of amorphous silica and high quality activated carbon*. Bioresour. Technol. 170, 132–137.

[5] Liu, D., et al., 2016. *A green technology for the preparation of high capacitance rice huskbased activated carbon*. J. Clean. Prod. 112, 1190–1198.

[6] Shafigh, P., Mahmud, H.B., Jumaat, M.Z., Zargar, M., 2014. *Agricultural wastes as aggregate in concrete mixturese A review*. Constr. Build. Mater. 53, 110e117.

[7] Pode, Ramchandra, 2016. *Potential applications of rice husk ash waste from rice husk biomass power plant*. Renew. Sustain. Energy Rev. 53, 1468e1485.

[8] Idris, U.D., Aigbodion, V.S., Abubakar, I.J., Nwoye, C.I., 2015. *Eco-friendly asbestos free brake-pad: using banana peels*. J. King Saud Univ.-Eng. Sci. 27 (2), 185–192.

[9] Yawas, D. S., Aku, S. Y., & Amaren, S. G. 2016. *Morphology and properties of periwinkle shell asbestos-free brake pad.* Journal of King Saud University Engineering Sciences, *28*(1), 103–109.

[10] Ikpambese, K. K., Gundu, D. T., & Tuleun, L. T. 2016. *Evaluation of palm kernel fibers (PKFs) for production of asbestos-free automotive brake pads*. Journal of King Saud University - Engineering Sciences, *28*(1), 110–118.

[11] Lagel, M. C., Hai, L., Pizzi, A., Basso, M. C., Delmotte, L., Abdalla, S., … Al-Marzouki, F. M. 2016. *Automotive brake pads made with a bioresin matrix*. Industrial Crops and Products, *85*, 372–381.

[12] Sisman, C.B., Gezer, E., Kocaman, I., 2011. *Effects of organic waste (rice husk) on the concrete properties for farm buildings*. Bulg. J. Agric. Sci. 17 (1), 40e48.

[13] Guo, Y., et al., 2002. *The preparation and mechanism studies of rice husk based porous carbon*. Mater. Chem. Phys. 74 (3), 320–323.

[14] Liou, T.-H., Wu, S.-J., 2009. *Characteristics of microporous/mesoporous carbons prepared from rice husk under base- and acid-treated conditions*. J. Hazard. Mater. 171 (1), 693–703.

[15] Xiao, Y., et al., 2014. *Porous carbon with ultrahigh specific surface area derived from biomass rice hull*. Mater. Lett. 116, 185–187.

[16] Huang, H., Gao, X., Wang, H., Ye, H., 2017. *Influence of rice husk ash on strength and permeability of ultra-high performance concrete*. Constr. Build. Mater. 149, 621-628. https://doi.org/10.1016/j.conbuildmat.2017.05.155.

[17] Guo, Y., et al., 2003. *Performance of electrical double layer capacitors with porous carbons derived from rice husk*. Mater. Chem. Phys. 80 (3), 704–709.

[18] Wang, G., et al., 2014. *Hierarchical porous carbon derived from rice husk as a low-cost counter electrode of dye-sensitized solar cells*. Renew. Energy 63, 708–714.

[19] Li, D., et al., 2015. *Preparation of porous carbons with high low-pressure CO2 uptake by KOH activation of rice husk char*. Fuel 139, 68–70.

[20] Ugheoke, Benjamin Iyenagbe, et al Leonardo, *Electronic Journal of Practices and Technologies 9* (2006): 167-178.

[21] Deiana, Cristina, et al, *Industrial & Engineering Chemistry Research*47.14 (2008): 4754-4757.

[22] M. Taufik Qurohman and Syarifudin. 2016. *Analisa Beban Pengereman Terhadap Kualitas Kampas Rem Tromolmobil Dengan Metode Oghosi*. Jurnal Matematika Vol. 19, No.1, April 2016 : 8 -12

[23] Emre G, Akkus A, et al. 2004. *Wear resistance of Polymethyl Methacrylate (PMMA) with the Addition of Bone Ash, Hydroxylapatite and Keratin*. 9th International Conference on Tribology (Balkantrib'17).

[24] Primaningtyas, E Widya, et al. 2018. *Pengaruh Ukuran Partikel Serbuk Bonggol Jagung Terhadap Sintesis Komposit Kampas Rem Non-Asbestos*. Jurnal IPTEK Volume 22 Nomer 1, Mei 2018, 45–52.

Acknowledgments

This study fully funded by Directorate of Research and Community Service - Directorate General of Research and Development Strengthening - Ministry of Research, Technology, and Higher Education of the Republic of Indonesia.

Design of Wheelchairs Robot Based on ATmega128 to People with Physical Disability

[1]Syahri Muharom, [2]Tukadi, [1]Tjahja Odinanto, [1]Syadidatul Fahmiah, [1]Diana Putri Permata Siwi.

[1]Electrical Engineering Department, Institut Teknologi Adhi Tama Surabaya
[2]Informatic Engineering Department, Institut Teknologi Adhi Tama Surabaya
Jl. Arief Rachman Hakim No 100, Surabaya, Indonesia

syahrimuharom@itats.ac.id

Abstract Physical disability is individual to a person having disorder movement, due to its abnormality muscular, bone structure, sick or accident. Wheelchairs are a tool to help people with physical disability. From it, the researcher proposed the creation of a wheelchair robot that was able to be used to assist in daily activities. This robot uses the Atmega128 as the primary system and the motor driver. To movement wheelchairs robot used button command, it gives commands to the Atmega128, and then the motor will drive the wheelchair movement. In the first test is motor rotation setting without load, from the data, obtained the most significant wheel rotation difference from the command is 3rpm, this different rotation because wheel slip at wheelchair and the next test with load people weight 72kg, the difference wheel rotation is 88 rpm, it because the load people at wheelchairs robot. The last test is the command button on wheelchair movement, from the tests performed get a success rate of 87.5%, this is because there was an error when testing the command to move right. From this research expected the wheelchair robot can help people with physical disabilities carrying out at daily activities a wider area.

1. Introduction

Disability is the inability body to do an activity such as ordinary people in general. Disability physic is a people with disorder movement, it's caused abnormality muscular authentic, like paralysis, or accident such as amputation of body parts. The level of disability is divided be three part, where is the first level, is the inability to do physical activity, but it still can be done with a tool. The second level is inability physic because motorists sensor has been injured and can't move the body.

Moreover, the third level is a total limitation in physical movements, and unable to control physical changes [1]. Wheelchairs are a type of tool to help people with physical disability of leg to carrying out at daily activities. At the wheelchairs having four-wheel, where two small wheels at the front of wheelchairs and two big wheels at the back [2]. From it, many disability physics used the wheelchairs to assist carrying out at daily activities [3].

Hashimoto has developed wheelchairs robot, the movement of wheelchair robot using heart rate and stress level of the user, where from this research prioritising comfort level of using a wheelchairs robot and can be weighed is 60 kg[4]. From this research, the wheelchairs robot can't reach the wider area, because, to navigated of wheelchairs robot is difficult to a user, and the wheelchairs can a movement with load is 60 kg. At the research developed by bong Keun Kim, the movement wheelchairs robot certain only at corridor and room. The study used a camera to navigate wheelchairs robot [5]. This

research prioritising navigate the system of wheelchair robot to find a room target, from it the wheelchairs robot can't reach another area.
Microcontroller Atmega128 having flash memory capacity is 128kb (kilobyte), and microcontroller 8-bit CMOS series created by Atmel, the architecture based on RISC (reduced instruction set computer). In general, the AVR having four class is Attiny, AT90sxx, AT-Mega and AT86RFxx family. To identify AVR class is the memory, peripheral and function, from architecture and instruction all the same AVR series[6]. To movement the wheelchairs robot using DC motor[6], where the DC motor converts electrical energy to be rotation mechanical energy, at general the DC motor working with hight rotation[7]. From it to robot application need gearbox to reduce hight rotation be a low rotation[8]. From the mechanism, it will increase torque[9]. A rotary encoder is a mechanic component having a function to angular position monitor, from an angular position convert be digital data pulse width, and then connected to a controller. Based on the data result from the position of angular, and processed by controller be speed data and position of the axis. Implementation of rotary encoder many used in robotics to control wheel speed left and right [10]. From the background, the researcher provides a breakthrough to created of wheelchairs robot which can be used wider area, and the researcher will build wheelchair robot with a load more than is 70 kg.

2. Design and Method

From the research, result has several methods, how to getting characteristic and analysis data. Moreover, then getting several parameters optimally wheelchairs robot. Figure 1 is a design of wheelchairs robot.

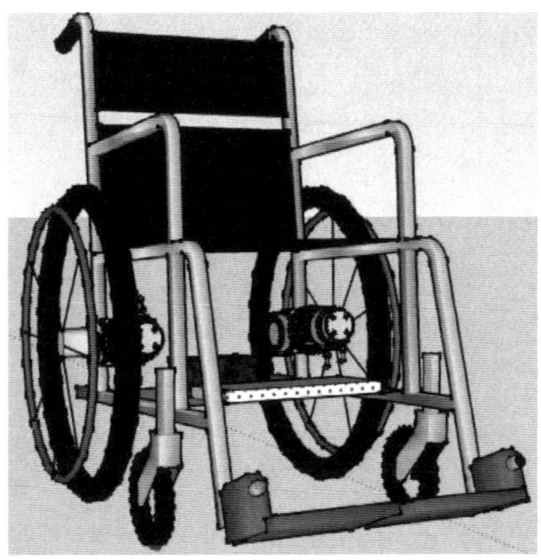

Figure 1. Design of Wheelchairs Robot

The design of wheelchairs robot, the wheelchairs having four wheels, two in front and two at the back. Moreover, the wheelchairs having two DC motor, the DC motor connected with the wheel of wheelchairs, the function of DC motor to wheel movement. To optimally the system of wheelchairs robot, the researcher makes a block system to movement wheelchairs. Moreover, figure 2 is diagram block system of wheelchairs robot.

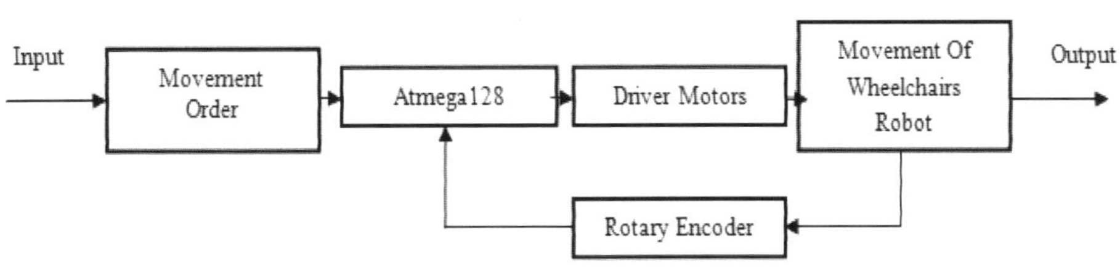

Figure 2. Diagram Block System

Block system at figure 2, the ordered movement giving a signal to Atmega128, and the Atmega128 processed it, and then providing a signal to driver motors, from driver motor giving the order to motors contained on wheelchairs, from the signal to movement of wheelchairs robot is forward, reverse, turns left and turn right. The encoder has a function to find out the wheel rotation and compared by signal order from main system Atmega128. Figure 3 is a flowchart system of wheelchairs robot

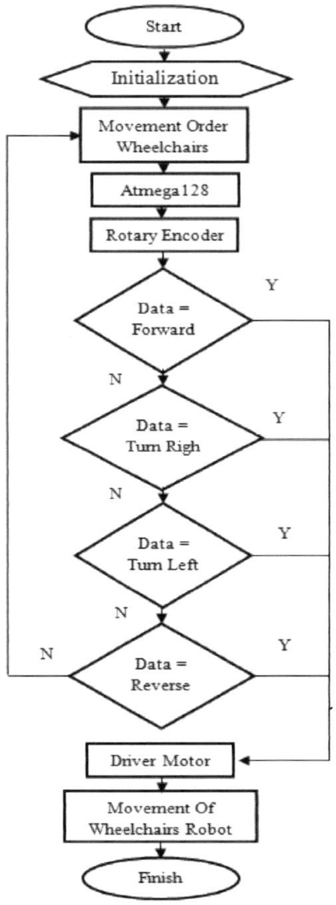

Figure 3. Flowchart System

From flowchart at figure 3, the movement of wheelchairs robot having four, forward, turn right, turn left and reverse, to movement order of wheelchairs robot using switch push button, from the flowchart system can know how to movement wheelchairs robot. Moreover, figure 3 is a design of wheelchairs robot.

3. Result and Discussion

This research having several findings, the first result is the set value of PWM (pulse width modulation) convert be rpm (rotation per minute), and then convert to be valued 0-100. Table 1 result of test rpm to wheelchairs robot without load.

Table 1. Result PWM Test Without Load

No	Input PWM (rpm)	Wheel Rotation (rpm)	Difference Rotation (rpm)
1	0	0	0
2	10	8	2
3	25	22	3
4	35	36	1
5	40	42	2
6	55	54	1
7	60	57	3
8	70	71	1
9	80	78	2
10	90	92	2
11	100	97	3

The result of this test is getting difference wheel rotation between input and output, this result influenced by several factors, among other wheel slipped. From the result, the most significant difference of wheel rotation is 3 rpm. Figure 4 is a graph of the relation between input PWM and output wheel rotation.

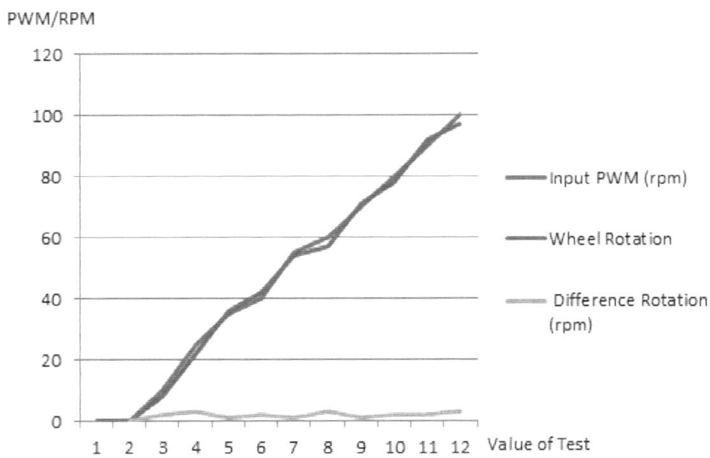

Figure 4. Graph Input Between Output Without Load Relation

The next test is the set value of PWM convert be rpm and then convert to be valued 0-100. Table 2 result of test rpm to wheelchairs robot with load weight people is 72kg.

Table 2. Result PWM Test With Load

No	Input PWM (rpm)	Wheel Rotation (rpm)	Difference Rotation (rpm)
1	0	0	0
2	10	0	10
3	25	0	25
4	35	2	33
5	40	2	38
6	55	3	52
7	60	5	55
8	70	8	69
9	80	9	71
10	90	10	80
11	100	12	88

From this test getting result difference input and output setting PWM, this result influenced by people load is 72kg. From this test the biggest difference wheel rotation is 88rpm. Figure 5 is the graph relation between input and output wheel rotation.

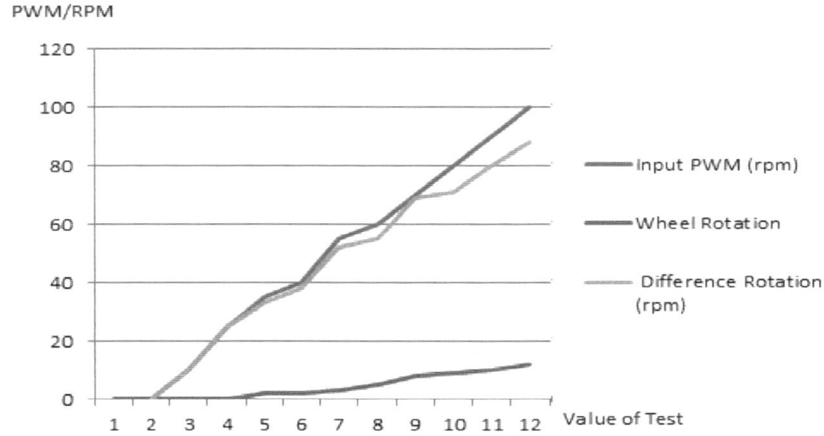

Figure 5. Graph of Relation Input Between Output With Load

The last discussion is movement order using the push button against the movement of wheelchairs robot, this test to getting a value of high success rate because it is having a function to the central navigation system. Furthermore, Table 2 is testing result of movement wheelchairs robot.

Table 3. The Result of Movement Wheelchairs Robot

No	Push Button Order	Movement of Wheelchairs Robot
1	Forward	Forward Moving
2	Turn Left	Left Moving
3	Turn Right	Little to Right Moving
4	Turn Right	Right Moving
5	Reverse	Reverse Moving

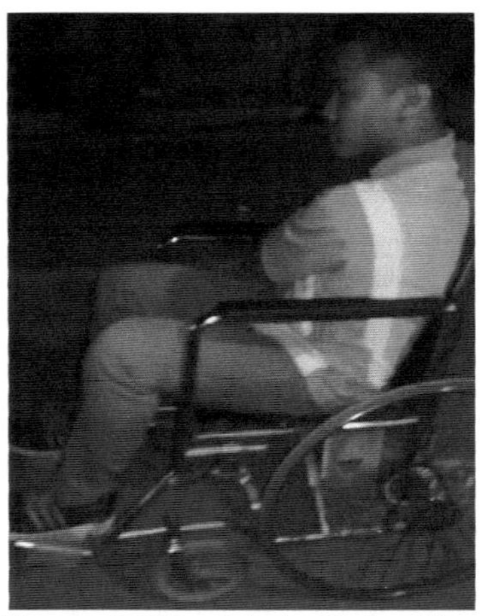

Figure 6. Wheelchairs Robot Test

The result at table 3, getting several data. From all test got one mistake, where contained at the data number 3, actually, the wheelchairs robot able to turn right is 90° degree, but the wheelchairs robot only turn right is 34° degree. It is because the system can't response order from push button to move the wheelchairs robot. From the data result, the system has a success rate is 87,5%. Moreover, at figure 6 is wheelchairs robot test.

4. Conclusion

From the result of a testing system, several data become point observation. The first test without load getting difference wheel rotation is 3rpm; it's because of slip at the wheel of the wheelchairs robot. Moreover, the second test with load weight people is 72kg, the difference of the set value of rpm input and output is 88rpm, because of a load of people at wheelchairs robot. The last test is button command to movement wheelchairs robot, from this test getting value of success rate is 87,5%. It is expected that with this wheelchairs robot can help people with physical disabilities activities at a wider area.

Reference

[1] Agus Riyanto, Mei 2013, United Nations Children's Found (UNICEF)"Rangkuman Eksekutif" UNICEF/HQ2007-0745/Noorani
[2] Syahri muharom, Tukadi,"Control of Wheelchair Robot Movement Using Flex Sensor Glove." Vol 3, No 2, (INFORM) Juli 2018.
[3] Iosif papadakis ktistakis and Nikolaos g. Bourbakis "Assistive intelligent robotic wheelchairs" Assistive Technology, IEEE PotEntIals, January/February 2017.
[4] Hashimoto R, Nomura R, Kanbara M, Ukita N, Nov-5-7, 2015 "Behavior Representation of Robotic Wheelchairs with Physiological Indices for Passenger Confort" Procedding IEEE, Yokohama, Japan.
[5] Bong Keun Kim, Tanaka H, Sumi Y, Oct, 13-16, 2015 "UML-Based Design of a Robotic Wheelchair System for Indoor Navigation Using a Visual Marker" ICCA.
[6] David Zier, 30 Maret 2003 "AVR Studio and Atmega 128 A Beginer's Guide" Oregon State University TekBots.
[7] Mariza Wijayanti, 2016 "Peralatan Energi Listrik, Motor Listrik"Pedoman Efisiensi Energi Untuk Industri se Asia.
[8] Djoko Purwanto, Ronny Mardiyanto, and Kouhei Arai , 2009 "Electric Wheelchair Control with Gaze Direction and Eye Blinking" The Fourteenth International Symposium on Artificial Life and Robotics (AROB 14th '09).
[9] Kohei Arai ,Ronny Mardiyanto, 2011 "Eyes Based Eletric Wheel Chair Control System" (IJACSA) International Journal of Advanced Computer Science and Applications, Vol. 2, No. 12.
[10] Widodo Budiharto, Djoko Purwanto, 2015 "Robot Vision" Teknik Membangun Robot Cerdas Masa Depan, Andi Publisher, ISBN: 9789792950199.

Feature Extraction Shape Kawi Numbers and Java Images Using The Zernike Moment

H Nugroho[1], W Widodo[1], R K Hapsari[1] and L A Hermanto[1]
[1]Department of Informatic Engginering, Faculty of Information Technology, ITATS, Jl Arief Rachman Hakim 100, Surabaya, Indonesia

dosh3ndro@itats.ac.id[1)]

Abstract. The writing of Kawi and Javanese numeric characters has different forms, in order to find out the extraction value of the image form features of Kawi and Javanese numerals, the Moment Zernike method is free from rotation. To prove the Zernike Moment method feature extraction, a number of rare research processes are carried out, the steps are (1) image input, (2) grayscale, (2) threshold and region growing segmentation process, (3) extraction of the Zernike Moment feature. The results obtained from the test from the nearest value approach using city block extraction feature Kawi and Java numeric images with extraction features of rotated features 90^0 and 180^0. The results obtained are the same as 0.01 for the Kawi 1 and 0.02 numbers for Java number 2.

1. Introduction

The development of ancient Javanese script writing originated from Pallawa script which experienced the improvement of the form of letters around the VIII century [1]. Early Javanese script writing is aksara kawi (750-925) [1]. During the Nusasantara kingdom there were Ddinoyo inscriptions from Malang, Sangkhara Inscriptions from Sragen, and the Plumpung Inscription from Salatiga. The development of Javanese Kawi characters ended in the days of the Medang kingdom, the kingdom of Kediri (925-1250 AD), and developed the Majaphit script (1250-1450 AD) for example the Kudadu inscription from Mojokerto.

From the ancient Javanese writings in the period to experience many changes from writing kawi to Javanese writing now. In this study the writing of ancient Javanese characters taken as research material is ancient Javanese numeric characters. Ten-based numerical writing already exists in our ancestors in the Kedukan Bukit inscription by reciting the number "sarivu tluratus sepuluh dua " [2].

The ancient Javanese digit numbers are taken as image data of ten of the Kawi characters and Javanese characters. To find out the writing pattern of ancient Javanese numerals (Kawi and Java), the image of ancient Javanese characters was used and the form feature extraction process was carried out in the image. The use of form feature extraction to get number arrangements that can be used to identify [3].

The form feature extraction is included among fourier descriptors, invariant moments, and Zernike moments[3]. The use of Zernike Moment is used for image processing for the first time by M.R. Teague in 1980 [4] [3]. The advantage of Zernike Moment which results in the form of Zernike Moment Descriptors (ZMD) which has advantages that are independent of the rotation, is not affected by noise and has a minimum of information redundation [5].

The development of Javanese numeric characters began with Pallawa script which experienced simplification of form around the eighth century. The ancient Javanese script called the Kawi script is a historical Brahmi script that was once used in Java and Bali. The kawi script in its development became the ancestor of traditional Javanese, Balinese, Sundanese characters, etc. [1]. In the period of development the Javanese script in writing was delivered in four parts, namely Javanese-Hindu script, Javanese-Islamic script, and Javanese-colonial script [6].

Javanese numeric script writing has the same number symbol as the Hindu-Arabic numerical system base, namely 0, 1, 2, 3, 4, 5, 6, 7, 8, and 9. For examples of writing the symbol of Jawa Kawi number can be seen in the picture 1. And for the symbols of Javanese numerals (Javanese-Hindu, Javanese-Islamic, and Javanese-Colonial) in Figure 2.

In Figure 1 and Figure 2 which will be input material for Javanese script pattern recognition research that is taken only one number to get an extracurricular pattern of form features from the Javanese numeric symbol.

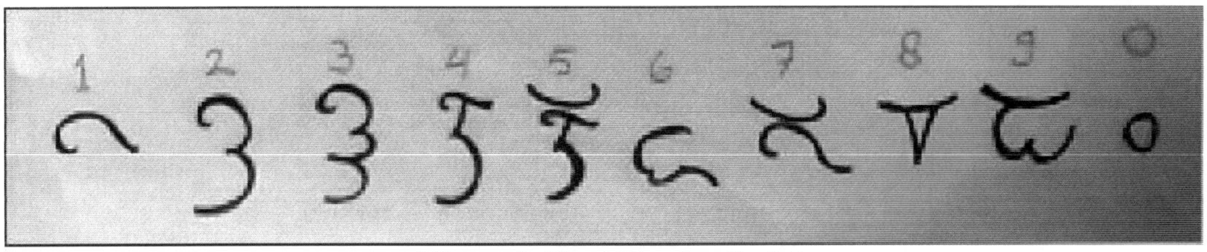

Figure 1. Kawi Number Script

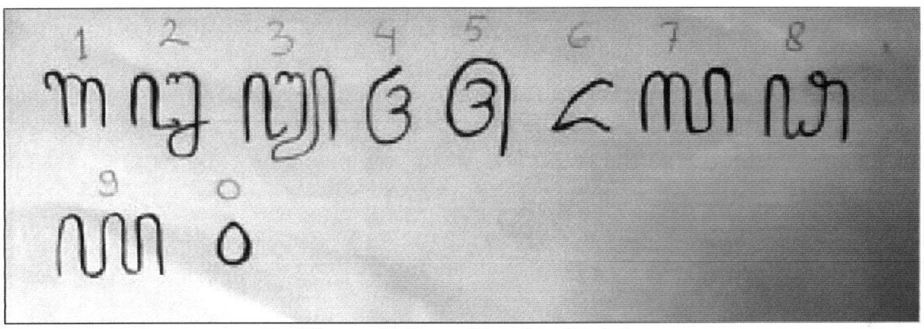

Figure 2. Jawa Number Script

2. Method

To get an introduction to the image pattern of ancient Javanese script writing, namely Kawi script and its developmental Javanese script (Figure 1 and Figure 2), the extraction feature of the Javanese script writing feature is needed. Before getting the Javanese numeral image feature extraction values the first step is segmentation of the Old Javanese numeral image image to get the binary value in the image and after that the form feature extraction process is performed. The segmentation used in this study is Region Growing and feature extraction used is the Zernike Moment.

2.1. Segements Region Growing

Image segmentation is a process that is shown to get objects contained in the image or divide the image into several areas with each object or area having similar attribute [3]. Separation between background and foreground images can use edge detector approaches or similarities between pixels in an area [7]. Based on segementation techniques can be divided into four categories, namely [8]. Threshold, edge-based, region-based, and hybrid methods.

The method used in this research is region-based because it only requires image segmentation in homogeneous regions of the Old Javanese numeric script written in black. The Growing Region method by starting several pixels (Seed) which represent different image regions and growing the region to fill the entire image.

For the Region Growing method, researchers have rules governing the mechanism for the growth of seeds and another rule that tests the homogeneity of the region after one stage of growth is complete.

The mechanism for growing seeds at each level and for each Region Ri(k), i=1,2,...,N and tested whether there are pixels that have not been classified in the 8-neighboring of each pixel on the edge of the region.

The allocation of a pixel x to the Region Ri (k), is tested whether the homogeneous region P P(Ri(k) ∪ {x}) = TRUE, is valid.

2.2. Metode Moment Zernike

Zernike's moment was introduced by F, Zernike (1934), the application of the Zernike moment for form feature extraction in the form of Zernike Moment Descriptors (ZMD). The Zernike moment is based on the Zernike polynomial that is orthogonal to the circle $x^2 + y^2 \leq 1$, which is stated as equation 1[3].

$$V_{pq}(x,y) = U_{pq}(r\,cos\theta, r\sin\theta) = R_{pq}(r).\exp(jq\theta) \tag{1}$$

With r is the radius of (x, y) the center of mass (centroid), θ is the range between r and x axis. $R_{pq}(r)$ is an orthogonal radial polynomial like equation 2

$$R_{pq}(r) = \sum_{s=0}^{(p-|q|)/2}(-1)^s \frac{(p-s)!}{s!\left(\frac{p+|q|}{2}-s\right)!\left(\frac{p-|q|}{2}-s\right)!}\rho^{p-2s} \tag{2}$$

In this case n = 0,1,2,...; $0 \leq |q| \leq n$, $j = \sqrt{-1}$, and p-|q| even value.
The Zernike moment codes p with the repetition of the continuous function f (x, y) as much as q is stated in equation 3.

$$Z_{pq} = \frac{p+1}{\pi}\int_y \int_x f(y,x).V^*{}_{pq}(y,x)\,dy\,dx;\; x^2 + y^2 \leq 1 \tag{3}$$

In this case, V * denotes the conjugate, while Vpq (x, y) is named as the function of the Zernike base with p-order with repetition of q. This function is in equation 4.

$$V_{pq}(y,x) = V_{pq}(\rho,\theta) = R_{pq}(\rho).\exp(jq\theta) \tag{4}$$

With p in the form of zero or positive integers and n = p- | q | even and | q | ≤ p.
If the digital image input f (x, y) using equation 4 can be used with equation 5.

$$Z_{pq} = \frac{p+1}{\pi}\sum_y \sum_x f(y,x).V^*{}_{pq}(y,x) \tag{5}$$

If the image is rotated at an angle of α Zernike moment functions in the form of equation 6.

$$Z'_{pq} = Z_{pq}.e^{-jq\varphi} \tag{6}$$

Equation 6 states that if only the magnitude of the Zernike moment used will be obtained the feature does not depend on the rotation.

Low-order Zernike polynomials are useful for obtaining global features of a form, while high-order polynomials can capture local form or feature details [9].

3. Result

The workmanship process in this study to get an introduction to the writing patterns of Old Javanese numeric numerals can be seen in the diagram of the workmanship steps that can be seen in Figure 3. In the figure 3 there are several steps in the process, namely the input of ancient Javanese numerals in the form of Kawi numerals and Javanese numeric characters, the process of grayscale image to get gray values in the image that will be used by the segemnetation process, the segmentation process uses the Threshold method to get binary values (black and white) and Growing Region to get the region of the number object from the binary image and to eliminate noise, and the last is extracting form features using the Zernik moment method.

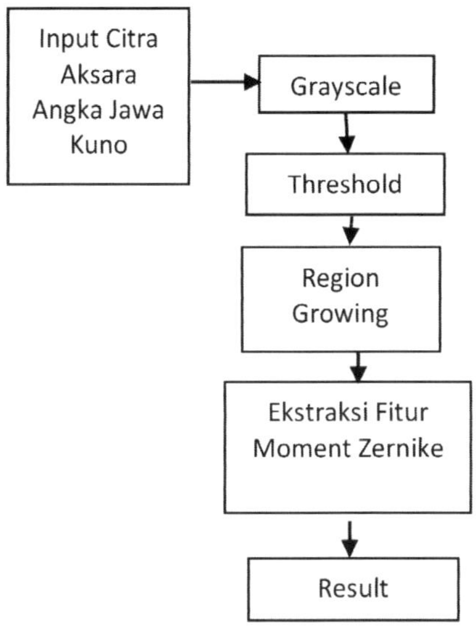

Figure 3. Introduction to Kawi and Javanese Literacy Patterns

The results obtained from the introduction of the Old Javanese Numerical Pattern pattern can be seen in figure 4. The results of the process begin the image input of one ancient Javanese numeric script, the grayscale process, the segmentation process of the Threshold, and Growing Region processes. The results of the Zernike Moment form extraction results in the form of tables resulting from Figures Kawi and Java as research material for Old Javanese writing patterns.

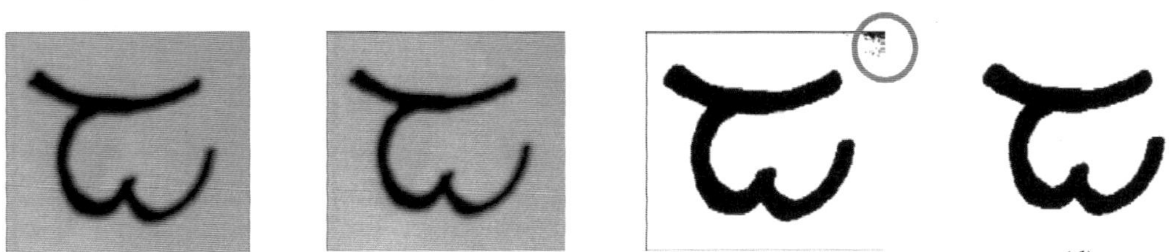

(a) (b) (c) (d)

Figure 4. (a) Input image of Kawi Number 9, (b) Image of Grayscale, (c) Image of binary binary (black and white) there is noise (red circle), (d) Citra Region Growing

The results of the extraction of the form feature of the Moment Zenike method have seven moment values based on the equation 1 to 6. The extraction features of the Moment Zenike method image of the Java Kawi number are presented in table 1, and for the Java numeral images are presented in table 2.

In table 1 extraction of the Moment shape feature Zenike numbers Kawi has seven moment values for the kawi image starting from number 0 to number kawi 9. From the extraction value the Zernike Moment form feature for the numerical image pattern recognition research can be compared to the extraction feature shape the image of the Java number presented in table 2.

Table 1. Zernike Moment Form Extraction Result Result of Kawi

Nama Citra	Moment Zernike							
	Z1	Z2	Z3	Z4	Z5	Z5	Z6	Z7
Kawi 0	0,000654	0,267173	0,000291	0,009712	0,117361	0,00283	0,207766	0,000654
Kawi 1	0,113842	0,501926	0,003291	0,008793	0,001037	0,254577	0,105368	0,113842
Kawi 2	0,026417	0,356537	0,003176	0,018925	0,093412	0,016929	0,092113	0,026417
Kawi 3	0,035137	0,361862	0,003596	0,03092	0,087044	0,034078	0,112283	0,035137
Kawi 4	0,063775	0,423388	0,001728	0,018131	0,055437	0,092623	0,028562	0,063775
Kawi 5	0,066467	0,396512	0,006539	0,027184	0,056722	0,09393	0,101055	0,066467
Kawi 6	0,053727	0,359792	0,003433	0,039916	0,078998	0,078066	0,123954	0,053727
Kawi 7	0,082643	0,419454	0,006833	0,015119	0,040018	0,1497	0,069585	0,082643
Kawi 8	0,072239	0,367871	0,010446	0,034853	0,063072	0,130536	0,149305	0,072239
Kawi 9	0,037403	0,318909	0,007814	0,027234	0,096436	0,050406	0,160168	0,037403

Table 2 results of extras extras on the Zernike Moment form feature on the Java Numbers image are also presented in the form of a table consisting of ten Javanese numeric images from numbers 0 to 9, and have seven moment values from the Moment Zenike method.

Table 2. Zernike Figures Java Moment Extraction Feature Features

Nama Citra	Moment Zernike							
	Z1	Z2	Z3	Z4	Z5	Z5	Z6	Z7
Jawa 0	0,024287	0,30313	0,001374	0,01975	0,10436	0,034309	0,190437	0,024287
Jawa 1	0,047861	0,360196	0,001293	0,037244	0,081885	0,007221	0,117963	0,047861
Jawa 2	0,036292	0,330865	0,003227	0,023391	0,097185	0,060691	0,125705	0,036292
Jawa 3	0,036497	0,330994	0,003341	0,023902	0,097208	0,061488	0,125056	0,036497
Jawa 4	0,038666	0,370691	0,002497	0,010592	0,088641	0,044132	0,088784	0,038666
Jawa 5	0,021895	0,319713	0,00489	0,036913	0,105536	0,021553	0,140585	0,021895
Jawa 6	0,010727	0,276985	0,005851	0,013508	0,11376	0,030554	0,168103	0,010727
Jawa 7	0,067674	0,400405	0,006638	0,02011	0,05912	0,100642	0,070168	0,067674
Jawa 8	0,047183	0,365176	0,002128	0,008097	0,081003	0,056895	0,098707	0,047183
Jawa 9	0,083648	0,449517	0,003811	0,015968	0,035971	0,154403	0,007435	0,083648

4. Discussion

To find out the success rate of the Moment Zenike method in the Kawi and Javanese numeric scripts, tests were carried out on the image of 90^0 and 180^0 rotation of the Kawi and Javanese numerals. This image is played to test the Zernike Moment free of rotation. Tests on rotation images of Kawi numerals can be seen in table 3.

IOP Conf. Series: Materials Science and Engineering **462** (2019) 012017 doi:10.1088/1757-899X/462/1/012017

Table 3. Example of Zernike Moment Shape Extraction Feature The Kawi 1 number is rotated

Citra Kawi	Moment Zernike							
	Z1	Z2	Z3	Z4	Z5	Z5	Z6	Z7
	0,113842	0,501926	0,003291	0,008793	0,001037	0,254577	0,105368	0,113842
	0,112789	0,502538	0,004833	0,010295	0,000399	0,251874	0,10648	0,112789
	0,112902	0,502563	0,006256	0,008231	0,00022	0,25236	0,107374	0,112902

In table 4 also presents the results of extraction in Javanese numeric script with the same the rotation process as the Kawi numerals, 90^0 and 180^0.

Table 4. Example of Zernike Moment Feature Extraction Results of Java 2 Numbers that are Rotated

Citra Kawi	Moment Zernike							
	Z1	Z2	Z3	Z4	Z5	Z5	Z6	Z7
	0,036292	0,330865	0,003227	0,023391	0,097185	0,060691	0,125705	0,036292
	0,034375	0,32837	0,005542	0,022631	0,098528	0,05826	0,131549	0,034375
	0,033614	0,329293	0,006774	0,027526	0,097911	0,055311	0,127374	0,033614

To find out the Zernike Moment method is free from the rotation, it is tested by approaching the feature extraction value of the Kawi numerals and Javanese characters which have not been rotated with rotated Kawi numerals presented in Tables 3 and 4.

Approach to knowing the similarity of form feature extraction values using the closest distance to the City Block method. The results that are pegged at the nearest value in the Kawi 1 numeral can be seen in table 5.

Table 5. Distance Result Nearest Zernike Moment Shape Feature Extraction Image of Kawi 1

Citra Kawi	Nilai	
	Rotasi (derajat)	Jarak Terdekat dengan citra Angka Kawi 1 Tanpa di Rotasi
	90	0,01
	180	0,01

Table 6. Distance Result Nearest Zernike Moment Shape Feature Extraction Image of Java 2

Citra Kawi	Nilai	
	Rotasi (derajat)	Jarak Terdekat dengan citra Angka Kawi 1 Tanpa di Rotasi
	90	0,02
	180	0,02

Table 5 and table 6 show the image of Kawi numerals and Javanese scripts which are manually rotated 90^0 and 180^0, calculated the proximity of Zernike Moment feature extraction value with the image that is not rotated has the same value of 0.01 (double digit) and 0 , 02 (rounded by two digits).).

5. Conculsion

The conclusion of the research extraction of the shape features using the Zernike Moment Method on the image of Kawi and Javanese numerals has different values in each Kawi numeral image and Javanese number. For images to be rotated by Kawi numerals and Java 900 and 1800 numbers manually as material for testing the Zernike Moment method free from the rotation shows the same value for the closest proximity using City Block method which is 0.01 for the image of Kawi 1 and 0 characters, 02 for Javanese numerals 2

6. References

[1] Wikipedia, "Wikipedia," Wikipedia, [Online]. Available: https://id.wikipedia.org/wiki/aksara kawi. [Accessed 14 Agustus 2018].

IOP Conf. Series: Materials Science and Engineering **462** (2019) 012017 doi:10.1088/1757-899X/462/1/012017

[2] A. Prabowo, "Goresan Angka Sang Citralekha," *Bersains,* vol. 1, no. 10, Oktober 2015.
[3] S. A. Abdul Kadir, Teori dan Aplikasi Pengolahan Citra, Yogyakarta: Andi Yogyakarta, 2013.
[4] Q. Y. X. a. Z. J. Chen, "Robust Image Watermaking with Zernike Moments," *on Proc. of the IEEE CCECE/CCGEI,* , pp. 1340-1343, 2005.
[5] Y. K. K. J. R. Mingqjang, "A Survey of Shape Feature Extraction Techniques on Pattern Recognition Technique," *Technology and Application,* pp. 43-90, 2008.
[6] S. A. P. S.S, "Kilas balik kelengkapan aksara Jawa dari masa ke masa," SMA 2 Wonosobo, Wonosobo.
[7] T. Acharya and A. Ray, "Image Processing Principles and Applications," John Wiley & Son, Inc, New Jersey, 2005.
[8] R. Rangyyan, "Biomedical Image Analysis," CRC Press, Boca Rotan, 2005.
[9] R. Choras, "Image Feature Extraction Techniques anf Their Applications for CBIR anf Biometrics System," *International Journal of Bilogy and Biomedical Engineering,* vol. 1, no. 1, pp. 6-16, 2007.

Acknowledgments
Thank you to the Chair of LPPM and colleagues of Informatics Engineering lecturers who supported this research.

IOP Conf. Series: Materials Science and Engineering **462** (2019) 012018 doi:10.1088/1757-899X/462/1/012018

Automatic Identification of Acute Lymphoblastic Leukemia on Blood Cell An image Using Geometric Features

Rahmi Rizkiana Putri[1], **Eka Prakarsa Mandyartha**[2], **and Annisaa Sri Indrawanti**[1]

[1]Institut Teknologi Adhi Tama Surabaya
[2]Universitas Pembangunan Nasional Veteran Jawa Timur

rahmirizkianaputri@gmail.com

Abstract. Acute Lymphoblastic Leukemia (ALL) is a human blood cancer which causes most deaths among other types of blood cancer. It commonly occurs in children and teenagers. ALL is recognised when there is a presence of lymphoblast cell in human blood. Haematologists diagnose ALL using visual microscopic observation from a human periphery blood sample. The inspection takes a long time and is limited to the subjective experience of the haematologist, and therefore, causes differences in the diagnostic results among haematologists. In this study, the researchers proposed automatic ALL Identification system to solve the problem. The method used an image processing technique that analysed a human blood cells morphology. The system identified the differences between normal white blood cells and lymphoblast cells by considering the shape and the size of the nucleus and also the cytoplasm. There are 33 microscopic an image data of a human blood cell that were used in this research. Geometric features used in the study were an area, perimeter, eccentricity, equivDiameter, solidity, roundness and circularity of a human blood the nucleus and cytoplasm and also the ratio of area and the perimeter of the cells and the nuclei. Firstly, the nucleus and cytoplasm were extracted from the other component of a human blood cell using Gram-Schmidt as an orthogonalisation-based segmentation method and then using Otsu threshold. Support Vector Machine was then used to classify whether the white blood cells were lymphoblasts or normal cells. Lastly, ALL identification was evaluated by the parameter of sensitivity and misclassification rate metrics using k-fold cross-validation.

1. Introduction

Acute lymphoblastic leukaemia (ALL) is a type of blood cancer caused by malignant blood cells [6,15,16,17] usually recognised by the presence of abnormal white blood cells called lymphocytes in the peripheral blood [3]. Those abnormal Lymphocyte cells are white blood cells which are imperfect (*immature*), also called leukemic cells or lymphoblast. ALL most commonly occurs in children aged 2 to 5 years old and adults aged 50 years old or older [13,12,9]. ALL diagnostic is challenging to do because ALL has quite unsuspicious symptoms such as fever, anaemia, lethargy, and pain in the joints [12]. Haematology specialist performs ALL diagnosis by examining a peripheral blood sample. The blood test includes complete blood count (with flow cytometry) and visual examination using a microscope [2]. Full blood count is performed to count the amount of each of the 5-type white blood cells. This automatic counting instrument uses flow cytometry to measure cytochemical contained in the blood and to observe the blood cells' physical characteristics using a light detector [9]. The light detector uses electric impedance or fluorescence to identify the white blood cells' type. Although the results given by the counting instrument is very accurate, the instrument itself is unable to detect the morphology of the abnormal white blood cells, which is why the microscopic visual examination is also necessary. Visual examination is performed to identify lymphoblast based on the morphology. As a visual examination of the blood sample takes a long time, and the diagnosis results among haematologists cannot be used as a standard since the accuracy still depends on the experts' proficiency [9,6,12,5]. To solve the problem above, ALL automatic identification system is proposed. ALL automatic identification system analyses the morphology of blood cell an image using peripheral blood sample imaging; the result is not subjective; with standardised accuracy [9,12]. The difference between normal lymphocytes and abnormal lymphocytes is determined by the changes in the shapes and the sizes of the nucleus and cytoplasm. Normal lymphocytes have regular and smaller the nucleus shape and have the nucleus ratio with smaller cytoplasm if compared to lymphoblast cells. Therefore, the shape and the size of cytoplasm can be used to differentiate normal lymphocytes from lymphoblast (ALL cell). Geometric feature descriptors used in this study were the width, perimeter, eccentricity, equivalent diameter, solidity, roundness, and circularity of white blood cells' the nucleus and cytoplasm, as well as the ratio of area and the nucleus's perimeter to the cell. The nucleus and cytoplasm had to be first extracted from other blood component using Gram-Schmidt orthogonalisation-based segmentation method and Otsu threshold. Classifier Support Vector Machine (SVM) was also used to classify whether a white blood cell was considered normal or lymphoblastic.

2. Literature Review

2.1. Gram-Schmidt Orthogonalization.

An image segmentation using Gram-Schmidt orthogonalisation theory was implemented in the study proposed by [5] and [10]. Gram-Schmidt orthogonalisation is a method used for vector set orthogonalisation of the inner product space. Set of orthogonal vector $S' = \{u1,..., un\}$ generated from Gram-Schmidt orthogonalisation of vector set $S = \{v1,..., vn\}$ has the same sub-space with Gram-Schmidt S. projection as shown in the equation (1).

$$proj_u v = \frac{(u,v)}{(u,u)} u = (u,v)\frac{u}{(u,u)} \quad (1)$$

With (u,v) as the multiplication in vector *u* and *v*. Vector *v* is orthogonal toward vector *u*. The Gram-Schmidt process is defined by equation (2) as follows:

$$u_k = v_k - \sum_{j=1}^{k=1} proj_{u_j} v_k, e_k = \frac{u_k}{\|u_k\|} \quad (2)$$

Vector $u_1,\dots, u_k$ is an orthogonal vector, and vector $e_1,\dots, e_k$ is a normalised vector that generates a set of orthonormal. Based on the method above, for vector $S = \{v_1,\dots, v_n\}$, it can be found a vector with maximum orthogonality toward desirable vector, and also a vector with minimum orthogonality toward other vectors in N-dimension space. Vector w_k can be determined by using equation (3):

$$w_k = v_k - \sum_{j=1}^{k-1} proj_{v_j} v_k, k = 1,2,3 \quad (3)$$

Therefore, the result of the inner product of set S with w_k is shown by equation (4) below:

$$\begin{cases} \langle \boldsymbol{v_j}, \boldsymbol{w_k} \rangle = 0 & j \in 1,\dots,n \; dan \; j \notin k \\ \langle \boldsymbol{v_k}, \boldsymbol{w_k} \rangle = K & K \neq 0 \end{cases} \quad (4)$$

2.2. Multilevel Thresholding Otsu

Otsu threshold method is used to find single threshold value which divides pixels into a bi-level value. The development of Otsu bilevel thresholding method for cases in which there are more than one threshold value in multilevel thresholding that has been proposed by [1], [8] and [5].

3. Research Methodology

The imaging data observed in the study were microscopic imaging of peripheral blood smear sample which contained lymphoblast (ALL) cell and normal cell. The microscopic imaging of peripheral blood smear sample itself included regions such as red blood cell, white blood cell, platelet, and background

(Figure 1). The white blood cell is comprised of the nucleus and cytoplasm. Error! Reference source not found. shows the structure of a white blood cell.

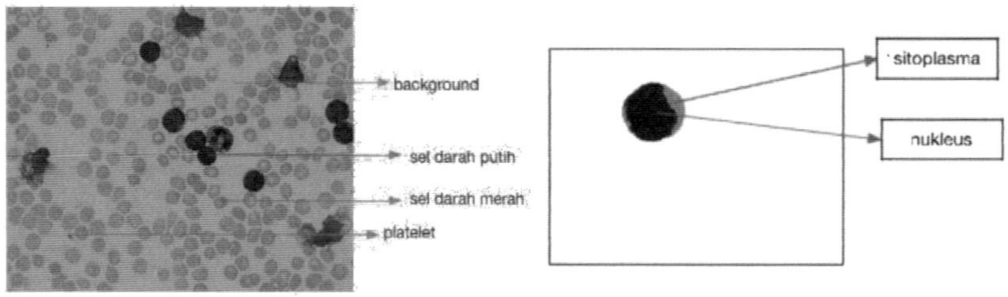

Figure 1. The physical structure of white blood cell

This study was divided into two parts. Those parts were segmentation process for extracting white blood cells' the nucleus and cytoplasm, which then followed by ALL identification. The white blood cells identification using Gram-Schmidt orthogonalisation was then followed by Otsu threshold. The identification was performed to divide the nucleus and cytoplasm. Geometric features used in the study were width, perimeter, eccentricity, equivDiameter, solidity, roundness, and circularity of the nucleus and cytoplasm, as well as the ratio of area and the edge of the cell and the nucleus. Those geometric features were used as attributes to classify white blood cells into lymphoblast or normal cells. Block diagram in Figure 2 represents the flow diagram of the complete process of automatic ALL identification.

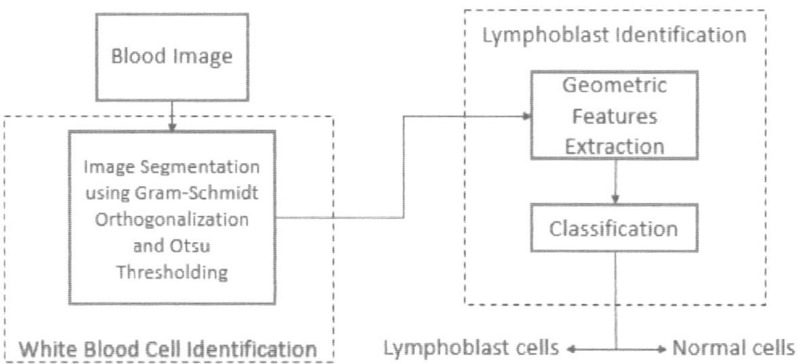

Figure 2. Flow diagram of the complete process of automatic ALL identification

3.1 White Blood Cell Segmentation

White blood cell segmentation was performed to extract white blood cells' nuclei and cytoplasm from the other blood components (red blood cell and platelet), including the background. In this study used segmentation method proposed by [5]. The algorithm steps of white blood cells segmentation are as the following:

1. Determine the feature vector for every pixel using imaging RGM component
2. Calculate weight vector w to strengthen the colour vector of the white blood cell *region* and to weaken unwanted colour vectors.
3. Calculate the inner product between weight vector w and the vector of input an images' pixels to produce composite an image with maximum intensity in the violet colour regions and minimum depth for the region of other colours, then continue with thresholding, and the result will be in binary an images

4. Cut input images into sub-an images, based on bounding box that surrounds connected component of regions in binary images.
5. For every sub-an image, do thresholding that finds out two threshold values using Otsu function: between class variance

3.2 Geometric Features Extractions

The ratio of area between the nucleus and cell (Figure 3) was determined by the surface width (the number of pixels) of the binary an image, Snu and Ssel, for the nucleus and cell area consecutively. The ratio of area between the nucleus and cell was defined by equation (5).

$$\mathrm{NCR}_{\mathrm{area}} = \frac{S_{nu}}{S_{sel}} \tag{5}$$

The perimeter ratio between the nucleus and cell was determined by the perimeter of the binary an image (the number of boundary pixel), Pnu dan Psel. The perimeter ratio between the nucleus and cell was defined by equation (6).

$$\mathrm{NCR}_{\mathrm{peri}} = \frac{P_{nu}}{P_{sel}} \tag{6}$$

Seven geometric feature values were counted. Those values were width, perimeter, eccentricity, equivDiameter, solidity, roundness and the nucleus circularity. Width value is the number of binary image pixels. Perimeter value is determined by the number of pixels surrounding (boundary) the binary an image (Figure 3(a)). Eccentricity value is determined by using the binary moment to count the ratio between major axis length and minor axis length from ellipse inertia as shown in equation (7).

$$\text{Eccentricity} = \sqrt{\frac{\text{major\_axis\_length}^2 - \text{minor\_axis\_length}^2}{\text{major\_axis\_length}}} \tag{7}$$

EquivDiameter is the diameter of a circle of the same width (region). EquivDiameter is determined using equation 8.

$$\text{EquivDiameter} = \sqrt{\frac{4 \times \text{width}}{\pi}} \tag{8}$$

Solidity is pixels proportion in the convex hull which was also in the region and was counted using equation (9). The convex hull is the smallest convex polygon which contains region. Convex area is the number of pixels on the convex an image (convex hull with all pixels in the hull filled), as shown by figure 3(b).

$$\text{Solidity} = \frac{\text{width}}{\text{convex\_area}} \tag{9}$$

Roundness was calculated using equation (10). Significant axis length is the pixel length of major axis ellipse (Figure 3(c)) which is the same with the second central moment from the region.

$$\text{Roundness} = \frac{4 \times \text{width}}{\pi \times \text{major\_axis\_length}^2} \tag{10}$$

Lastly, Circularity or the form factor is calculated using equation (11).

$$\text{Circularity} = \frac{\text{perimeter}^2}{4 \times \pi \times \text{luas}} \tag{11}$$

Therefore, the number of the element from geometric feature vectors was fourteen, seven features each for the nucleus and cytoplasm.

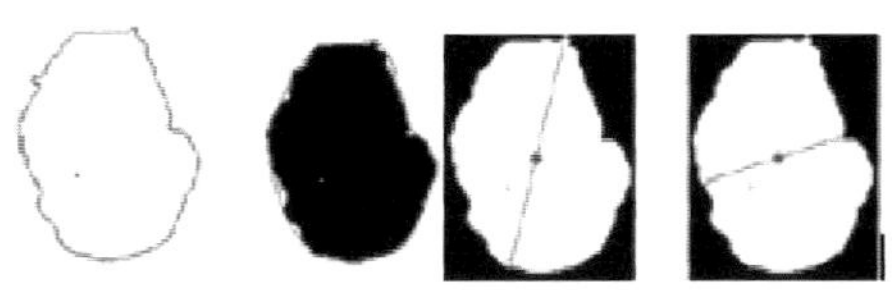

(a) (b) (c) (d)

Figure 3. Geometric features [14]: (a) perimeter, (b) convex area, (c) major axis length, (d) minor axis length.

3.3 Classification

For the purpose of training and classifying, the algorithm of classifier Support Vector Machine (SVM) with linear kernel function was used. Training algorithm took training data to be trained by classifier, so that training model could be determined. Training model was then used in classifying new data into two classes: ALL cells and normal cells.

4. Results and Discussion

The *Dataset* of the microscopic peripheral blood an image (blood-smear images) used in the study was acute lymphoblastic leukaemia an image database ALL-IDB1 provided by Donida Labati, *et al.* [4]. ALL-IDB1 consists of 108 24-bit resolution JPG images. For system trial in this study, the researchers used 35 blood cell images comprised 269 white blood cells, taken by Olympus C2500L camera (1712 x 1368 pixel resolution) under the same lighting setting. The number of the successfully segmented white blood cell was 263 white blood cells with an average accuracy of 98%. Each segmented white blood cell had its 16 geometric features calculated. To evaluate the suggested method, k-fold cross-validation (k = 10) method was used. Since the problem of the study was the binary classification, which was the identification of ALL cells and normal cells, the classifier algorithm Support Vector Machine (SVM) with linear kernel was used. The classification result was analysed using a confusion matrix. This matrix, as shown in Table 1, contains information about the actual value and image prediction testing obtained from SVM. True Positive (TP) is the number of ALL cells correctly classified as ALL cells. False Negative (FN) is the number of ALL cells classified as normal cells. False Positive (FP) is the number of normal cells classified as ALL cell, and True Negative (TN) is the number of cells correctly classified as a normal cell.

Table 1. Confusion Matrix

	Actual Class	
Prediction Class	ALL Cell	Normal Cell
ALL Cell	True Positive (TP)	False Positive (FP)
Normal Cell	False Negative (FN)	True Negative (TN)

The result was then analyzed using two evaluation measures which were sensitivity and misclassification rate. Sensitivity is the probability of ALL cell being identified correctly defined by the equation (12) below:

$$\text{Sensitivity} = TP/(TP+FN) \quad (12)$$

Misclassification determines the number of cells that are classified incorrectly; Misclassification is defined by the equation (13) below:

$$\text{Misclassification rate} = (FP+FN)/(TP+FP+FN+TN) \quad (13)$$

The calculation results are shown in Table 2.

Table 2. The Evaluation of the Proposed Method Performance

Evaluation Matrix	Value
TP	247
TN	6
FP	6
FN	4
Sensitivity	0.984
Misclassification	0.061

Table 2 shows that most ALL cells were identified as normal cells which means that ALL cells were correctly classified, proven by 98.4% Sensitivity and 6.1% misclassification rate.

5. Conclusion

In this study proposed the use of automatic acute lymphoblastic leukaemia (ALL) cells identification method by processing and analysing peripheral blood-smear an images. The technique provided an automated procedure to support medical examination, especially to help acute lymphoblastic leukaemia screening. The process was initialized by identifying white blood cell using threshold technique, which then continued by geometric features extraction for classification using SVM Linear. The results of the experiment showed that the proposed method was able to classify all extracted white blood cells; both lymphoblastic cells (ALL cells) and normal cells, with 98.4% sensitivity and 6.1% misclassification rate. The focus for further development of the proposed method was on the phase of separating overlapped blood cells, which was very important in calculating all individual white blood cells on the microscopic peripheral blood an image. In addition, the multi-class classifier algorithm was also needed in identifying the types of acute lymphoblastic leukaemia.

6. References

[1] Huang, D C, Hung, K D, & Chan, Y K 2012. *A computer-assisted method for leukocyte the nucleus segmentation and recognition in blood smear images. Journal of Systems and Software, 85*(9), 2104-2118.

[2] Inaba, H, Greaves, M, & Mullighan, C G 2013. *Acute lymphoblastic leukaemia. The Lancet, 381*(9881), 1943-1955.

[3] Isnani, N, Perwitasari, D A, Andalusia, R, & Mahdi, H I 2014. *Evaluasi Toksisitas Hematologi Akibat Penggunaan 6-Merkaptopurin Dalam Fase Pemeliharaan Pada Pasien Pediatri Kanker Leukimia Limfoblastik Akut Di RS Kanker Dharmais Jakarta.* Media Farmasi, 11(1), 90-97.

[4] Labati, R D, Piuri, V, dan Scotti, F 2011, "*All-IDB: The Acute Lymphoblastic Leukemia An image Database for An image Processing*", *Proceedings of the 18th IEEE ICIP International Conference on An image Processing*, Eds: Macq, B., dan Schelkens, P., IEEE Signal Processing Society, Brussels, 2045-2048.

[5] Mandyartha, E P, & Fatichah, C 2016. *Three-level Local Thresholding Berbasis Metode Otsu untuk Segmentasi Leukosit pada Citra Leukemia Limfoblastik Akut. Jurnal Buana Informatika, 7*(1).

[6] Mohapatra, S, Patra, D2, & Satpathy, S 2014. An ensemble classifier system for early diagnosis of acute lymphoblastic leukemia in blood microscopic an images. Neural Computing and Applications, 24(7-8), 1887-1904.

[7] Nanni, L, Lumini, A, & Brahnam, S 2010. *Local binary patterns variants as texture descriptors for medical an image analysis.* Artificial Intelligence in Medicine , 49, 117–125.

[8] Oliva, D, Cuevas, E, Pajares, G, Zaldivar, D, & Perez-Cisneros, M 2013. *Multilevel thresholding segmentation based on harmony search optimization. Journal of Applied Mathematics, 2013.*

[9] Putzu, L, Caocci, G, & Di Ruberto, C 2014. *Leucocyte classification for leukaemia detection using an image processing techniques. Artificial intelligence in medicine, 62*(3), 179-191.

[10] Rezatofighi, S H, & Soltanian-Zadeh, H 2011. *Automatic recognition of five types of white blood cells in peripheral blood. Computerized Medical Imaging and Graphics, 35*(4), 333-343.

[11] Salsabila, K 2016. *Perbandingan Nilai Ureum Dan Kreatinin Sebelum Dan Sesudah Kemoterapi Fase Konsolidasi Pada Pasien Leukemia Limfoblastik Akut Anak Di RSUP H. Adam Malik Medan* Tahun 2011-2013.

[12] Singhal, V, & Singh, P 2016. *Texture Features for the Detection of Acute Lymphoblastic Leukemia.* In *Proceedings of International Conference on ICT for Sustainable Development* (pp. 535-543). Springer Singapore.

[13] Widjajanto, P H, Veerman, A J P, & Sutaryo, S 2016. *Apoptotic cell identification: An in-vivo study during induction treatment of childhood acute lymphoblastic leukemia*. Paediatrica Indonesiana, 46(5), 195-8.

[14] Singhal, V, & Singh, P 2014. *Local Binary Pattern for automatic detection of Acute Lymphoblastic Leukemia*. In 2014 Twentieth National Conference on Communications (NCC) (pp. 1–5). IEEE. doi:10.1109/NCC.2014.6811261

[15] Thanh, T.T.P, Vununu, C., Atoev, S., Lee, H-S., Kwong, K-R 2018. *Leukemia blood cell image classification using convolutional neural network*. International Jpurnal of Computer Theory and Engineering,

[16] Shafique, S & Tehsin S 2018.*Computer-aided diagnosis of acutelLymphoblastic leukemia*. Computational and Mathematical Methods in Medicine

[17] Taleb, A S T A & Atiya A F. 2017. *A new approach for leukemia identification based on cepstral analysis and wavelet transform*. International Journal of Advanced Computer Science and Applications.

Acknowledgement

The study was sponsored by DRPM DIKTI Ministry of Research, Technology, and Higher Education of The Republic of Indonesia.

The Application of Mobile Geographic Information System (MGIS) for Android-based Mapping of Micro, Small and Medium Enterprises

B D Meilani[1], Subianto[1]

[1] Information Systems Department, Institut Teknologi Adhi Tama Surabaya
budanis@itats.ac.id.

Abstract. Micro, Small and Medium Enterprises (MSMEs) have great potential in driving community economic activities and at the same time become a source of income for most people in improving their welfare. Therefore, to improve the efficiency and effectiveness of independent and developing MSMEs, Information Technology (IT) is needed to promote enterprises and their products. For this purpose, Mobile Geographic Information System (MGIS) application was developed for android based mapping of MSMEs. In data collection process, the authors employed observation and interview as methods. The purpose of developing this application is as an alternative promotion media and to make it easy to find the location of MSMEs in Surabaya. Black box test was conducted to test the appropriateness of the application. In addition, testing was also conducted using a questionnaire involving 50 respondents with 15 questions representing 5 stakeholder needs. From the questionnaire results, it was concluded that 4 needs, namely, the need to promote products, to spread information, community service, and information of MSMEs have been highly fulfilled by the application, while the need to increase gross domestic product was sufficiently fulfilled by the application.

1. Introduction

The small industry sector is one form of alternative strategies to support economic growth in long-term economic development. Currently, the condition of small and medium industrial products in Indonesia is still a concern. Products from other countries easily penetrate Indonesian market causing the local products to be less competitive [1].

Since the monetary crisis in 1997/1998, the authorities have begun to pay attention to the development of MSMEs both in terms of quantity and quality. This is because MSMEs hae relatively good resilience compared to large-scale businesses. In addition, compared to large-scale businesses, the role of MSMEs is very important due to their main characteristics [2].

MSMEs stands for Micro, Small and Medium Enterprises. MSMEs is regulated based on Law Number 20 of 2008 concerning Micro, Small and Medium Enterprises. In the regulation, it is explained that a company classified as a MSME is a small company owned and managed by a person or small group of people with a certain amount of wealth and income [3].

One of the problems faced by MSMEs is the lack of promotion and the difficulty in finding types of MSMEs are in a region. Problems and research on MSMEs led to the idea of registering MSMEs to the Department of Cooperatives and MSMEs Surabaya and making an application to search for MSMEs based on Android with appropriate software as well as using

SQLite database to store MSMEs location. This application can be used by the user with the selection of a place or guide to the MSMEs' location. The activity in question is like seeing a route to MSMEs. While MSMEs players can register their business easily by filling out the form available in the application then waiting for the verification results from the Department of Cooperatives and MSMEs Surabaya.

The inherently complex field of mine action, with its many political, financial, and physical consid - erations, is also a spatial, data-driven field; and as a result, geographic information systems (GIS) stand to play a major role [4].

The model used in designing this system is the Prototype model. Software engineering establishes a foundation for a complete software process by identifying a number of framework activities applied to all software projects, regardless of size and complexity [5]. The key to making this prototype model work well is to define the rules of the game at the beginning, namely the customer and developer must agree that the prototype was built to define needs. Prototype will be partially or completely removed and the actual software engineered with determined quality and implementation [6].

The expected results of this application are helping MSMEs players and consumers, particularly Surabaya community and generally Indonesian people. The main objective of this research is to develop application for MSMEs in Surabaya which can be used to provide information about MSMEs and their location so as to facilitate them in delivering their promotion.

Android is a collection of software for mobile devices that includes the operating system, middleware and main mobile application [7]. SQLite is a public domain software package that provides a relational database management or RDBMS. The relational database system is used to store user-defined records on large table size and process complex query commands as well as combining data from various tables to produce reports and data summaries. The word "Lite" in SQLite does not refer to its ability, but refers to the nature of SQLite which is light when associated with settings complexity, administrative overhead and resource usage [8].

Database Management System is a software system that allows users to define, create, manage, and control the access to the database [9]. While Global Positioning System (GPS) is a radio navigation system to determine the position using satellite. GPS can provide position of an object on earth accurately and fast (three-dimensional coordinates x,y,z) and provide information on time and speed continuously all over the world[10].

This research discusses the Location Based Service (LBS). LBS technology is a part of Mobile Geographic Information System (MGIS) implementation which is more likely to provide daily applied function such as displaying city directories, vehicle navigation, address searching, and social networking than the functionality of popular MGIS technology for Field Based GIS [10].

2. System Design

The location used as a research study was generally Surabaya city area. The data collected in this study came from the Department of Cooperatives and MSMEs Surabaya in April 2017. Application development method used were object-oriented method with Prototype model as system development method which can be used to connect client misunderstandings about technical matters and clarify the client's desired specification needs to software developers. System analysis was carried out by identifying users between MSMEs players, Department of Cooperatives, and public. The problems are presented in table 1:

Table 1. Needs Analysis

Actors' Needs Analysis	**Needs**	**System**
MSMEs (User)	Promoting products	Using Map/ LBS as location guide
	Spreading information about MSMEs	Detailed information on MSMEs is easier to obtain

Department of Cooperatives and MSMEs Surabaya (Administrator)	Increasing Gross Domestic Product	Data visualization
	Community service	MSMEs information that is easy to access
Public (User)	MSMEs information	MSMEs data is based on location by using Map/ LBS as a location guide

The following is the use case diagram in Figure 1, a system development carried out by the admin and user in carrying system functionality that is running on the application.

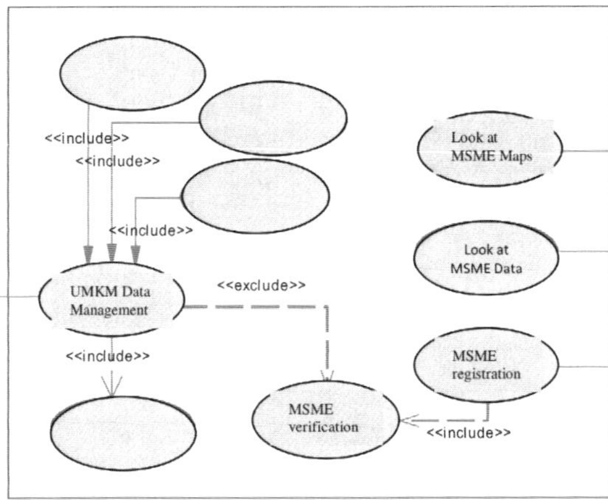

Figure 1. *Use Case of System Functionality*

The flow process of activities carried out by system users is the administrator accessing the laptop/ PC media. While the user accesses the electronic media (smartphone). First, the user installs the application on android smartphone. The flow process starts when the user wants to see the information they want. Then, the application will send the data to the server via web service. The server will process to display information that the user needs, shown in figure 2:

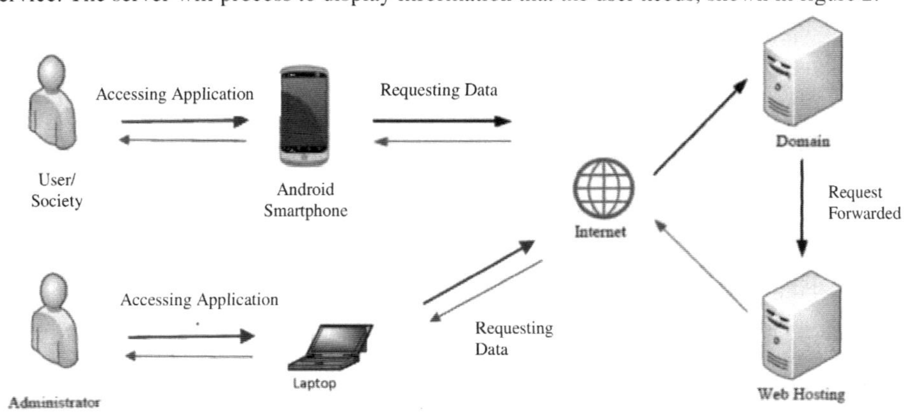

Figure 2. *Deployment Diagram*

3. Result

3.1. System Testing

The system that has been completed is built through coding stage, then tested. This is done to ensure the software made can be used properly and is suitable for use or it still needs to be refined. Black Box testing is a method of designing test data based on software specifications. The results obtained after testing are presented in table 2 and table 3:

Table 2. The Test Result of Android Application Functions

Test ID	Objective	Input	Expected Result	Status
1	Opening the application's main menu	Choosing application icon on Android	Showing the main menu	[√] Suitable [ ] Not suitable
2	Opening MSMEs map	Choosing "maps" menu	Showing MSMEs map	[√] Suitable [ ] Not suitable
3	Opening MSMEs details	Choosing "MSMEs details" menu	Showing MSMEs details	[√] Suitable [ ] Not suitable
4	Looking at MSMEs route map	Choosing navigation button	MSMEs map is shown	[√] Suitable [ ] Not suitable
5	Calling MSMEs	Choosing call button	Confirmation message for calling is shown	[√] Suitable [ ] Not suitable
6	Sharing MSMEs	Choosing share button	Smartphone application in which MSMEs information will be shared is shown	[√] Suitable [ ] Not suitable
7	Looking at the list of registered MSMEs	Choosing "MSMEs list" menu	Showing MSMEs data	[√] Suitable [ ] Not suitable
8	Registering MSMEs	Choosing registration menu and filling out registration form	Showing registration form and the data were sent to the server	[√] Suitable [ ] Not suitable
9	Application information	Choosing info on the list of menu	Showing application information	[√] Suitable [ ] Not suitable
10	Closing the application	Back button on the smartphone	The application is closed	[√] Suitable [ ] Not suitable

Table 3. The Test Result of Administrator Functions

Test ID	Objective	Input	Expected Result	Status
1	Opening the application	Inputting username and password	Showing the main menu	[√] Suitable [ ] Not suitable
2	Looking at MSMEs data	Choosing "MSMEs data" menu	Showing MSMEs data	[√] Suitable [ ] Not suitable

3	Adding MSMEs data	Choosing "add" menu and fill out the form	Saving MSMEs data	[√] Suitable [ ] Not suitable
4	Editing/ deleting MSMEs data	Choosing "edit/delete" menu on "action" column	Renewing the data	[√] Suitable [ ] Not suitable
5	Printing the new MSMEs data	Choosing "MSMEs request" menu and choose "print" icon	Printing registration form	[√] Suitable [ ] Not suitable
6	Closing the application	Choosing "log out" on the application	The application is closed	[√] Suitable [ ] Not suitable

3.2. System Evaluation and System Use

The system that has been tested was then evaluated and used. After the questionnaire was given to the participants, the data obtained were processed to get the assessment results. The statement of needs analysis and observation results that have been distributed to respondents are as follows:

3.2.1. The need to promote products.

Does MGIS application help to promote products?

Answer Code	**SS**	**S**	**TS**
Answer Frequency	14	30	6
Score	14 x 3 = 42	30 x 2 = 60	6 x 1 = 6
Total Score	108		
Percentage	108 / 150 x 100% = 72 %		

Continuum Line

Score

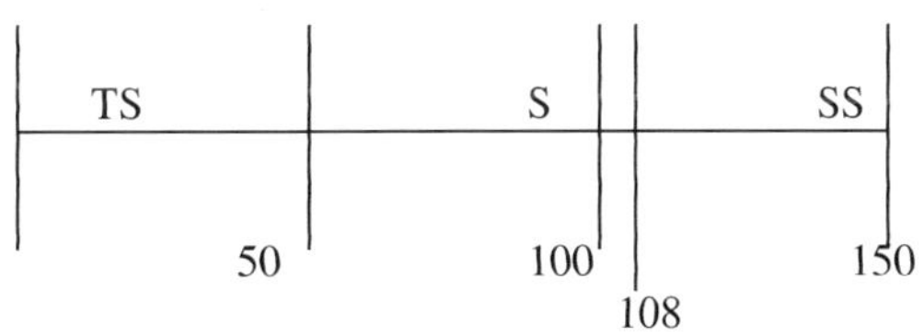

Based on the aforementioned percentage results, it can be concluded that the assessment of the need to promote products was 108 of which was expected to be 150, or can be categorized as strongly agree.

3.2.2. The need to spread information

Does MGIS application help to spread the information of MSMEs in Surabaya?

Answer Code	**SS**	**S**	**TS**
Answer Frequency	24	23	3
Score	24 x 3 = 72	23 x 2 = 46	3 x 1 = 3
Total Score	121		
Percentage	121 / 150 x 100% = 80 %		

IOP Conf. Series: Materials Science and Engineering **462** (2019) 012019 doi:10.1088/1757-899X/462/1/012019

Continuum Line

Score

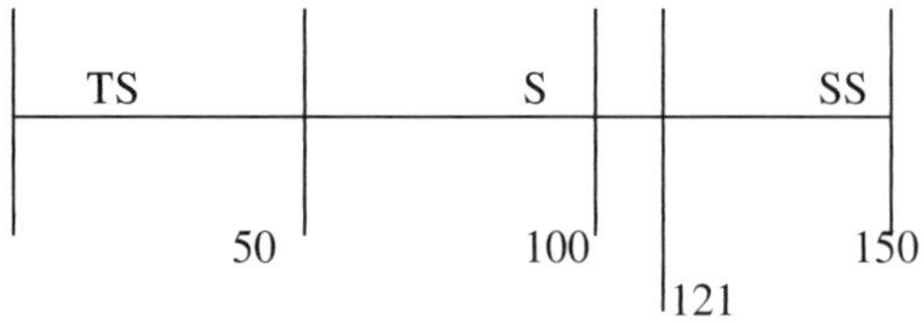

Based on the aforementioned percentage results, it can be concluded that the assessment of the need to spread information was 121 of which was expected to be 150, or can be categorized as strongly agree.

3.2.3. The need of community service

Does MGIS application give information services on MSMEs to the society?

Answer Code	**SS**	**S**	**TS**
Answer Frequency	12	29	9
Score	12 x 3 = 36	29 x 2 = 58	9 x 1 = 9
Total Score	103		
Percentage	103 / 150 x 100% = 68 %		

Continuum Line

Score

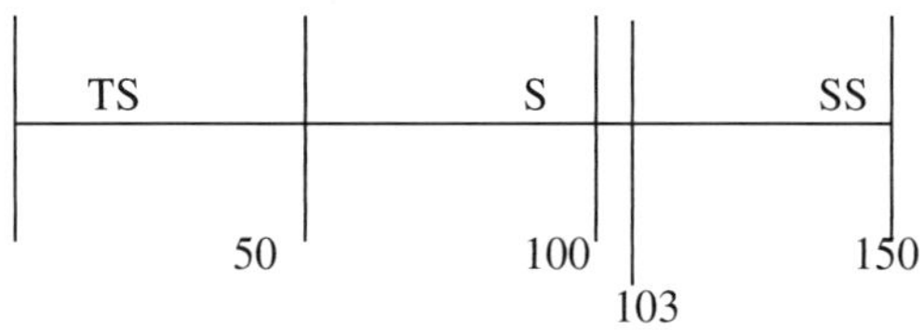

Based on the aforementioned percentage results, it can be concluded that the assessment of the need of community service was 103 of which is expected to be 150, or it can be categorized as strongly agree.

3.2.4. The need to increase gross domestic product

Can MGIS application increase gross domestic product in the Department of Cooperatives and MSMEs Surabaya?

Answer Code	**SS**	**S**	**TS**
Answer Frequency	4	27	19
Score	4 x 3 = 12	27 x 2 = 54	19 x 1 = 19
Total Score	85		
Percentage	85 / 150 x 100% = 56 %		

Continuum Line

Score

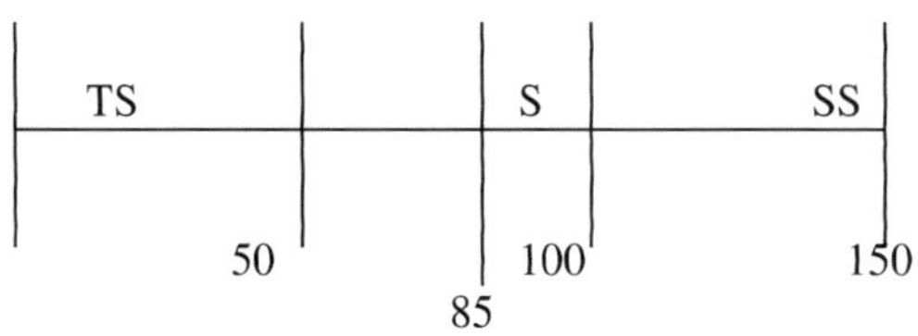

Based on the aforementioned percentage results, it can be concluded that the assessment of the need to increase gross domestic product was 85 of which is expected to be 150, or can be categorized as agree.

3.2.5. The need of MSMEs information

Does MGIS application show the need of data desired by consumers?

Answer Code	**SS**	**S**	**TS**
Answer Frequency	20	25	5
Score	20 x 3 = 60	25 x 2 = 50	5 x 1 = 5
Total Score	115		
Percentage	115 / 150 x 100% = 76 %		

Continuum Line

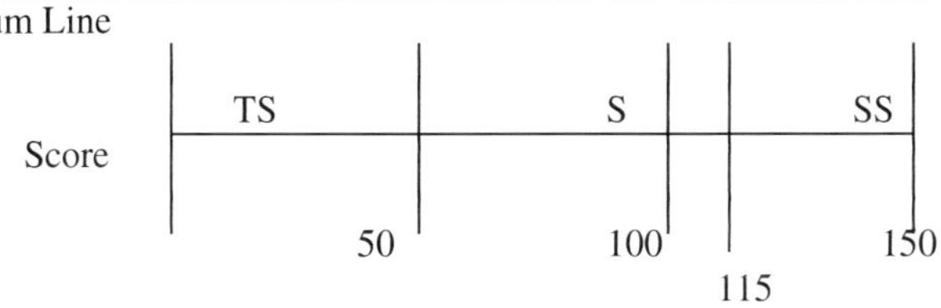

Based on the aforementioned percentage results, it can be concluded that the assessment of the need of MSMEs information was 115 of which expected to be 150, or can be categorized as strongly agree.

4. Conclusion

1. Based on the questionnaire result of 50 respondents, the Mobile Geographic Information System application for mapping MSMEs in Surabaya functioned well in accordance with the expected needs analysis with the following achievements:
    - Assessment of the need to promote products was 108 of which was expected to be 150, or can be categorized as strongly agree.
    - Assessment of the need to spread information was 121 of which was expected to be 150, or can be categorized as strongly agree.
    - Assessment of the need of community service was 103 of which is expected to be 150, or it can be categorized as strongly agree.
    - Assessment of the need to increase gross domestic product was 85 of which is expected to be 150, or can be categorized as agree.
    - Assessment of the need of MSMEs information was 115 of which expected to be 150, or can be categorized as strongly agree.

    It can be concluded that the application developed was sufficiently fulfilled the initial development purpose.
2. With this application, it is expected that it can help the government of Surabaya in increasing public interest in MSMEs in order to increase the independence of an area.

5. References

[1] Suara Merdeka(2012), (http://www.suaramerdeka.com)
[2] Tambunan, Tulus T.H. (2009). *UMKM di Indonesia*. Jakarta : Ghalia Indonesia.
[3] Kementerian Negara Koperasi dan UKM (2009). *Undang-Undang Usaha Mikro Kecil dan Menengah (UMKM) UU RI No. 20 Th. 2008*. Jakarta : Sinar Grafika.
[4] Ittenhouse, Paul and Aldrich, Lindsay (2017), "*Using Mobile Geographic Information Systems to Improve Operational Efficiency, Data Reliability, and Access in Mine Action*", Journal of Conventional Weapons Destruction: Vol.21:Iss. 2, Article 5.
[5] Pressman, R.S. (2010). *Software Engineering : a practitioner's approach, McGraw-Hill*. New York.
[6] McLeod Jr. Raymond, (2001). *Sistem Informasi Edisi 7 Jilid 2*. Jakarta : Prenhallindo.
[7] Sistem Operasi *Android*, (*http://www.android.com/about/*, diakses 24 Februari 2011)
[8] Kreibich. Jay A, (2010), *Using SQLite*, O'reilly

[9] Connolly, Thomas & Begg, Carolyn. (2002). *Database Sistems :A Practical Approach to Design, implementation and management. Third Edition.* Addision Wesley, England.

[10] Riyanto. (2010). *Membuat Sendiri Aplikasi Mobile GIS : Platform JAVA ME, Blackberry & Android.* Yogyakarta : Andi Offset.

High-Performance Computing (HPC) design to improve the quality of Introduction of Parallel Computing lectures

E Alfianto[1], A Sa'diyah[2], F Rusydi[3], I Puspitasari[4],

[1] Dep. of Computer System, Institut Teknologi Adhi Tama Surabaya, Jl Arif Rachman Hakim 100, Surabaya, Indonesia. 60111
[2] Dep. Of Marine engineering, Politeknik Perkapalan Negeri Surabaya, Jl Jalan Teknik Kimia, Kampus ITS-Sukolilo, Surabaya, Indonesia, 60117
[3] Theoretical Physics Research Group, Dep. of Physics, Fac. of Science and Technology, Airlangga University, Jl. Mulyorejo, Surabaya, Indonesia 60115
[4] Dep. of the information system, Fac. of Science and Technology, Airlangga University, Jl. Mulyorejo, Surabaya, Indonesia 60115

Email: enggar@itats.ac.id

Abstract. High-Performance Computing (HPC) has been created to improve the quality of *Pengantar Komputasi* Parallel lectures. HPC consists of two PCs with AMD Ryzen 7 processor connected by high-speed LAN. HPC is used to solve problems that students can't practice Parallel Computing because the computer used is not adequate. So it takes a machine that can used together that has many processors, so the problems encountered are resolved. The results obtained, participants can use more than four processors to solve tree problems directly.

1. Introduction

The research on Course Introduction to *Pengantar Komputasi* Parallel is a new course in the Department of Computer Systems, ITATS. These courses discuss how to use multiple processors on a PC for simultaneous use. The contents of the route start from the history of processor development, to how to use multiple processors simultaneously. Participants are generally college students who have learned to code using code using C language, so the programming language used in this course using the C language

Participants of this lecture, have a different laptop. The number of processors used they can classify into two. Laptop has two processors and it has more than two processors. It is a problem because there are students who cannot practice programming using more than two processors. So that learning becomes obstructed. Therefore, it takes a solution so that all students can follow the lecture well with support facilities.

From the problems that exist, we try to find a solution that can solve the problem. Namely using a computer that has some processors more than 4, so that one class can use it with the same specifications. However, if it realized by buying some computers for each student, then the funds needed are substantial.

Another solution is to utilize a multi-user PC that can share simultaneously. The alternative is to use Computer-Cluster. With the computer-cluster is expected to be a solution to solving the implementation of numerical computing courses. Computational method and Implementation.

2. Related Work

2.1. Processor Development

The processor is one of the fastest growing electronic components. The first processor was released by Intel in 1971, with the name of a 4004 microprocessor, used as the brain for the Bascom counter machine. This chip is the starting point of the development of current generation processors, in that same year (1971) called the era of integrated electronics (Era of Integrated Electronics) [1].
Development issues the next processor is at increased speed [1]. Evident from the next generation processor raises the speed only. In 1985, the 32-bit processor was first launched by Intel with the name of Intel 30386. In the same year, AMD made a similar processor. The birth of this processor supports the early of Windows operating system that has helped the multitasking process. Since then the competition between Intel and AMD began to tighten [2].
In 1997, Intel created the Pentium MMX (P55C) which is an early generation of Intel Pentium. In the same year, AMD built AMD K6 which is a rival of the Pentium MMX. Pentium generation continues to perfect itself until born a processor with many cores. Currently, Intel has created Intel core i7 generation to 8, and answered with AMD is also issued AMD Ryzen 7 that its performance is almost the same as Intel core i7 [3].

2.2. Development of Cluster-Computers

Linux users in the 1960s first introduced Cluster-Computer. In that era, Cluster-Computers utilise a second-hand computer which is then coupled with a LAN so that it can work together. The purpose of cluster making is that the performance of used computers can increase. Increased production is needed to meet the increasingly complex calculation requirements [1] [4].
Computer Performance-Cluster obeys Amdahl's law, which was formulated by Gene Amdahl of IBM who published his writings in 1967. Datapoint Corporation commercially commenced the use of Computer-Cluster in 1977 [4].
To date, computer clusters have widely developed. The development followed by the development of communities that participate in developing this Computer-Cluster. From that formed, was born Rock Cluster, Open HPC, Beowulf and so forth [5].

2.3. Development of parallel computing research

Parallel computing usually utilizes Message Passing Interface (MPI) as a means to run programs in parallel. MPI built by a consortium consisting of several elements, Education, Industry and several developers [6]. Although MPI does not use IEEE or ISO standards, MPI has used as an industry standard in parallel program writing [7] [8]. In this study, MPI used as the main engine in running parallel programming.

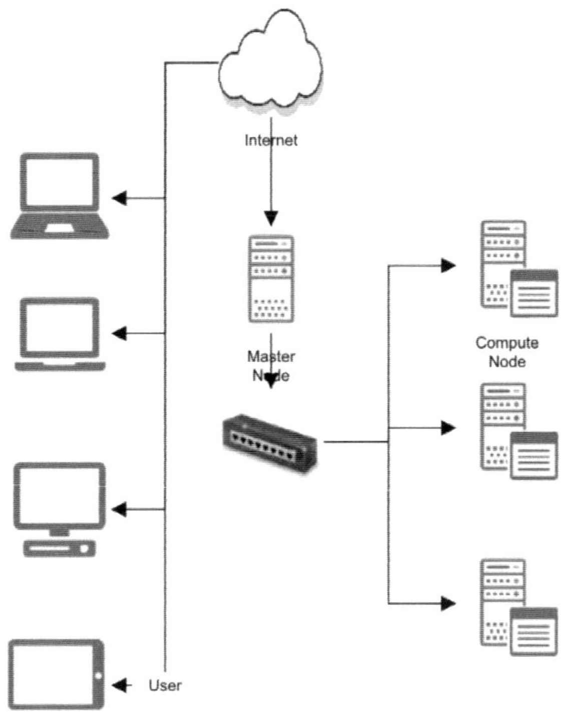

Figure 1. Cluster-Computer scheme

3. Research Method

This research begins with the preparation of computer clusters which will be utilised to support the Parallel Computing lectures. The Cluster Setup is

1. Analyze the PC to used as a computer cluster
2. Installing the operating system used (Centos 7-x86-64-DVD-1804).
3. Filtering unnecessary services for numerical computing lectures.
4. Create a local repository (Centos 7-x86-64-Everything-1804.iso)
5. Build system environment for node installation with PXE.
6. Performing the Compute Node installation
7. Setting up the Master Node to enter the Compute Node without using a password.
8. MPI Installation.

The computer cluster has a master node, and three compute nodes (**Figure 1**), between the master node and the compute node connected by a hub switch. The master node directly related to the internet, so the master node can be accessed by various devices (Laptops, PCs, Tablets). To access Clusters using the SSH client that is on the user's device. So that it can be used to assign tasks to computers anytime [8][9].

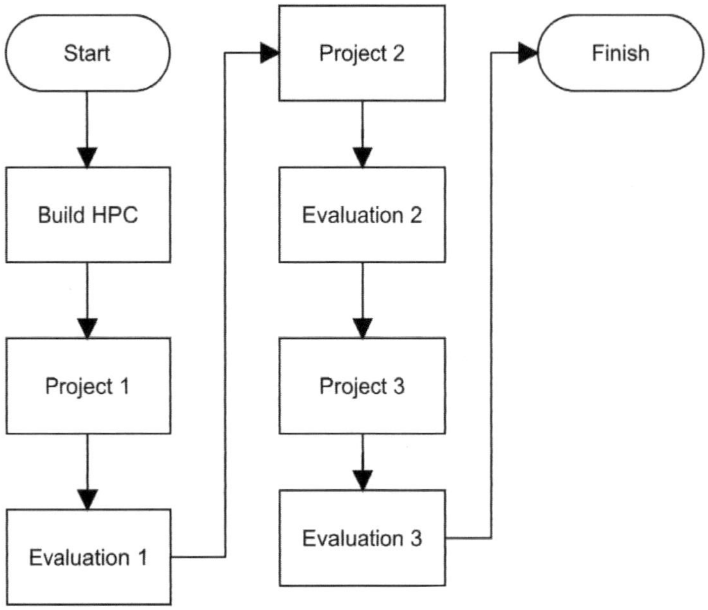

Figure 2. Flowchart of the research

The study continued with the introduction of the use of MPI in students. Then students are taught to run programs remotely using ssh. In a test, it was carried out by giving three projects which contained examples of cases in parallel programming. The flowchart is shown in **Figure 2**.

4. Project

4.1. Project 1

The first project that given was serial programming to find the root of the Gauss elimination method and the backward method. The matrix used is 100x100 with a random number generated by generating a random number. Students must record the program performance time.

4.2. Project 2

The second project is to create a program by assigning tasks to each processor to do different jobs at the same time. So that at the same time, the program can work on various functions. The goal is for students to be able to define tasks for each processor.

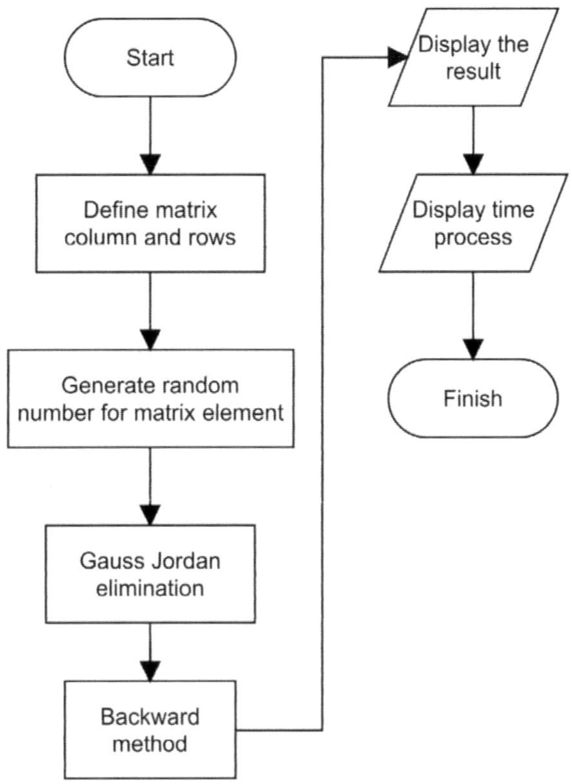

Figure 3. Flowchart of Project 2

4.3. Project 3

The third project is to make a program based on the elimination of Gauss to find the value of xi with some 1000x1000 matrices. Programming is done using a parallel paradigm. Here, students are expected to be able to understand how its performance entirely. How useful algorithms divide processor tasks in parallel Algorithm.

5. Result and discussion

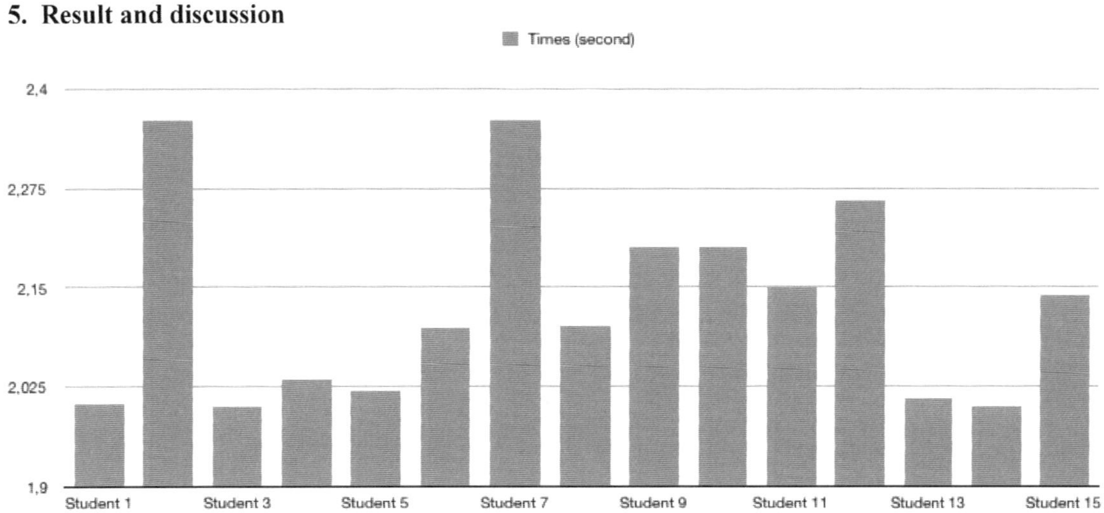

Figure 4. Running time of project 1

From the project, then the calculation time obtained by students to do the project. From the assignment given, students are free to do coding. Then, the program that has completed run in computer clusters. The results show in **Figure 4**.

6. Conclusion

From this, it can conclude that computer clusters that are made to facilitate students in computer parallel introductory courses are a solution to answer existing problems. By using HPC, students can run parallel programs with more than 4 processors. So that they can understand how parallel computers performance works. One of these understandings obtained from the difference in time between parallel and serial. Other parameters can be seen with the student's score of 70% getting an A.

7. References

[1] Ahmad, Ashari, and Riasetiawan Mardhani. "High Performance Computing on Cluster and Multicore Architechture." *Telkomnika*, 2015: 1408-1413.

[2] Sibaroni, Yuliant. "The Optimal High Performance Computing Infrastructure for Solving High Complexity Problem." *Telkomnika*, 2016: 1545

[3] Zhang, Cuilian. "Parallel Program Design Based on MPI." *Computer Technology and Development 8*, 2006: 24

[4] Carry, R, and Stevlana G Sasharina. "Comparison of C++ and Fortran 90 for Object-Oriented scientific programming." *Computer Physics Communications*, 1997: 20 - 36.

[5] Ye, Yun, and D,A Bader. "GPUMemSort: A high performance graphics co-processors sorting algorithm for large scale in-memory data." *GSTF Journal on Computing (JoC)*, 2018: 1-2

[6] Mergen, M. F., V. Uhlig, O., Krieger, and J. Xenidis. "Mergen, Mark F., et al. "Virtualization for high-performance computing." *ACM SIGOPS Operating Systems Review*, 2006: 8-11.

[7] Alfianto, Enggar. "Implementasi Metode Teori Fungsional Kerapatan pada bahasa C untuk menemukan energi keadaan dasar berbagai atom." *Jurnal Arus Elektro Indonesia*, 2015: 1

[8] Hsu, Chung-hsing, and Wu-chun Fen. "A power-aware run-time system for high-performance computing." *Proceedings of the 2005 ACM/IEEE conference on Supercomputing*. IEEE Computer Society, 2005.

[9] Tang, Qinghui, Sandeep Kumar S. Gupta, and Georgios Varsamopoulos. "Energy-efficient thermal-aware task scheduling for homogeneous high-performance computing data centers: A cyber-physical approach." *IEEE Transactions on Parallel and Distributed Systems*, 2008: 1458-1472.

Acknowledgement

This calculation of this work supported by computer research facility of Computational Material Design laboratory, Institut Teknologi Bandung, Indonesia.

IOP Conf. Series: Materials Science and Engineering **462** (2019) 012021 doi:10.1088/1757-899X/462/1/012021

Analysis of Channels Impulse Response Due to Transducer Movements in Underwater Acoustic Communication

I P A I Kusuma[1] and M Margareta Z B[1]

[1] Dept. of Naval Architecture and Shipbuilding Engineering,
Adhi Tama Surabaya Institute of Technology, Indonesia

Email: indira@itats.ac.id

Abstract. This study discussed the effect of transducer movement in underwater acoustic communication based on simulation analysis. A proposed underwater communication testing platform which is a 10 m x 2.5 m x 1.5 m mini towing tank is used as the communication channel. In the simulation analysis, channels boundary interaction is simulated using image method consists of 6 ray element and the transducer movements is assumed to have relative velocity between each other. Due to the relative velocity, Doppler frequency shift is occurred and gives amplitude variation in Channels Impulse Response (CIR). Several Doppler spread value was given to analyze the CIR. The simulation result indicates that, Relative velocity Doppler effects contributes big amplitude variation in the arrival structure of channels impulse response (CIR). When relative velocity or Doppler spread increases, the amplitude variation and the phase variation of arrival structure also increases.

1. Introduction

The simple ping signal which is used by sonar that operates in audible band was developed as the fundamental improvement for underwater communication [1]. Recently, numerous developments had been introduced in this field i.e. simulation or measurements approach in ocean, point to point acoustic link until the applications of autonomous network for ocean observations [2]. Besides the numerous applications, underwater acoustic communication (UAC) has many challenges and problems. Due to strong absorption of high frequency sounds by sea water the available bandwidth was very limited [3]. Small scale fading usually appears in digital communication, which is because of the signal time spreading and time variant nature from the channel [4]. It is a result of the motion within channel and specifically movement of source and receiver. Doppler spread and coherence time were the parameters which are usually used to describe the channel's time-varying nature [5]. Several inter symbol interference (ISI) and fluctuations in the received signal occur due to fluctuation of random amplitudes and phases of multiple arrivals. Multipath and time-varying effect can be characterized by channel impulse response (CIR). It was considered varying over time if the channels itself are not in stationary condition [6]. Time-varying effect will reduce the communication quality and become very challenging point for researchers to solve this problem. In 2007, Chitre [7] investigated the variations of arrival time lag and amplitude which are determined by transducer movement and Doppler effects. Assigning the time-varying multipath effect as the main topic, the purpose of this study is to simulate underwater acoustic communication in testing platform. A 10 m x 2.5 m x 1.5 m mini towing tank is proposed as the suitable and very challenging testing platform to test the underwater communication process under time-varying condition. Utilizing the benefit of wave making towing tank, Kusuma [8] presented simulation analysis of underwater acoustic communication using mini towing tank as channel of communication which the result indicates that this channel has a characteristic that multipath effect is the main challenging problem. However, in this simulation study discussed the effect of transducer movements in underwater acoustic communication testing platform. The transducer movement is assumed to have relative velocity between each other. Due to the relative velocity, Doppler frequency shift is occurred and gives amplitude variation in Channels Impulse Response (CIR). All of the simulation process is conditioned in accordance with the measurement in real condition. Channels impulse response is analyzed with considering several values of Doppler frequency shift and comparing it with the original impulse response. The aim of this analysis is to understand the characteristics and sensitive variables of time-varying multipath effect in the testing platform.

2. Methodology

2.1. Channel Modeling

In the simulation process, Binary phase–shift keying (BPSK) signals with 2 ms in length and centered at 10 kHz was transmitted from source to the receiver. Furthermore, the interactions between acoustic signals and boundaries of towing tank may give multipath effects. To analyze channel's boundary interactions, the method of image was applied with considering 8 reflection paths as shown in Figure 1. The 6 ray elements which shown in Figure 1. i.e. Direct path *(dp),* Surface path *(sp),* Surface bottom path *(sb),* Bottom surface path *(bs),* Bottom path *(bp),* Left wall path *(lw),* Right wall path *(rw)* and back wall path *(bw).* The interaction between signal and boundaries generates multiple travel paths in the signal transmission. The channel impulse response (CIR) of multipath arrival can be written as

$$H(\omega) = \sum_{n=1}^{L} a_n e^{-j\omega\tau_n} \tag{1}$$

Here, a_n and τ_n are the amplitude and time delay of the nth ray respectively. Doppler spread is one of parameter which is used to describe channel's time-varying nature. The time variance itself is a result from the movement of transducer or channel itself. For Doppler frequency shift due to moving transducer, multipath signals are emitted from source at different angles and arrive at receiver with different angles [17]. Maximum Doppler frequency shift *(f_D)* can be written like the following equation

$$f_D = f \frac{v}{c} \cos(\theta) \tag{2}$$

Where, f is carrier frequency, v is moving speed of transducer, c is sound speed and θ is the emission or arrival angle. If we consider the range of cosine is within -1 and +1, then f_D becomes

$$\pm f_D = f \frac{v}{c} \tag{3}$$

Furthermore, the maximum Doppler angular frequency shift of the nth ray is

$$\omega_n = 2\pi f_n, |f_n| \leq f_D \tag{4}$$

Thus, the impulse response affected by Doppler shift becomes:

$$H(\omega, t) = \sum_{n=1}^{N} a_n e^{-j\omega\tau_n + j\omega_n t} \tag{5}$$

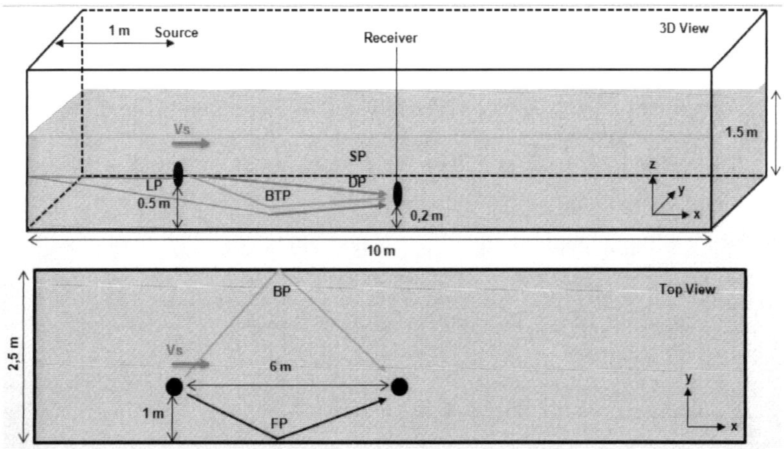

Figure 1. The interaction of ray element with channel boundary in the testing platform.

2.2 Time Reversal Mirror

Time Reversal Mirror (TR) or Passive Phase Conjugation (PPC) in frequency domain were proposed to focus energy at desired depth thus to mitigate the effects of multipath dispersion and inter-symbol interference (ISI) from time-varying condition. TR can be classified into active time reversal and passive time reversal process [9]. The active time reversal process must implement directly in real environment, on the other hand, passive TR procedure for transmitting signals was implemented via computer simulation. Suppose that *s(t)* is the source signal time domain, *p(t)* is the receive signal time domain, *h(t)* is the channel impulse response (CIR) between the source and the receiver. Furthermore, *p(t)* in time domain can be expressed as the following equation:

$$p(t) = h(t) \otimes s(t) \tag{6}$$

Where $\otimes$ denotes the convolution integration. According the process of TR, if source location is interchanged with receiver location and source signal is changed to time reversal signal of receiver, then the received signal *z(t)* and it's time reversal signal *z(-t)* at original source location can be expressed as the following equations respectively.

$$z(t) = h(t) \otimes p(-t) \tag{7}$$

$$TRM(t) = z(-t) = h(-t) \otimes h(t) \otimes s(t) \tag{8}$$

The output of *TR (t)* is similar with original source signal.

2.3 Image Method

In this technique, the Green's function is decomposed into the free-space Green's function plus a sum of contributions from image sources, which fall outside the region of interest [10]. From the benefit of this method, it can be explored into underwater communication especially in the shallow water to simulate the boundary condition. As we know that the main problem for shallow-water communication channel, especially in the underwater communication testing platform is the reflection from the boundary. The interaction between signals and boundaries will be extended multipath and refractive properties and induced severe inter-symbol interference in the received signal. The solution of the problem is where the upper interface is soft and the lower interface is hard. Here, the channel surface and bottom are considered as two mirrors. The rays which hit the surface and bottom are then starting exactly at the images of the actual sources of origin. With this logic the whole image or mirror method is developed and thereby it is easy to provide mathematics for multipath propagation [11]. A schematic representation of the contributions from the physical source at depth z_s and the first three image sources, leading to the first four terms in the expression for the total field can be described by the following equation:

$$P(r,z,\omega) \cong A(\omega)\left\{\frac{e^{-jkL_{01}}}{L_{01}} + \hat{R}_1(\varphi_{02},\omega)\frac{e^{-jkL_{02}}}{L_{02}} + \hat{R}_2(\varphi_{03},\omega)\frac{e^{-jkL_{03}}}{L_{03}} + \hat{R}_1(\varphi_{04},\omega)\hat{R}_2(\varphi_{04},\omega)\frac{e^{-jkL_{04}}}{L_{04}}\right\} \tag{9}$$

Where $A(\omega)$ is amplitude of the signal and i = 1,2. Then the length's of all rays can be write as

$$L_{m1} = \sqrt{r^2 + (2Dm - z_s + z)^2}, \tag{10}$$

$$L_{m2} = \sqrt{r^2 + (2Dm + z_s + z)^2} \tag{11}$$

$$L_{m3} = \sqrt{r^2 + (2D(m+1) - z_s - z)^2} \tag{12}$$

$$L_{m4} = \sqrt{r^2 + (2D(m+1) + z_s - z)^2} \tag{13}$$

3. Simulation Process

In the simulation, first analysis is simulated stationary channel condition, this condition means that the transducer position was fixed and without the movements of transducer occur during the transmission process. The stationary channel was simulated as the original condition of the channels impulse response (CIR). After that, the second analysis is considering the movement of transducer whit the assumption of relative velocity between each transducer. Due to the relative velocity, Doppler frequency shift is occurred and gives amplitude variation in Channels Impulse Response (CIR). Several value of Doppler spread is analyzed to understand the effect of each Doppler spread value to CIR. With adding random analysis in CIR allowed us to provide time-varying condition for this case. Furthermore, the CIR for each case will be compared between original condition and time-varying condition. The simulation input data and diagram for the interaction of ray element with channel boundary in the testing platform can be shown in Table 1 and Figure 1 respectively. These cases will be discussed respectively as the following sections in order to understand clearly the influence of each case in underwater communication.

Table 1. Simulation parameter

Parameter	Symbol	Value
Range	*mx*	6 m
Water Depth	*D*	1.5 m
Source Depth	*sz*	0.5 m
Receiver Depth	*mz*	0.2 m
Sound Speed	*C*	1500 m/s
Carrier Frequency	*Fc*	10 kHz
Sampling Frequency	*Fs*	200 kHz
Transmitted Time	*Ts*	0.02 s
Transmitted Symbol	*Bs*	100 bits BPSK Signal
Doppler Spread	f_D	5 Hz, 10 Hz, 15 Hz, 25 Hz, 40 Hz, 50 Hz, 80 Hz, 100 Hz
Ray Element Number	*N*	6

4. Simulation Analysis

4.1 Analysis of Channel Impulse Response at Original Condition

In this paper, it only shows the simulation results for one-time signal transmission at 6 m distance of communication. The Binary phase-shift keying signaled with consists of 100 bits symbol was transmitted through the channel. The transmission time is 0.02 s with 200 kHz carrier frequency. The communication result is characterized in the arrival structure to analyze the arrival time of CIR as shown in Figure 2 (a). Furthermore, the received signal shown to analyze the frequency shifting due to Doppler effects as shown in Figure 2 (b). The result indicates that 6 ray element of image method describes the interaction between the signal and boundary of the channel and contributes the multipath condition. Based on previous research, this condition will reduce the communication quality and bit error rate was high [8]. However, this condition will more complicates if the transducer have relative velocity between each other. In order to understand how big the movement of transducer affects this CIR, the original condition is given random analysis of Doppler frequency shift.

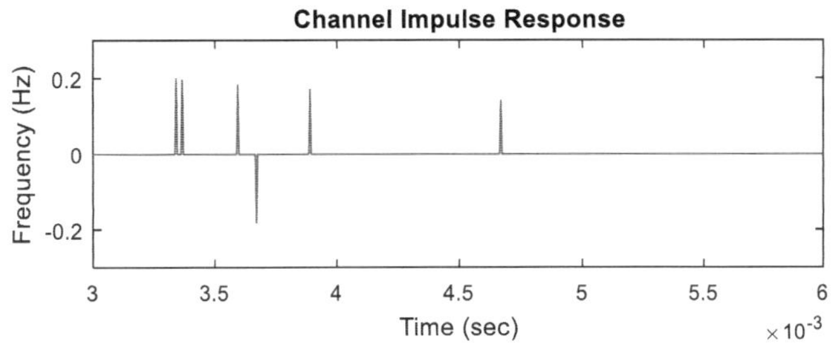

(a)

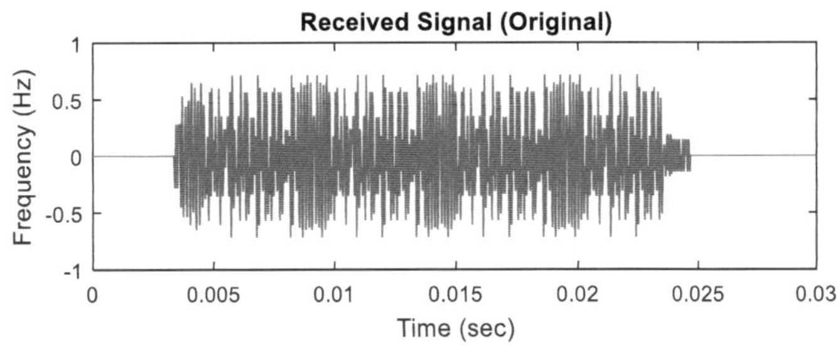

(b)

Figure 2. (a) Channels impulse response and **(b)** Received signal of 6 m distance of communication without the movement of transducer (Original)

4.2 Analysis of Channel Impulse Response Affected by Doppler Frequency Shift

In this case, the transducer movement is assumed to have relative velocity between each other. Due to the relative velocity, Doppler frequency shift is occurred and gives amplitude variation in the arrival structure of CIR. Eight variations of the Doppler spread were evaluated i.e. f_D=5 Hz, f_D=10 Hz, f_D=15 Hz, f_D=25 Hz, f_D=40 Hz, f_D=50, f_D=80 Hz and f_D=100 Hz. Doppler Spread is determined by the relative velocity of the transducer based on equation (2-5) respectively.

Table 2. Comparison of Frequency shifting at Ray element

Condition of CIR	**Frequency Shifting at Ray Element (Hz)**					
	1	**2**	**3**	**4**	**5**	**6**
Original	0.1996	0.1981	0.1854	-0.1817	0.1713	0.1427
Doppler Shifts (f_D=5 Hz)	0.1992	0.1975	0.1853	-0.1809	0.1704	0.1417
Doppler Shifts (f_D= 10 Hz)	0.1982	0.1958	0.1848	-0.1792	0.1689	0.1419
Doppler Shifts (f_D= 15 Hz)	0.1975	0.1919	0.1853	-0.1806	0.1702	0.1365
Doppler Shifts (f_D= 25 Hz)	0.1882	0.1745	0.1800	-0.1609	0.1488	0.1085
Doppler Shifts (f_D= 40 Hz)	0.1869	0.1928	0.1395	-0.1712	0.0960	0.0655
Doppler Shifts (f_D= 50 Hz)	0.1855	0.1285	0.1845	-0.1784	0.1221	0.0558
Doppler Shifts (f_D= 80 Hz)	0.1502	0.1771	0.0244	-0.1409	-0.06355	-0.0826
Doppler Shifts (f_D= 100 Hz)	0.1876	-0.0179	-0.0232	0.0848	0.02695	-0.1034

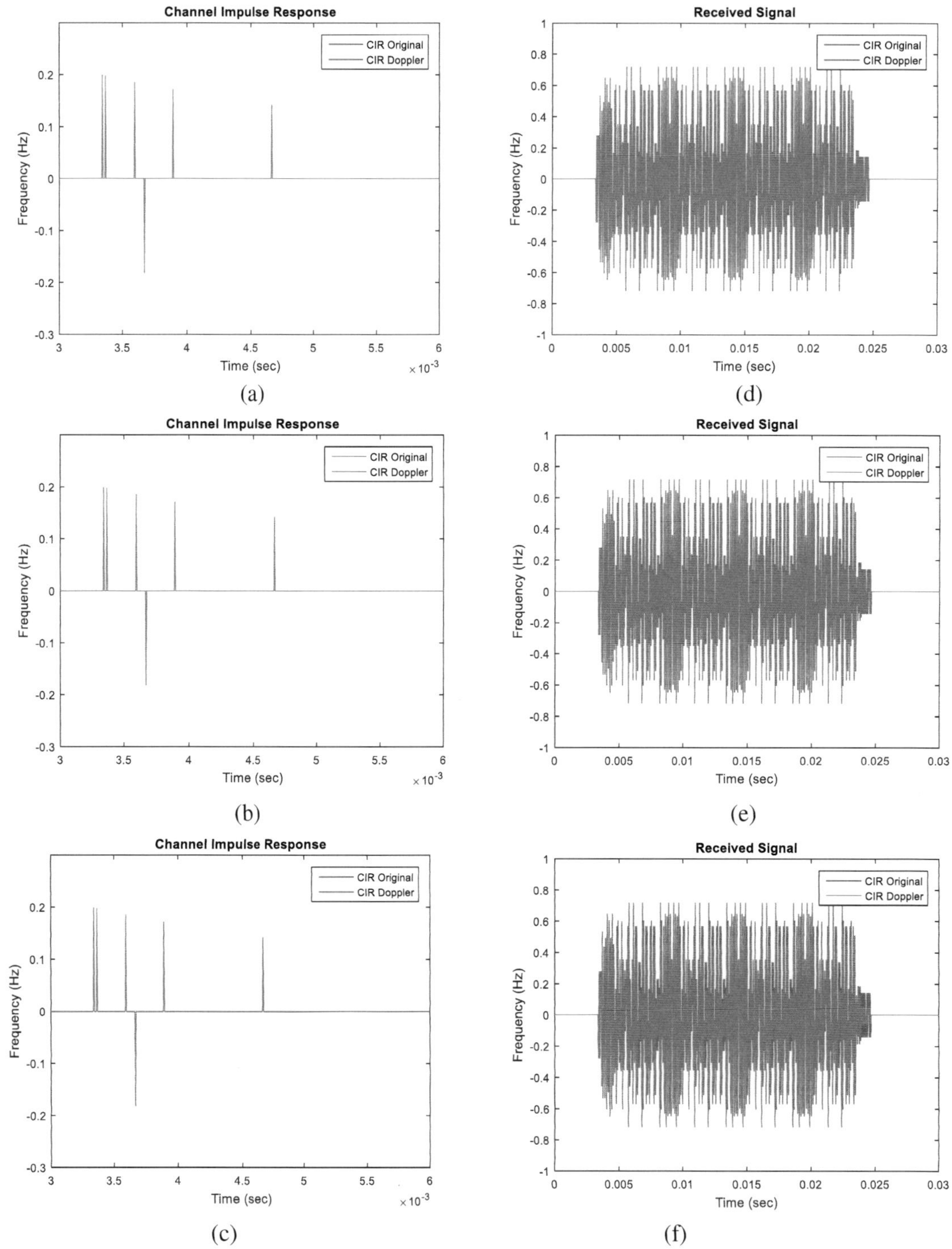

Figure 3. (a,b,c) Channels impulse response of 6 m distance of communication with 15 Hz, 25 Hz, 40 Hz Doppler Spread and **(b)** Received signal of 6 m distance of communication with 15 Hz, 25 Hz, 40 Hz Doppler Spread

With the similar procedure and process from previous case, the Doppler spread effect and frequency shift can be analyzed in the arrival structure and received signal figure 3 (a, b, c) and (d, e, f) above. These figures indicate that, Relative velocity Doppler effects contributes amplitude variation in the arrival structure of channels impulse response (CIR). When relative velocity or Doppler spread increases, the amplitude variation and the phase variation of arrival structure also increases. Relative velocity Doppler effects leads to a significant error rate when Doppler spread (f_D) is larger than15 Hz. The full comparison between stationary condition and Relative velocity Doppler effects can be shown in table 2 above. The comparison allows us to understand the characteristic of amplitude variation at each Doppler spread.

5. Conclusion

This simulation study discussed the effect of transducer movement in underwater acoustic communication. The transducer movement is assumed to have relative velocity between each other. Due to the relative velocity, Doppler frequency shift is occurred and gives amplitude variation in the arrival structure of CIR. The results can be concluded that amplitude variation increases when the Doppler spread increases and gives a significant frequency shifting when Doppler frequency spread above 15 Hz. However, the current study was limited by frequency domain cases. Further investigation, such as, combination effect between multipath effect and relative velocity Doppler effect is need to be consider for further parametric study as the real condition approach.

6. References

[1] Citree M, Shahabudeen S, Freitag L, and Stojanovic M 2008, *Proc. Int. Conf. Oceans* (Quebec) (Canada: IEEE)

[2] Trujillo R P 2010, *Interference cancellation and network coding for underwater communication systems*, (Denmark: Department of Telecomunication Technology, West Aalborg University Press)

[3] Duke P S 2002, *Direct-sequence spread-spectrum modulation for utility packet transmission in underwater acoustic communication networks*, (USA: Master Thesis, Naval Post Gradute School, Monterey-California)

[4] Stojanovic M, and Preisig J 2009, *Proc of IEEE Communication Magazine*, p 84

[5] Aik T B, Sen Q S, and Nan Z 2007, *Proc of Oceans Asia Pacific*, (Singapore), p 1

[6] Eggen T H 1997, *Underwater acoustic communication over Doppler spread channels*, (Massachusetts: Ph.D Dissertation, Department of electrical Engineering and Ocean Engineering, Massachusetts Institute of Technology-Woods Hole Oceanographic Institution)

[7] Citree M 2007, A high frequency warm shallow water acoustic communications channel model and measurements, *J. Acoust. Soc. Am*, 122, p 25

[8] Kusuma I P A I 2017, Aplikasi Metode Passive Time Reversal Mirror Untuk Mengurangi Pengaruh Multipath Pada Komunikasi Akustik Bawah Air, *J. Inovtek Polbeng*, 07, p 1

[9] Wu B H, Too G P and Lee S 2010, Audio signal separation via a combination procedure of time-reversal and deconvolution process, *J. Mech Syst and Signal Proc*, 24, p 1431

[10] Kuperman W A, Hodgkiss W S, Song H C, Akal T, Ferla C, and Jackson D 1998, Phase conjugation in the ocean: Experimental demonstrationof an acoustic time-reversal mirror, J. Acoust. Soc. Am, 103, p 25

[11] Frisk G V 1994), *Ocean and Seabed Acoustics: A theory of wave propagation*, (New Jersey: Prentice Hall)

Acknowledgement

This work was financially supported by Direktorat Riset dan Pengabdian Masyarakat, Direktorat Jendral Penguatan Riset dan Pengembangan Kementrian Riset Teknologi dan Pendidikan Tinggi in accordance with the 2018 fiscal year research contract.

The Influence of Safety Posters on Employee Compliance in Creating a Culture of Workplace Safety and Health in the Construction of Serbaguna Community Centre Sidoarjo Building

F Harianto[1], M F N Aulady[1], F T Nuciferani[1], and Hariyadi S[1]
[1]Adhi Tama Institute of Technology, Surabaya

Abstract. Multi-story building construction project may pose a high rate of risks especially to worker safety, thus developing workplace health and safety is an urgent matter to take notice of. Workplace safety is an effort in protecting employees from any harm from a workplace accident, and in avoiding violation against government laws and regulations. The objective of this study was to examine the influence of safety posters on employees' effort and compliance in developing workplace health and safety at the construction of "Serbaguna Community Centre Sidoarjo" building. This study used the experimental method in which the respondents were the aluminium composite panel (ACP) and scaffolding installation workers. The data was collected based on observation employing a checklist. The data were analysed using paired sample t-test. The result of the study was that the safety posters had a significant impact on the employee's compliance (t-score =11.50 > t-table =1.771) with α=5%. This result indicates that safety posters are able to improve employees' consistency in behaving safely within the work environment.

1. Introduction

The multi-story construction project is subject to a high rate of risks since the activities are oftentimes restricted by the agreed cost, time and quality [1] [3]. As a consequence, contractors are required to work professionally, when in fact, their concern about workplace health and safety is relatively low. This is so unfortunate since human resources are the company's valuable asset. The statistics of work-related accidents is suspected not to represent the actual incidents happened, which is more. Work-related accidents can impact the various significant economic loss. Among construction activities, working at height is one of the most dangerous works because it is more likely to cause permanent disability or even death [2]. The said risks are typically prompted by the negligence of using personal protective equipment stated in the occupational health and safety regulations. There were several infelicities found during observation, some of which the workers were rarely seen in the safety gears provided. Also, they also seem to neglect the workplace sanitary. Therefore, the objective of this study serves as an effort to put workplace health and safety into practice within the construction site of "Serbaguna Community Center Sidoarjo" Building. Poster, as the treatment, was put up in order to raise awareness amongst the workers working at height specifically the ones installing aluminium composite panel (ACP).

2. Literature Review

2.1. Poster

The poster is a visualisation displayed as an announcement, caution, or persuasion typically in forms of graphics [5]. It is an image combining visual elements, such as lines, graphics, and texts intended for drawing attention and covying a message concisely [9]. It is a visual combination of strong design using colours and messages designed for catching the attention of passersby long enough to put emphasis to and instil a meaningful idea in their memory [7][8].

It can be concluded from the variety of opinions stated earlier that poster is a type of media containing written message, either in the form of graphics or texts intended for drawing people's attention so that the message can be delivered and accepted with ease. The functions of the poster are as follows [9]:

- Motivating employees. Poster serves a purpose as a reminder to encourage and motivate employees compliance.
- Warning. It contains reminders to implement the rules.
- Creative experience. Producing an idea, story, or composition of the poster to display can lead to more creative activities.

According to Sadiman, the characteristics of a good poster are simple, serving one idea to achieve one primary goal, and using vivid colours, clear slogan, clear prints, a variety of patterns and texts, in a relatively large size: A3 or bigger [5].

2.2. Workplace Health and Safety

Workplace safety is a series of attempts to foster a safe and healthy work environment for employees of the company [11]. It is human attempts to prevent the company, employees, society, and natural environment from any harming incidents.

There are two approaches in workplace safety: modern safety approach and operational approach. Modern workplace safety approach is a combination of both. Industrial safety approach emphasises on the obedience to the occupational health and safety regulations and the use of personal protective equipment, while operational approach emphasises more on the operations involving hazardous materials which are processed using specific operational parameters, such as temperature, pressure, flow, and others [12].

Occupational health, according to ILO and WHO, is an attempt to maintain and improve employee's physical, mental, and social well-being to the highest degree. Every work component can be the source of harm to employee's health either in the form of injury or physical or mental health problems. The primary purposes of occupational health are :

- Maintenance and improvement of employee's health and working capacity
- Increase in the working environment to become conducive for occupational health and safety
- Developing work organisation and working culture toward the ones which are supportive of workplace health and safety.

2.3. Compliance

Compliance is an attitude, an act to always follow the rules. Employee compliance is the key to preventing work-related accidents. Non-compliant attitude has to be curbed by giving warning to the workers who violate the regulations or to the workers who disrupt others [11]. Good compliance is reflected on workers' spirited behaviour and responsibilities for the work; also on having high solidarity among co-workers.

3. Research Methodology

The study was done in the construction of "Serbaguna Community Centre Sidoarjo" building, especially in engineering facade using Aluminium Composite Panel (ACP) and scaffolding installation. The experimental method was utilised, with work safety posters being the treatment variable and the worker's compliance being the dependent variable. The themes of the posters were appeals for prioritizing safety (poster 1), using safety gears (poster 2), reporting any damage (poster 3), checking safety harnesses (poster 4), protecting their valuable possessions (poster 5), believing that the risk is not worth their life (poster 6), checking protective equipment (poster 7), and putting on helmet (poster 8). The display location is detailed in Figure 3.
Observation as the data collection technique used in this study employed a checklist to measure the worker's attitude. The dependent variables were the use of personal protective equipment (PPE), work method and the working environment. Likert scale was employed to measure workers compliance with four response categories: 1 - never, 2 - rarely, 3 - often, and 4 - always. The indicators of workers compliance are described in Table 1. The research respondents were 25 workers selected by simple random sampling. The observation of workers attitude was done three times before and after the treatment, with time interval of 14 days (refer to Figure 1 for more details). The validity and reliability of the instrument was assessed before it was put to use. The validity value with α=5% was t-score> t-table (11.281 > 1.782) which is generally taken to indicate that the instrument was valid. The reliability value with α=5% was r-score> r-table (0.989 > 0.514) which is usually taken to show that the instrument was reliable. Equation 1 is the statistical formula used in the analysis of the paired t-test, in which the symbol 'n' represents the total number of individuals in the sample, the symbol 'di' represents the difference in post-test and pretest scores, and the symbol 't' represents the value of t score.

$$t = \frac{\sum d_i}{\sqrt{\frac{n \sum d_i^2 - (\sum d_i)^2}{n-1}}} \qquad (1)$$

Table 1. Observation Indicators

FACTOR	INDICATOR
Personal Protective Equipment	Wearing helmet, goggles, waistcoat, shoes, mask, gloves, and body harness.
Work Method	Carrying out work procedures well on the installation of ACP and panels.
Working Environment	Reporting any damage in the work area
	Maintaining cleanliness of the work area
	Care to the safety of the work area
	Feeling while working

Source: company document

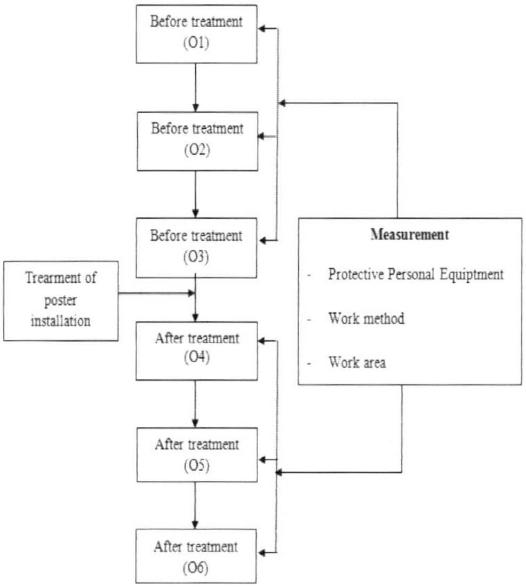

Figure 1. Research frame.

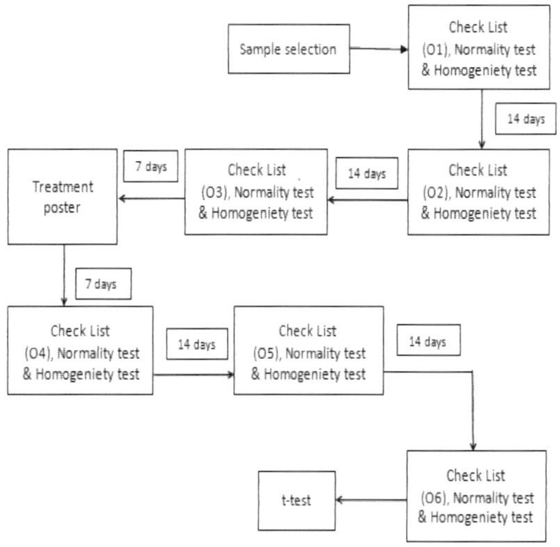

Figure 2. Research procedure.

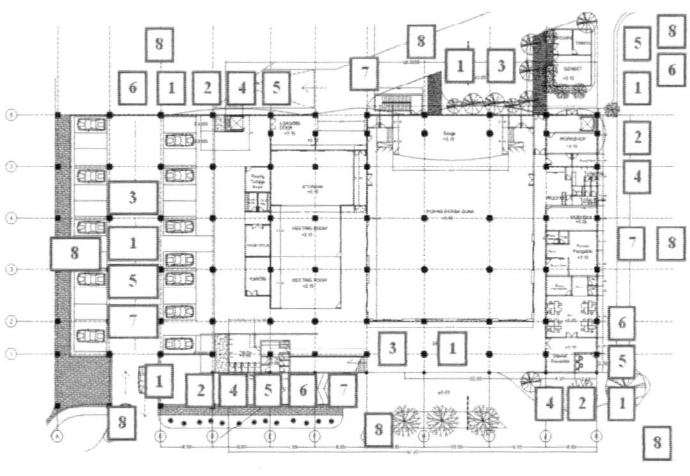

Figure 3. Display location of the posters.

4. Result and Discussion

The characteristics of the respondents in this study are presented in Table 2, 3, and 4, which in general indicate that the respondents are of sufficient age, have an adequate level of educational attainment and working experience. 96% of the respondents are under 45 years old which falls in the working age population. By the statement of the Department of Labor, younger workers are more productive than the older ones. When it comes to the awareness of workplace safety and health, the level of education plays an important role [6]. With 64% of the respondents have attended high school (SMA) / vocational high school (SMK), raising the awareness is assumed not to be a problem. 68% of them have more than five years of working experience. Workplace safety and health is closely tied to the duration of working experience [11]. Work-related accidents in Hong Kong which resulted in injuries on hands happened to the workers with less than one year of experience [6].

Table 2. Age

Age (years)	Percentage (%)
18-25	32
26-35	52
36-45	12
46-55	4

Table 3. Education

Education	Percentage (%)
SMP	36
SMA/SMU	64

Table 4. Experience

Experience (years)	Percentage (%)
More than 5	68
Less than 5	32

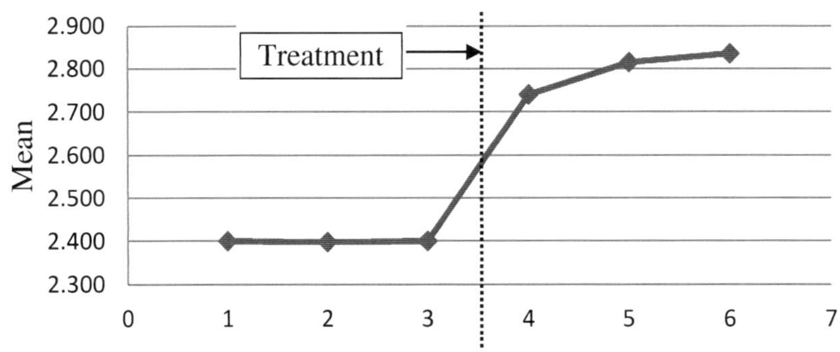

Figure 4. The Mean of Pretest and Post-test Scores

As presented in Figure 4, the sample was given occupational health and safety test. Comparing Observation 1 to Observation 2 with $\alpha=5\%$, the value of t-score was less than t-table ($-0.44<1.771$). This means that there was no significant difference between their safety attitudes on both occasions. Comparing Observation 2 and Observation 3, the value of t-score was once again less than t-table ($0.569<1.771$). Since there was no significant difference between the worker's safety attitudes on the first and third observation, which was 42-day apart, treatment in the form of the poster was given. Another observation was done seven days afterwards. The value of t-score was higher than t-table ($11.5>1.761$) resulted from comparing the post-test scores to the mean of the pretest scores on Observation 3 and 4. The mean of Observation 4 being higher than the mean of Observation 3 ($2.74>2.40$) also indicated that poster as the treatment had given significant impact on the worker's safety attitude. Comparing Observation 4 and Observation 5, the value of t-score was also higher than t-table ($3.565>1.771$), and so was on Observation 5 and 6, with t-score = 1.995 and t-table = 1.771. The significant difference between t-test means of Observation 4, 5, and 6 (2.74, 2.815, and 2.835) also indicated that the worker's safety attitudes kept getting better on every observation after the treatment. This was a sign of positive impact the posters gave the workers.

5. Conclusion

This study concludes that the provision of work safety posters in the construction of "Serbaguna Community Centre Sidoarjo" building had a positive impact on workers regarding occupational safety behaviour ($11.5 > 1.771$). It is expected that the construction project will continue to provide posters as an attempt to create a culture of safety at work. In addition, supervision of the implementation of workplace health and safety which includes safety patrol, safety supervisors, safety meetings and reporting the handling of work-related accidents are still carried out as planned.

6. References

[1] Soeharto, Iman. (1995). *Manajemen Proyek Dari Konseptual Sampai Operasional*. Jakarta : Erlangga.
[2] Wirahadikusuma, Reini. W. (2007). *Tantangan Masalah Keselamatan Dan Kesehatan Pada Proyek Konstruksi di Indonesia*. Bandung : Fakultas Teknik Sipil dan Lingkungan, Institut Teknologi Bandung.
[3] Ervianto, Wulfram. (2002). *Manajemen proyek konstruksi*. Yogyakarta : Andi.
[4] Riadi, Edi. (2015). *Metode Statistik : Parametrik Dan Nonparametrik*. Tangerang : Pustaka Mandiri.
[5] Mustiqon. (2012). *Pengembangan Media Dan Sumber Pembelajaran*. Jakarta: PT. Prestasi Pustakarya.
[6] Helda, et.al, 2007. *Hubungan Karakteristik Tenaga Kerja dan Faktor Pekerjaan Dengan Kecelakaan Kerja Di Perusahaan Meuble Kayu Kelurahan Oesapa Kota Kupang*. MKM Vol. 02 No. 01 Juni 2007.
[7] Sujana, Nana &Rivai, Ahmad. (2010). *Media Pengajaran*. Bandung: Sinar Baru Algensindo.
[8] Susilana, Rudi & Riyana, Cepi. (2009). *Media Pembelajaran Hakikat, Pengembangan, Pemanfaatan Dan Penilaian*. Bandung: CV. Wacana Prima.
[9] Anitah, Sri. (2008). *Media Pembelajaran*. Surakarta : UNS Press.
[10] Kurniawidjaja, L. Meily. (2012). *Teori Dan Aplikasi Kesehatan Kerja*. Jakarta : UI Press.
[11] Suma'mur. (2014). *Higiene Perusahaan Dan Kesehatan Kerja*. Jakarta : Sagung Seto.
[12] Gunawan. F.A. [2012]. *Safety Leadership*. Jakarta: Dian Rakyat.

Acknowledgement

The writers sent their gratitude to the contractors of PT. Bangun Kubah Sarana – PT. Indokon Raya, KSO for permitting the study to be carried out in the construction of "Serbaguna Community Centre Sidoarjo" building.

Triglyceride of Kapok seed Oil to biofuel over a synthesised Cu-Mo supported HZSM-5 catalyst

Y W Mirzayanti[1,2], A Roesyadi[2], D H Prajitno[3*]**

[1]Chemical Engineering, Faculty of Industrial Technology, Institute Technology of Adhi Tama Surabaya, Indonesia.
[2]Department of Chemical Engineering, Industrial Technology Faculty, Institute Technology of Sepuluh Nopember, Surabaya, Indonesia, 60111.
[3]Department of Industrial Chemical Engineering, Vocational Faculty, Institute Technology of Sepuluh Nopember, Surabaya, Indonesia, 60111.
*corresponding author: dana@chem-eng.its.ac.id
**corresponding author: yustiawulandari@yahoo.com

Abstract. In this study, triglyceride of kapok seed (*Ceiba pentandra*) oil was catalytic hydrocracking (T = 350°C, P = 10-20 bar) to biofuel production using Cu-Mo supported HZSM-5 (Cu-Mo/HZSM-5_51) as the catalyst. The catalyst was synthesized using incipient wetness impregnation. The obtained catalysts were characterized by X-ray diffraction (XRD), and Brunauer-Emmett-Teller (BET) method. The catalyst performance on catalytic hydrocracking of kapok seed oil using CuMo/HZSM-5 with ratio molar for Cu:Mo is 1:1 and loading 5%wt. The catalytic evaluation results indicate that for the employed conditions gasoil was the major product. The biofuel fraction formed consisted of 0.98 area% of gasoline, kerosene range of alkanes as much as 1.55 area% and gas-oil range of alkane as much as 1.14 area% at 350°C over Cu-Mo/HZSM-5_51. Triglycerides of kapok seed oil can be cracked by catalytic hydrocracking process to carboxylic acid as much as 78.78 area% by hydrogenation reaction using Cu-Mo/HZSM-5_51 catalyst. The catalyst is proved to be effective for the conversion triglyceride of kapok seed oil to corresponding biofuel gasoil range.

Key words: catalytic hydrocracking, kapok seed oil, biofuel, Cu-Mo metal transition, HZSM-5 catalyst

Introduction

Energy consumption continues to increase along with increasing population growth. The dominated energy consumption area is the transportation sector that demands the amount of fuel and it is unavoidable in the future. The fossil fuels are the vital source for the transportation sector, but it had caused global issues such as increasing CO_2 emission in the air and that is responsible for the greenhouse effect and global warming. The use of vegetable oil as an alternative source of transportation energy can be promoted by national or international legislation, sustainability, and economic growth. Fuels derived from the biomass lipid fraction are an alternative that is very suitable to replace carbon from fossil fuels for the industrial and transportation sectors [1-3]. The first generation developed from renewable fats from edible oil such as coconut oil, palm oil, soybean, sunflower etc. But, edible oil has the disadvantage which is a food commodity so that currently renewable fuels from non-edible oil are developed. They are cannot be haven as a food. In general, transesterification is used to produce fatty acid methylester (FAME), but from this process also produced as the by-product, glycerol [4].

Vegetable oil that can be used as biofuel one of them is kapok seed oil (KSO). KSO can be used for biofuel production by trans-esterification and, it has high free fatty acid content for the general trans-esterification process using basic homogeneous catalysis. Catalytic hydrocracking of triglyceride of kapok seed oil is an alternative route to makes biofuel. The catalytic hydrocracking process consumes a significant amount of hydrogen. Hydrogen consumption can be reduced by the use of catalysts during the process of catalytic hydrocracking [5].

Most researches work had reported on the production of biofuel from the catalytic hydrocracking of vegetable oils using zeolites [6]. The type of zeolite that is widely used is HZSM-5. HZSM-5 zeolite has a crystalline porous structure, high cracking activity, large surface area and strong acid [6,7]. Zeolite also has hydrophobic properties, namely the ability to crack vegetable oil containing triglycerides [8]. The main product using zeolite HZSM-5 as a catalyst is the dominant aromatic compounds and some poly-aromatic [9]. The acidic site can support the hydrogen transfer reaction. Transition metal (Ni, Cu, Co, Mo etc) on zeolite can support modify the acid and textural properties of the support [6, 10].

The combination of metal site and acid site of HZSM-5 gives a bi-functional characteristic of the catalyst. Cu metal catalyst can provide active hydrogen and is in over of promoting hydrogenation [11,12]. Mo metal catalyst is excellent hydrotreating catalyst [13]. Combination of Cu and Mo transition metals loaded into zeolite can form the bi-functional acidic c'atalyst for catalytic hydrocracking of triglyceride of kapok seed oil to biofuel.

Most research in single promoted using Cu or Mo metals [12]. In the previous study, Cu-HZSM-5 is used for continuous processes for hydrogenation of aromatic aldehydes in fixed-bed reactors. Cu can be relied upon as the main active species for the hydrogenation process to biofuel produce [12]. In this study, a dual promoter focused on the use of Cu-Mo/HZSM-5 catalysts was chose to study triglycerides of kapok seed oil for biofuel with gasoil-range alkanes.

Experimental methods

1.1. Material and chemicals

Kapok seed oil content unsaturated fatty acids in the form of linoleic acid and palmitic acid were 78.85% and 21.12% as a majority composition in the previous study [14]. The HZSM-5 zeolite synthesis is made from NH_4-ZSM-5, which was purchased from Zeolyst International. Cu from 98% copper nitrate and Mo from 98% ammonium heptamolybdate tetrahydrate purchased by Merck were used as a metal precursor.

1.2. Catalyst preparation and characterization

NH_4-ZSM5 zeolite (SiO_2/Al_2O_3 = 80), in powder form, was calcined under air at 550°C for 5 hours to obtained HZSM-5, as reported in the previous study [14,15]. The ratio of Cu to Mo used is 1:1 by loading 5% by total weight on the HZSM-5 catalyst. Before impregnation, the incipient wetness volume of support was determined by slowly adding water to 0.5 g of copper nitrate to a solution. A solution of copper nitrate was prepared as the copper precursor. The molybdenum precursor was prepared as same as a copper precursor. The copper precursor solution was slowly adding to HZSM-5 zeolite by the incipient wetness impregnation method while stirring using a magnetic stirrer at 30°C for 3 hours. The catalyst was dried in the oven at 120°C for 12 hours. The catalyst called by Cu/HZSM-5. After that, the molybdenum precursor solution was added to the Cu/HZSM-5 catalyst while stirring using a magnetic stirrer at 30°C for 3 hours too. The catalyst sample was dried in the oven at 120°C for 12 hours. After impregnation, the catalyst was calcinated at 500°C for 5 hours in the air and reduction with hydrogen at 550°C for 3 hours. It is designated as CuMo/HZSM-5_51 catalyst.

The catalyst was characterized by X-ray diffraction (XRD) using XPERT-Pro system apparatus operated at 40 kV, 30 mA, and Cu Kα radiation. The diffraction angle was varied from 10° to 90° at a scan rate of 10.16 s. The Brunauer-Emmett-Teller (BET) surface area, average pore diameter, and total pore volume were determined from nitrogen adsorption or desorption.

1.3. Catalytic hydrocracking of triglyceride of kapok seed oil

Triglyceride of kapok seed oil via catalytic hydrocracking was carried out in a slurry pressure batch reactor system with a stirrer (PARR 4563) as reported by the previous study [14-16]. As much as 200 ml of kapok seed oil is put into the reactor with a Cu-Mo/HZSM-5 catalyst of 1 g. The next step is flowed N_2 gas into the reactor for 5 minutes and stopped the flow to replace hydrogen gas for 1 hour with 10-20 bar. The reaction was carried out at a temperature of 350°C for 2 hours. Liquid products from conversion of kapok seed oil triglycerides were analyzed using Gas Chromatography-Mass Spectrometry (GC-MS) as reported by previous studies. [14,15].

Results and discussion

1.1. Characterization of catalyst

The characterization of catalyst was done by X-ray diffraction (XRD) and Brunauer-Emmett-Teller (BET) method. Figure 1 shows the X-ray diffraction pattern of Cu-Mo/HZSM-5 catalyst. The main peak of HZSM-5 zeolite observed at 2θ of 22° – 25° with framework type MFI [16,17]. After the impregnation process of Cu and Mo metal becomes HZSM-5, the structure of HZSM-5 on the catalyst does not experience significant changes. Cu metal detected a peak at 2θ of 43.18° while for Mo metal was detected at peaks at 2θ of 37.19° and 41.25°. The XRD pattern also found Cu in CuO detected a peak at 2θ of 35.65° and 35.97° while Mo metal in MoO_2 form was detected a peak at 2θ of 36.63°, 36.87°, 37.19°, 53.00°, 53.39° and 53.83°.

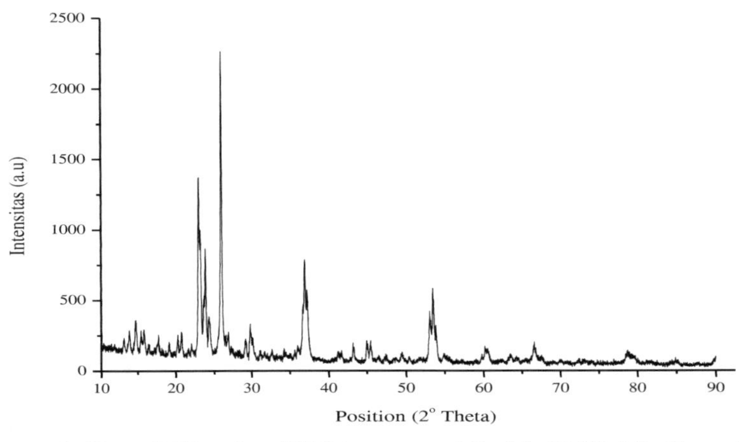

Figure 1. X-ray Diffraction (XRD) pattern of Cu-Mo/HZSM-5_51 catalyst

Table 1 shows the properties of Cu-Mo/HZSM-5 catalyst with 5% metal loading and 1: 1 metal ratio to the total weight of HZSM-5 catalyst (Cu-Mo/HZSM-5_51). On the HZSM-5 catalyst, a surface area of 375.121 $m^2.g^{-1}$ was obtained as reported in a previous study [15]. Then, after the impregnation Cu and Mo, the wet impregnation method into HZSM-5 has reduced the surface area to 281.140$m^2.g^{-1}$. This is due to the presence of Cu and Mo metal which cover the pores on the catalyst surface.

Table 1. The properties of catalyst Cu-Mo/HZSM-5

Catalyst	surface area, $m^2.g^{-1}$	Total pore volume, $cm^3.g^{-1}$	Average pore diameter, nm
HZSM-5*	375.121	0.2712	2.89
Cu-Mo/HZSM-5_51	281.140	0.1618	1.15

*[15]

1.2. Catalytic hydrocracking of triglyceride of kapok seed oil

Figure 2 shows the GC-MS spectrum of catalytic hydrocracking of triglyceride products from kapok seed oil using CuMo/HZSM-5_51 catalyst at 350°C. The fatty acid content of kapok seed oil is dominated by linoleic acid then after catalytic hydrocracking occurs, the carbon chain triglyceride linoleic acid is cracked into palmitic acid by 26.67 area%. At 350°C, the formation of carboxylic acids in hydrocarbon liquid products shows that triglycerides found in kapok seed oil are crack to free fatty acids (etc. palmitic acid, oleic acid). The formation of free fatty acids in the catalytic hydrocracking process is caused by hydrogenation reactions in the crack chain of triglycerides [16-18]. In addition to free fatty acids, 14.56% of the hydrocarbon compounds have been formed. Olefin detection shows that in the hydrogenation reaction there has been a cyclization reaction simultaneously as has been reported in previous studies [15].

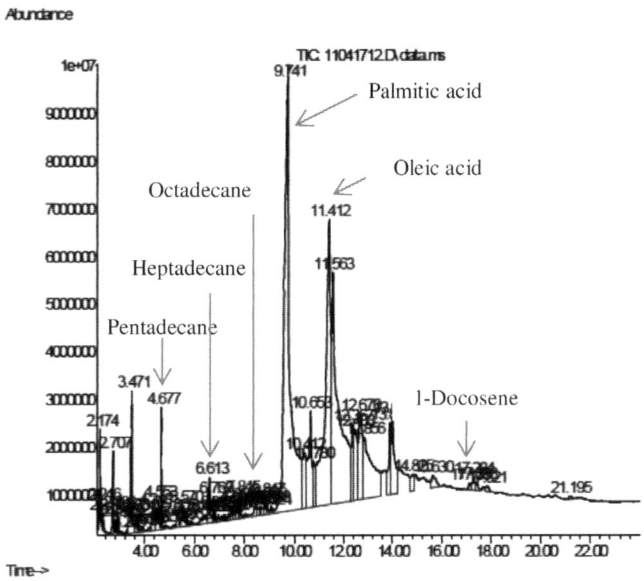

Figure 2. GC-MS spectra of liquid hydrocarbon products using Cu-Mo/HZSM-5_51 catalyst at 350°C

Figure 3 showed that the compounds of liquid hydrocarbon products using Cu-Mo/HZSM-5_51 catalyst at 350°C. The liquid hydrocarbon products are dominated by carboxylic acid as much as 78.78 area%. This shows that unsaturated fatty acids have broken the chain into saturated fatty acids. Cu metal is more selective for hydrogenation reactions when combined with HZSM-5 and has the role of adding acidic side so it is more effective for hydrogenation reactions [12]. And Mo metal leads to the hydrodeoxygenation reaction. Hydrocarbon compounds formed consist of 1.5 area% of n-paraffin, 2.4 area% of the cycloparaffin, 9.79 area% of olefins and 0.76 area% of the aromatic area. This shows that at 350°C it has succeeded in breaking the double chains on the long chain of fatty acids. The presence of 9.79 area% olefins in hydrocarbon liquid products shows that there has been a dehydrogenation reaction from n-paraffin.

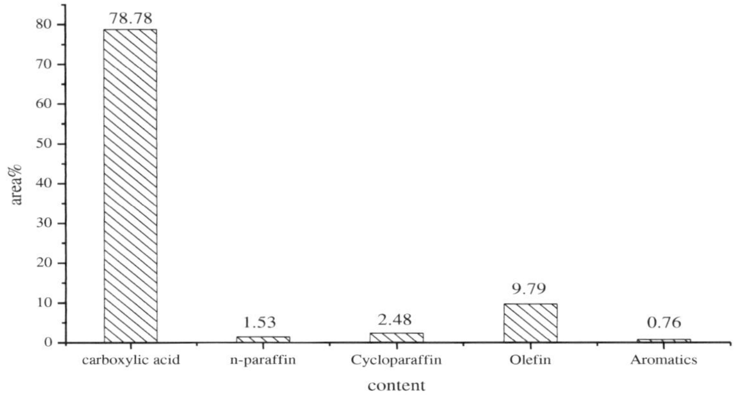

Figure 3. The compounds of liquid hydrocarbon products using Cu-Mo/HZSM-5_51 catalyst at 350°C

Table 2 shows the number of variations of biofuel in liquid products in the catalytic hydrocracking process of triglycerides of kapok seed oil using Cu-Mo/HZSM-5_51 catalyst at 350°C. The biofuel fraction formed consisted of 0.98 area% gasoline, the kerosene alkane range was 1.55 area% and the gas-oil alkane range was 1.14 area%. Whereas for the other products formed were classified into cycloparaffin, olefin and aromatic as much as 11.07 area%.

Table 2. Contents of various biofuel in liquid product over Cu-Mo/HZSM-5_51 catalyst at 350°C

Contents of various hydrocarbon in liquid product (area%)				
nC_5-nC_9	nC_{10}-nC_{14}[a]	nC_{15}-nC_{18}[b]	$nC_{>18}$	Cycloparaffin, olefin, and aromatic
0.98	1.55	1.14	0.00	11.07

[a]kerosene range alkanes
[b]gas-oil range alkanes

Conclusion

A catalytic hydrocracking process for the produced biofuel from triglyceride of kapok seed oil over Cu-Mo/HZSM-5_51 in the slurry pressure batch reactor was established. Cu-Mo/HZSM-5_51 can support hydrogenation reactions. Where HZSM-5 is a catalyst that has strong acidic properties and the acidic side participates in hydrogenation reactions. From the analysis of the liquid product content of biofuel alkane gas-oil range is formed as much as 1.14 area%. Whereas the hydrocarbon content of liquid products is dominated by carboxylic acid as much as 78.78 area%.

References

[1] Ameen, M., Azizan, M.T., Yusupp, S., Ramli, A., and Yasir, M., 2017, "Catalytic hydrodeoxygenation of triglycerides: An approach to clean diesel fuel production", Renewable and Sustainable Energy Reviews, vol. 80, pp. 1072-1088.

[2] Aslam, M., Konwar, L.J., Sarma, A.K., and Khotiyal, N.C., 2015, "An investigation of catalytic hydrocracking of high FFA vegetable oils to liquid hydrocarbons using biomass derived heterogeneous catalysts", Journal of Analytical and Applied Pryrolysis, vol. 115, pp. 401-409.

[3] Li, J., Xia, H., Wu, Q., Hu, Z., Hao, Z., and Zhu, Z., 2016, "Hydrocracking of the crude oil from thermal pyrolysis of municipal wastes over bi-functional Mo-Ni catalyst", Catalysis Today, vol. 271, 172-178.

[4] Ishihara, A., Fukui, N., Nasu, N., and Hashimoto, H., 2014, "Hydrocracking of soybean oil using zeolite-alumina composite supported NiMo catalysis", Fuel, vol. 134, pp. 611-617.

[5] Kubicka, D., and Kikhyanin, O., (2015), "Opportunities for zeolites in biomass upgrading—Lessons from the refining and petrochemical industry", Catalysis Today, 1 April 2015, Vol. 243, pp. 10-22.

[6] Teixeira, C.M., Frety, R., Barbosa, Barbosa, C.B.M., Santos, M.R., and Bruce, E.D., 2017, "Mo influence on the kinetics of jatropha oil cracking over Mo/HZSM-5 catalysts", Catalysis Today, Vol. 279, p. 202-208.

[7] Chen, L., Li, H., Fu, J., Miao, C., and Lv, P., 2015, "Catalytic hydroprocessing of fatty acid methyl esters to renewable alkane fuels over Ni/HZSM-5 catalyst", Catalyst Today, vol. 259, pp. 266-276.

[8] Sandstede, G and Lehmann, T., 2014, "Catalyst with supplement component for hydroprocessing of bio-feedstock", Pub. No. US 20140058182 A1, Pub date: Feb 27, 2014

[9] Rabaev, M., Landau, M.V., Vidruk-Nehemya, R., Koukouliev, V., Zarchin, R., and Herskowitz, M., 2015, "Conversion of vegetable oils on Pt/Al2O3/SAPO-11 to diesel and jet fuels containing aromatic", Fuel, vol. 161, pp. 287-294.

[10] Ewald, S., Standl, S., and Hinrichsen, O., 2018, "Characterization of nickel catalysts with transient methods", Applied Catalysis A: General, vol. 549, pp. 93-101.

[11] Li, Y., Zhang, C., Liu, Y., Tang, S., Chen, G., Zhang, R., and Tang, X., 2017, "Coke formation on the surface of Ni/HZSM-5 and Ni-Cu/HZSM-5 catalysts during bio-oil hydrodeoxygenation", Fuel, vol. 189, pp. 23-31.

[12] Kong, X., and Chen, L., 2014, "Hydrogenation of aromatic aldehydes to aromatic hydrocarbons over Cu-HZSM-5 catalyst", Catalysis Communications, vol. 57, pp. 45-49.

[13] Sousa, L.A., Zotin, J.L., and Teixeira da Silva., V., 2012, "Hydrotreatment of sunflower oil using supported molybdenum carbide", Applied Catalysis A: General, 449, p. 105-111.

[14] Mirzayanti, Y.W., Kurniawansyah, F., Prajitno, D.H., and Roesyadi, A., 2018, "Zn-Mo/HZSM-5 Catalyst for Gasoil Range Hydrocarbon Production by Catalytic Hydrocracking of *Ceiba pentandra* oil", Bulletin of Chemical Reaction Engineering & Catalysis, 13 (1): 136-143. (doi:10.9767/bcrec.13.1.1354.136-143)

[15] Mirzayanti, Y.W., Prajitno, D.H., and Roesyadi, A., 2017, "Catalytic Hydrocracking of kapuk seed oil (*Ceiba pentandra*) to produce biofuel using Zn-Mo supported HZSM-5 catalyst", 7th International conference on Environment and Industrial Innovation, IOP Conf. Series: Earth and Environment Science 67 (2017) 012023, doi:10.1088/1755-1315/67/1/012023

[16] Parahita, I.G.A.A., Mirzayanti, Y.W., Gunardi, I., Roesyadi, A., and Prajitno, D.H., 2018, Production of Biofuel via catalytic Hydrocracking of Kapuk (*Ceiba pentandra*) seed oil with Ni-Mo/HZSM-5 Catalyst", Matecweb of Conferences, 156, 06001 (2018), doi: 10.1051/matecconf/201815606001.

[17] Hemalatha, P., Lee, B, J., Lee, J, G., and Jang, T, H., 2013, "Synthesis, characterization of ceria modified ZSM-5 and its performance towards NO oxidation", International journal of control and automation, vol. 6, No. 5, pp. 83-90.

[18] Al Dughaither, A.S and de Lasa, H., 2014, "Neat dimethyl ether conversion to olefins (DTO) over HZSM-5: Effect of SiO_2/Al_2O_3 on porosity, surface chemistry, and reactivity", Fuel, vol. 138, pp. 52-64.

Power Loss Classification on Shifts Based on SMS (Singlemode-Multimode-Singlemode) Structured Fiber Optic Using Gaussian Naïve Bayes Method

D H Sulaksono[1] and A C P Siregar[1]

[1]Computer Science Department, Institut Teknologi Adhi Tama Surabaya, Arief Rahman Hakim 100, Surabaya, Indonesia

danang_h_s@yahoo.com

Abstract. Singlemode-multimode-singlemode based optical fiber can be used as a good communication medium, but energy or power carried by light will be weakened (losses) due to leakage or due to lack of clarity or shift in optical fibers. In this study, power losses in fiber optics based on SMS will be classified based on changes in the values of power losses to shifts and divided them according to three classes, there are good, average, and bad. The shift will be used as a classification variable that is between 0 μm to 450 μm with an increment of 50 μm for every interval. The SMS optical fiber structure used is 5.5 with 25 attempts on different optical fibers. The classification method used is Naïve Bayes with a Gaussian distribution. Gaussian distribution is used in Naïve Bayes because the dataset will be processed in the form of continuous values. From the results of testing based on TP+TN=6, FP=6 FN =6 on confusion matrix, the classification accuracy value was 42.86%. This indicates that this classification method is still less effective for classifying fiber optic power losses with an SMS structure. For further study, another classification methods can be used in the power loss classification to get better results.

1. Introduction

An optical fiber is a flexible, transparent fiber made by making the diameter of the glass (silica) or plastic material smaller than human hair. There are several types of optical fibers, one of which is that the structure has two singlemode fibers (SM) which are axially connected to the end of a fiber optic parabolic multimode (MM) core. Optical fiber with this structure is also called SMS (singlemode-multimode-singlemode). The feasibility of optical fiber is a power loss. Power loss can occur for numbers of reasons, such as the presence of a leak or lack of speed when the light is delivered. A good power loss change is continuous, in other words, when the power loss value goes up and down it will interfere with the optical fiber to communicate when it's used for transmission.

Fiber optic quality against power loss can be determined by a decision-making system, one of which is data mining. Data mining itself is a process for finding patterns from a data set and involves methods that bring machine learning, statistics and databases together. One branch of data mining is classification, where this branch compares a sample data compared to training data. In this study, a classification method with a Naïve Bayes with Gaussian distribution is used to make decisions on the quality of an optical fiber based on power loss. This method is used because the dataset used is a continuous data value.

The dataset itself will be divided into three classes, there are good, average, and bad. The division of these classes is a representation of the value of the change in power loss value at each shift. For the shift variables used in this classification use a distance between 0 µm, 50 µm, 100 µm, 150 µm, 200 µm, 250 µm, 300 µm, 350 µm, 400 µm and 450 µm with a multimode width of 5.5 cm which will be tested on 25 different optical fibers.

2. Methods

2.1 Naïve Bayes Classifier

Naive Bayes is a simple probabilistic classification that calculates a set of probabilities by adding up the frequency and combination of values from a given dataset. The algorithm uses Bayes theorem and assumes all independent or non-interdependent attributes given by values in class variables [8]. Another definition says Naive Bayes is a classification with probability and statistical methods proposed by British scientist Thomas Bayes, which predicts future opportunities based on previous experience [1].

Naive Bayes is based on the assumption of simplification that attribute values are conditionally free when given output value. In other words, given the output value, the probability of observing together is a product of individual probabilities [8]. The advantage of using Naive Bayes is that this method only requires a small amount of training data to determine the parameter estimates required in the classification process. Naive Bayes often works much better in most complex real-world situations than expected [2]. Naïve Bayes formula is written as follows :

$$p(C_k|x) = \frac{p(C_k)p(x|C_k)}{p(x)} \tag{1}$$

where $p(x|C_k)$or can be called posterior probability is the probability of the C_k hypothesis based on condition x. $p(C_k)$ is also called prior, the probability of a hypothesis in class C_k. In this study, prior uses three classes determined from the results of fiber optic power loss values, which are good, medium and bad. For $p(x|C_k)$ called likelihood, the likelihood value is the result of the vector between all variables with each class. There are 10 variables used, where all of these variables are the result of shifting optical fibers, namely 0 µm, 50 µm, 100 µm, 150 µm, 200 µm, 250 µm, 300 µm, 350 µm, 400 and 450 µm. Whereas for $p(x)$ is usually referred to as marginal likelihood or evidence, which is the sum of all $p(C_k)p(x|C_k)$.

2.2 Gaussian Distribution

As the value of features, the power loss data used is continuous, so it can also be assumed that the continuous values associated with each class will be distributed by using Gaussian distribution. This Gaussian distribution is used to find the value of $p(x|C_k)$. with the following equation:

$$p(x = v \mid Ck) = \frac{1}{\sqrt{2\pi\sigma_k^2}} e^{-\frac{(v-\mu_k)^2}{2\sigma_k^2}} \tag{2}$$

Where, x is a training data set that has a continuous attribute, in this case are the values of power loss. Whereas, µk is the average value of the x value that is assosiated in each class of C_k, with the following equation:

$$\mu = \frac{1}{N}\sum_{i=1}^{N} x_i \tag{3}$$

For σ_k^2 is a variance value that calculates how far a set of numbers is scattered from the average value. The result σ_k^2 is obtained from the value of x which is assosiated by class C_k, the variance formula is as follows :

$$\sigma^2 = \frac{1}{N}\sum_{i=1}^{N}(x_i - \pi)^2 \quad (4)$$

In Gaussian distribution, v value itself is the sample data that is tested against training data, so the probability of the distribution of the x value to the class C_k that is $p(x = v \mid Ck)$, can be calculated by entering the value v into a normal distribution which is the value of μ_k and σ_k^2 are used as parameters.

3. Results And Discussions

3.1 Training Data

There was a set of tests to get the power loss values which are used as training data. The variables used in this study are the value of shifting in optical fibers with a distance between 0µm to 450µm at intervals of 50µm at each shift.

Table 1. Power Loss Training Data

0µm	50µm	100µm	150µm	200µm	250µm	300µm	350µm	400µm	450µm	C_k
3.5704	3.7543	3.8642	3.8789	4.1258	4.3584	4.3698	4.4568	4.477	4.632	good
3.629	3.6185	3.7541	3.7642	3.5575	3.5499	3.4864	3.5815	3.6914	3.5889	bad
3.3302	3.469	3.407	3.4536	3.4017	3.5543	3.5106	3.6344	3.5129	3.6598	bad
3.2584	3.2687	3.2989	3.3112	3.3189	3.3338	3.3476	3.3491	3.4589	3.5686	average
3.6777	3.7123	3.7521	3.8195	3.8721	3.9013	4.0781	4.012	4.1445	4.152	good
3.6734	3.6772	3.6851	3.7121	3.7431	3.7486	3.8112	3.9112	3.9211	3.9389	average
3.6471	3.6511	3.6576	3.7214	3.7354	3.7231	3.7395	3.7415	3.7613	3.7781	average
3.715	3.7421	3.7561	3.7588	3.8471	3.851	3.8563	3.8761	3.8845	3.895	average
3.5671	3.621	3.7612	3.8121	3.8812	4.0214	4.135	4.2158	4.2568	4.3541	good
3.5891	3.5196	3.5527	3.5247	3.5595	3.638	3.4998	3.5931	3.4094	3.4135	bad
3.3164	3.1236	3.4187	3.7392	3.7658	3.6905	3.655	3.7828	3.7823	3.8705	bad
3.6143	3.6517	3.6778	3.6989	3.7124	3.731	3.7342	3.7421	3.748	3.7681	average
3.5761	3.5833	3.6122	3.6816	3.7151	3.792	3.9511	3.9621	3.978	4.0321	good
3.514	3.5251	3.5326	3.5387	3.5443	3.5124	3.5161	3.5312	3.5452	3.571	average
3.5925	3.6834	3.9877	4.2511	3.9734	3.9553	3.9355	3.9549	4.1345	3.9981	bad
3.6122	3.7011	3.7612	3.7732	3.8122	3.8418	3.912	4.0181	4.081	4.1225	good
3.512	3.587	3.4126	3.4137	3.523	3.5112	3.5145	3.6541	3.6127	3.5124	bad
3.7127	3.7124	3.701	3.8212	3.7816	3.7986	3.7812	3.7932	3.7986	3.8124	bad
3.8112	3.8162	3.821	3.841	3.8472	3.851	3.8612	3.8713	3.8811	3.8826	average
3.6315	3.6608	3.543	3.5337	3.5911	3.6138	3.662	3.2142	3.3336	3.4867	bad
3.512	3.5262	3.5413	3.5512	3.584	3.6121	3.6268	3.6321	3.6326	3.6413	average
3.5126	3.5268	3.6171	3.6577	3.7124	3.7223	3.7431	3.7851	3.8781	3.9511	good
3.6471	3.6321	3.721	3.8454	3.8732	3.9413	4.351	4.4588	4.5126	4.5324	good
3.6788	3.7246	3.8412	3.8962	3.9131	3.9463	3.9821	4.0511	4.121	4.1322	good
3.5172	3.611	3.621	3.6289	3.7031	3.7358	3.7481	3.7921	3.797	3.813	good

3.2 Mean And Variance For Each Class

After the training data is obtained, the calculation of the mean μ_k and variance σ_k^2 is done on the x values that associated with the values of each class C_k. These values are used to provide parameters to the set of training data x against the values of the test data v. The results of the mean and variance in each class C_k are shown in table 2, table 3, and table 4.

Table 2. Mean And Variance for C_k=good

	0µm	50µm	100µm	150µm	200µm	250µm	300µm	350µm	400µm	450µm
mean	3.595	3.6518	3.7279	3.777	3.8453	3.9178	4.030	4.083	4.138	4.191
variance	0.00786	0.0113	0.0178	0.0193	0.0354	0.0748	0.1040	0.1242	0.1205	0.1432

Table 3. Mean And Variance for C_k=average

	0µm	50µm	100µm	150µm	200µm	250µm	300µm	350µm	400µm	450µm
mean	3.5931	3.607	3.621	3.641	3.6665	3.6703	3.686	3.706	3.7290	3.7554
variance	0.0492	0.0496	0.0462	0.04901	0.05513	0.0548	0.0568	0.06607	0.0504	0.0379

Table 4. Mean And Variance for C_k=bad

	0µm	50µm	100µm	150µm	200µm	250µm	300µm	350µm	400µm	450µm
mean	3.539	3.5467	3.597	3.6876	3.6441	3.663	3.6306	3.6510	3.6594	3.667
variance	0.0365	0.0629	0.0737	0.135	0.0812	0.0388	0.0457	0.0561	0.0517	0.0747

3.3 Test Data

After getting the mean μ_k and variance values σ_k^2, the next step is to calculate the likelihood value $p(x|C_k)$ for each $C_k$class using Gaussian distribution. A set of power loss test data is substituted into the variable v in the Gaussian distribution equation. The likelihood value $p(x|C_k)$ of each class that produced by Gaussian distribution is used to classify the sample data to produce a classified test data. Test data that will be used for classification testing are shown in table 5.

Table 5. The Results Of Classified Test Data

0µm	50µm	100µm	150µm	200µm	250µm	300µm	350µm	400µm	450µm	C_k	Actual
3.6471	3.6511	3.6576	3.7214	3.7354	3.7231	3.7395	3.7415	3.7613	3.7781	average	average
3.6016	3.6014	3.601	3.7101	3.6705	3.6875	3.6701	3.6821	3.6875	3.7014	average	bad
3.5360	3.5210	3.6209	3.7343	3.7621	3.8302	4.2408	4.3477	4.4015	4.4214	good	good
3.2053	3.0125	3.3077	3.6281	3.6547	3.5811	3.5454	3.6717	3.6712	3.7641	average	bad
3.6016	3.6015	3.610	3.7101	3.6705	3.6875	3.6701	3.6832	3.6875	3.7011	average	bad
3.604	3.6310	3.6450	3.6477	3.7360	3.7401	3.7452	3.7650	3.7634	3.784	average	average
3.5704	3.7543	3.8642	3.8789	4.1258	4.3584	4.3698	4.4568	4.477	4.632	good	good
3.5623	3.5661	3.5740	3.6510	3.6320	3.6374	3.7001	3.8001	3.8122	3.8278	bad	average
3.5172	3.611	3.621	3.6289	3.7031	3.7358	3.7481	3.7921	3.797	3.813	good	good
3.7127	3.7124	3.701	3.8212	3.7816	3.7986	3.7812	3.7932	3.7986	3.8124	bad	bad

3.4 Classification Accuracy and precission

Confusion matrix is used for measuring the effectiveness of classification. By using Gaussian Naïve Bayes classification, Classification Accuracy and precission calculation is performed for 10 test data against training data.

Table 6. Confusion Matrix

		Prediction			
		Good	**Average**	**Bad**	**FN**
Actual	**Good**	3	0	0	0
	Average	0	2	1	1
	Bad	0	3	1	3
	FP	0	3	1	

From the table above we have obtained the value of TP+TN=6, FP=6 FN =6, so that the accuracy can be calculated as :

$$Accuracy = \frac{6}{6+4+4} * 100\% = 42.86\% \quad (5)$$

4. Conclusions

The classification accuracy calculation using Gaussian Naïve Bayes classifier method is less effective when it used for automatic classification of power loss data on optical fiber. The accuracy is still below 50% using the effectiveness measurement Confusion Matrix. For further study, another classification methods can be used in the power loss classification to get better results.

5. References

[1] Bustami 2013 Penerapan Algoritma Naive Bayes Untuk Mengklasifikasi Data Nasabah Asuransi. TECHSI : Jurnal Penelitian Teknik Informatika. Vol. 3. No.2. Hal. 127-146.

[2] Pattekari S A and Parveen A 2012 Prediction System for Heart Disease Using Naive Bayes. International Journal of Advanced Computer and Mathematical Sciences. ISSN 2230-9624. Vol. 3. No3, 290-294.

[3] Lou W, Wang X, Chen F, Chen Y, Jiang B and Zhang H 2014 Sequence Based Prediction of DNA-Binding Proteins Based on Hybrid Feature Selection Using Random Forest and Gaussian Naïve Bayes. PLoS ONE 9(1): e86703.

[4] Ghaith A, Siti S K and Husniza H 2018 Conceptual Framework for Stock Market Classification Model Using Sentiment Analysis on TwitterBased on Hybrid Naïve Bayes Classifiers. International Journal of Engineering & Technology. 7(2.14) (2018) 57-61.

[5] Dishant K and Arunima S 2018 Kernel-Based Naive Bayes Classifier for Medical Predictions. Advances in Intelligent Systems and Computing. Volume 695.

[6] Raizada RDS and Lee Y-S 2013 Smoothness without Smoothing: Why Gaussian Naive Bayes Is Not Naive for Multi-Subject Searchlight Studies. PLoS ONE 8(7): e69566.

[7] Ziemann and Olaf et al 2008 POF Handbook-Optical Short Range Transmission Systems. Springer : Berlin.

[8] Nian Z, Jiang X, Jing Z and Keenan L 2018 Gaussian Process Regression Method for Classification for High-Dimensional Data with Limited Samples. IEEE. ISSN: 2573-3311. DOI: 10.1109/ICIST.2018.8426077.

[9] Aslam C P S and Danang H S 2017 Perancangan Sensor Suhu Menggunakan Metode Interpolasi Lagrange Berbasis Serat Optik Berstruktur Sms (Singlemode-Multimode-Singlemode). JEEE UMSIDA. Vol. 1 No. 2. DOI: https://doi.org/10.21070/jeee-u.v1i2.976.

[10] Aslam C P S and Danang H S 2017 Perancangan Sensor Pergeseran Menggunakan Metode Interpolasi Lagrange Berbasis Serat Optik Berstruktur Sms (Singlemode-Multimode-Singlemode). SNTEKPAN. C-123. ISBN : 978-602-98569-1-0.

IOP Conf. Series: Materials Science and Engineering **462** (2019) 012025 doi:10.1088/1757-899X/462/1/012025

Application of Saving Matrix Methods and Cross Entropy for Capacitated Vehicle Routing Problem (CVRP) Resolving

Lukmandono[1], M Basuki[1], M J Hidayat[1] and F B Aji[1]

[1]Industrial Engineering Department, Faculty of Industrial Technology, Institut Teknologi Adhi Tama Surabaya, Jln. Arief Rahman Hakim 100, Surabaya, Indonesia

lukmandono@itats.ac.id

Abstract. Distribution activity is one part of Supply Chain Management and very important component. It can improve profit company with a decrease in transportation cost. This activity has many problems like capacity of vehicle, difference of consumer demand, and difference of consumer location. These problems must be handled by determine the route to minimize distribution cost. My research is the company engaged in printing service where the product has been printed will be sent to several customers with different locations. In determining the route will be used Saving Matrix and Cross Entropy method. The purpose of them is to form an optimal route where saving matrix method is processed manually and Cross Entropy method by Matrix Laboratory software. The result obtained by calculation of Saving Matrix method are four sub routes and Cross Entropy method are three sub routes. The comparison of two methods used is Cross Entropy method 5,81% more efficient than Saving Matrix method based on total transportation cost.

1. Introduction

Distribution is the activity of moving and storing goods from the level of suppliers to the level of consumers in the supply chain. The distribution will be at each level of the supply chain where the material flow moves from supplier to factory, then the finished product from the factory moves to the consumer so this will directly affect the supply chain cost [1]. This research takes the object of a company that has goods distribution activity between producers and end users (users) located in Surabaya City, East Java. This company is engaged in printing services that produce based on orders from consumers.

The problem in this study is how to determine distribution routes that are able to make optimal distance and costs. One thing that often happens in distribution activities is Vehicle Routing Problem. namely the problem of determining the vehicle route to serve several customers with different locations. The VRP is to find the minimum cost routes to be traveled by a fleet of vehicles. All of the vehicles must start and terminate at a central depot. Each customer must be visited once, by one vehicle only [2]. The total demand for each route does not exceed the vehicle's carrying capacity which is commonly called a capacitated vehicle routing problem (CVRP).

In this study, the optimal distribution of products will be determined to solve distribution problems. This is so that the product can meet market demand well, simplify and accelerate distribution to reach consumers quickly so that it can minimize the total cost of transportation. The approach used through the saving matrix method and the Cross Entropy method where both have the ability to solve transportation and distribution problems. Transportation in companies produces the highest costs in the distribution system, which is one third to two thirds of the total distribution costs [3].

2. Theory

2.1 Distribution

Distribution is an activity to move products from a factory or distribution center to retailers or consumers. The distribution process is generally carried out by transportation systems and modes or can also be through other media [4].

2.2 Capacitated Vehicle Routing Problem (CVRP)

Capacitated Vehicle Routing Problem (CVRP) is the basic model of VRP problems. This is consistent with the constraints experienced by having a vehicle with a limited capacity. In its implementation, CVRP is more difficult to solve than TSP (Traveling Salesman Problem). All requests from customers are known at the beginning, for each customer and carried out on the same route, so all customer requests are entered on the same route [5].

The classic CVRP can be described as follows: N customers geographically dispersed in a planar region must be served from a unique depot. Each customer asks for a quantity qi (i=1,2,….,n) of goods. The transport cost from node i to node j is Cij. m vehicles with a fixed capacity Q are available to deliver the goods stored in the depot. Each customer must be visited just once by only one vehicle. The objective of the problem is minimizing the total cost of all routes without violating the individual capacity of each vehicle. The depot is denoted by i=0 [6].

The model of the vehicle routing problem is established based on the following assumptions [7]:

1. The vehicles in each warehouse are of the same type (the same load), and the number of vehicles varies with customer demand.
2. There is a limit to the load of the vehicle, which cannot be overloaded.
3. Each customer must be accessed and can only be serviced once; no repeat access is allowed.
4. After completion of the vehicle distribution task, the vehicle fleet must return to the original warehouse.

3. Saving Matrix

The saving matrix method is one method that is easy to apply for complex distribution problems quickly. The advantage of this method lies in its ease of modification. If there are obstacles such as vehicle capacity, number of vehicles, delivery time or other constraints, it can provide a quick and practical solution even though the results do not guarantee optimal solutions [8]. In the application of the saving matrix method there are steps that must be taken, namely as follows;

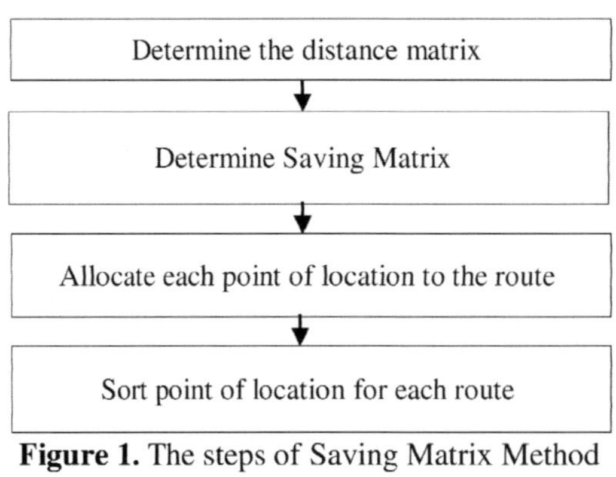

Figure 1. The steps of Saving Matrix Method

IOP Conf. Series: Materials Science and Engineering **462** (2019) 012025 doi:10.1088/1757-899X/462/1/012025

4. Cross Entropy Method (CE)

Cross-Entropy (CE) is general stochastic optimization technology based on a fundamental principle of information theory called cross entropy [8]. The CE method can be used to solve estimation and optimization cases. The advantage of the CE method is that it is very reliable in optimization problems if it is formulated as an estimation problem. In its development, there has been a modification to the CE method so that this method can be used also to solve optimization problems that have been successfully applied in the completion of VRP, TSP, orienteering problems, job scheduling, and others.

In addition, the advantage of this method lies in its relatively small iteration, the CE algorithm is also a very concise algorithm, and the larger the resulting sample the better the resulting solution [10].Once the associated stochastic optimization is defined, then CE method follow these two phases [11]:

1. Generate a solution sample randomly (vectors, trajectories, etc.) according to a specified random mechanism.
2. Update of the parameters of the random mechanism, on the basis of the data, in order to produce a "better" sample in the next iteration.

The following is the CE algorithm for the case of Capacitated Vehicle Routing Problem [12]:

1. Determine the supporting parameters, namely the percentage of elite sample parameters (ρ), smoothing parameters (α), stop tolerance (β), and the number of generation samples (N).
2. Determine the intial value of the P. transition matrix.
3. Determine generating N paths from the P transition matrix and calculating the probability of a node being served by the following mechanism.
4. Calculate and select the path with the optimum value of ρ x N sample.
5. Updating the transition matrix from the P matrix based on the generation of the track.
6. If the stop criteria are met, then the iteration stops, if not then the process returns to step 2.

In the steps that must be carried out in this study as well as identifying problems, conducting a literature review, collecting and processing data and analyzing the results of a study. and the last is to conclude the results of this research. The research method flowchart used in this study can be seen in Figure 2.

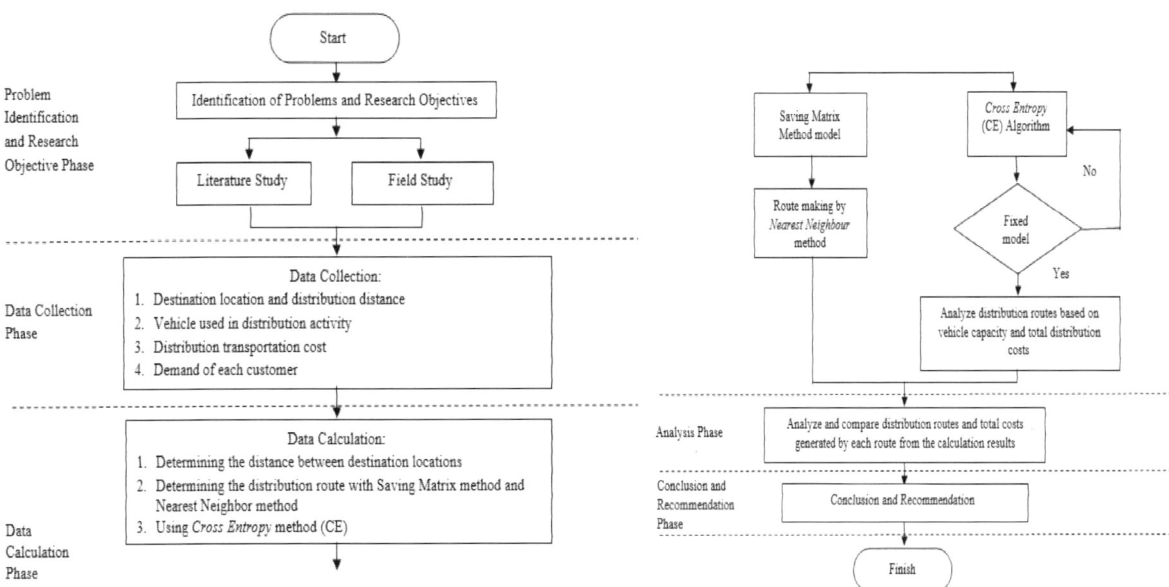

Figure 2. Research Methodology

IOP Conf. Series: Materials Science and Engineering **462** (2019) 012025 doi:10.1088/1757-899X/462/1/012025

5. Results

5.1 Sub Route Using Saving Matrix Method

From the results of route savings, several proposed routes were found in accordance with the vehicle's carrying capacity so that the value of distance savings from the route obtained by using the nearest neighbor method was obtained. After obtaining the sub route, the calculation of distribution costs is done by multiplying the distance of each route with the price of the fuel of the vehicle used to distribute the product which can be seen in Table 1.

Table 1. Sub Route of Product Distribution with the Saving Matrix Method

No	Sub Route	Vehicle Capacity (Unit)	Mileage (Kilometers)	Fuel Ratio (Liter)	Fuel Cost (Rupiah)	Vehicle Used	Cost (Rupiah)
1	P-PSRN-PRBLG-LMJG-JMBR-KPNJN-BLT-TLG-KDR-MDN-PNRG-BJNGR-TBN-LMGN-P	37.380	858	39	5.150	Truck	200.850
2	P-PUSAT-SIDOARJO-MOJOKERTO-JOMBANG-MALANG-P	32.981	271	6,82	5.150	Truckl	35.114
3	P-SURABAYA- P	22.180	14	2 x 0,56	5.150	2 x L300	5.768
4	P-GRESIK-P	8.824	70	7	8.000	Pick Up	56.000
	TOTAL		1213				297.732

By using the saving matrix method is expected to be able to complete and make distance savings to get the route to be formed. In determining the route also must pay attention to the transport capacity of the vehicle used which will be filled by the product requested by the consumer. The routes formed from the results of the Saving Matrix include:

1. Factory – Pasuruan – Probolinggo – Lumajang – Jember – Kepanjen – Blitar – Tulungagung – Kediri – Madiun – Ponorogo – Bojonegoro – Tuban – Lamongan – Factory
2. Factory – Pusat – Sidoarjo – Mojokerto – Jombang – Malang – Factory
3. Factory – Surabaya – Factory
4. Factory – Gresik – Factory

From the results of the sub route that was formed obtained by using the Saving Matrix method, the total distance was 1213 kilometers with a total cost of Rp. 297.732 every month

5.2 Sub Route Using Cross Entropy Method

By using the Cross Entropy method with a value of $\rho = 0.6$, $\alpha = 0.9$, $N = 30$ which was carried out with the help of Matlab software version 2018, a sub route was obtained based on the best results, as follows:

Table 2. Sub Route of Product with Cross Entropy Method

No.	Sub Route	Distance (Kilometer)	Fuel Ratio (Liter)	Fuel Cost (Rupiah)	Cost (Rupiah)
1	1 – 7 – 2 – 21 – 16 – 5 – 14 - 10 – 18 – 1	613	27,86	5.150	143.498
2	1 – 19 – 17 – 13 – 8 – 1	514	23,36	5.150	120.323
3	1 – 12 – 9 – 20 – 3 – 1	71	3,23	5.150	16.620
	Total	1198			280.441
	computing time (detik)		0,9375		

By using the Cross Entropy method which is done with the help of the 2018 version of Matlab software, sub routes are obtained based on the best results including:

1. Factory – Kediri – Blitar – Tulungagung – Probolinggo – Jember - Pasuruan – Lumajang – Sidoarjo – Factory
2. Factory – Surabaya – Pusat – Mojokerto – Kepanjen – Factory
3. Factory – Malang – Lamongan – Tuban – Bojonegoro – Factory

From the results of the sub route formed, the total distance is 1198 kilometers with a total cost of Rp. 280.441 every month.

5.3 Comparison of Saving Matrix Method and Cross Entropy Method

Based on the results of data piloting, a comparison was made between the initial sub routes that became company policies with the proposed sub routes obtained using the Saving Matrix method and the Cross Entropy method. The following is a comparison table which can be seen in Table 3.

Table 3. Comparison of Saving Matrix and Cross Entropy Method

Indicator	Method		
	Saving Matrix	*Cross Entropy*	Company Policy
Distance (Kilometers)	1213	1198	1677
Cost (Rupiah/Month)	297.732	280.441	875.932

From the table above it can be seen that the comparison between company policy with the use of the Saving Matrix method and the Cross Entropy method has a considerable difference that is by using the Saving Matrix method, obtained route savings of 464 kilometers or 27.67%. Whereas by using the Cross Entropy method, obtained route savings of 479 kilometers or 28.56%.

In terms of the total costs incurred for the initial distribution route of Rp. 875,932.00 per month, the total cost spent on the proposed route using the Saving Matrix method is Rp. 297,732 every month or obtained a savings of 66.01%, while the total costs incurred by using Cross Entropy amounted to Rp. 280,441.00 or obtained savings of 67.98%.

6. Conclusion

From the results of the analysis, using the Cross Entropy method obtained three proposed sub routes with the first sub route namely Factory - Kediri - Blitar - Tulungagung - Probolinggo - Jember - Pasuruan - Lumajang - Sidoarjo - Factory, the second sub route is Factory - Surabaya - Pusat - Mojokerto - Kepanjen - Factory, and the third sub route is Factory - Malang - Lamongan - Tuban - Bojonegoro - Factory with a total distance of 1198 kilometers with a total cost of Rp. 280,441.00. To save total costs, the Cross Entropy method is more efficient at 5.81% from the Saving Matrix method.

From the comparison of the two methods, it can be said that the use of the Cross Entropy method is better than the use of the Saving Matrix method in this case. In addition, in terms of computing the Cross Entropy method has a very fast time with the help of Matrix Laboratory software, but depending on the number of samples generated (N), the more N, the longer the computation time. Whereas the Saving Matrix method takes a very long time because it still uses the manual method. So that the more points visited, the longer the computation time.

7. References

[1] Chopra S. & Meindl P. 2004. *Supply Chain Management : Strategy, Planning, and Operation.* New Jersey: Prentice-Hall, Inc.

[2] Du, L. & He R. *Combining Nearest Neighbor Search with Tabu Search for Large-Scale Vehicle Routing Problem.* 2012.Physics Procedia. 25, 1536 – 1546.

[3] Ballou, Ronald H. 1998. *Business Logistic Management.* New Jersey: Prentice Hall International Edition.

[4] Siahaya, Williem. 2013. *Sukses Supply Chain Management.* Jakarta: In Media

[5] Amri M. Rahman A. Yuniarti R. 2014. *Penyelesaian VRP Dengan Menggunakan Metode Nearest Neighbour (Studi Kasus: MTP Nganjuk Distributor PT. Coca Cola).*

[6] Qi, Chengming. Application of Improved Discrete Particle Swarm Optimization in Logistics Distribution Routing Problem. 2011. Procedia Engineering. 15, 3673 – 3677.

[7] Zhao Y. Leng L. Qian, Z. Wang W. *A Discrete Hybrid Invasive Weed Optimization Algorithm for The Capacitated Vehicle Routing Problem.* 2016.Procedia Computer Science. 91, 978 – 987.

[8] Yuniarti, Rahmi & Murti Astuti. 2013. *Penerapan Metode Saving Matrix Dalam Penjadwalan Dan Penentuan Rute Distribusi Premium Di SPBU Kota Malang.* , 4(1), pp.17–26.

[9] Bian Li. 2012. *Parameters Analysis to Pattern Null Placement Based on the Cross Entropy Method.* 2012. Physics Procedia. 24, 1444 – 1451.

[10] Soenandi, Iwan A., Budi Marpaung, & Meriastuti G. 2017. *Optimasi Vehicle Routing Problem Dengan Pendekatan Metaheuristik (Studi Kasus: Distribusi Bahan Baku Makanan).*

[11] Santosa Budi. *Multiclass Classification with Cross Entropy-Support Vector Machines.*2015. Procedia Computer Science. 72, 345 – 352.

[12] Rera, Gladiez F. 2010. *Penerapan Metode Cross Entropy dalam Penyelesaian Capacitated Vehicle Routing Problem*, Jurusan Teknik Industri Institut Teknologi Sepuluh Nopember: Surabaya.

Acknowledgement
The author would like to thank the financial support from The Ministry of Research Technology and Higher Education (no kontrak) for Applied University Research scheme (2018).

IOP Conf. Series: Materials Science and Engineering **462** (2019) 012026 doi:10.1088/1757-899X/462/1/012026

The Effect of Coal Powder Addition to Asphalt Concrete - Wearing Course (AC-WC) Mixture to Increase Road Surface Hardness Quality

K H Putra[1], **M Firdausi**[1] **and M Rubbyanto**[1]

[1]Department of Civil Engineering, Institut Teknologi Adhi Tama Surabaya, Indonesia

adyputra.putra@gmail.com

Abstract. The most common type of structural asphalt pavement in Indonesia is asphalt concrete-wearing course (AC-WC). This type of layer is the most vulnerable type, with damages mostly caused by weather and tire pressure of heavy vehicles. In this study, the test object was produced based on standardized general specification called '*Spesifikasi Umum Bina Marga*' 2010 (Division VI). The mixture of the test object contained varied amount of charcoal powder: 10%, 20%, and 30%. From Marshal Test data obtained in the study, the mixture which met the qualification of '*Spesifikasi Umum Bina Marga*' 2010 was the one containing 30% charcoal powder. The mixture had 1183 kg stability value, 3.73 mm flow value, 316.9 kg/mm Marshal quotient, 4.47 % VIM, 17.08% VMA, and 73.82% void filled. Therefore, it was concluded that 30% charcoal powder content met the Marshal test standard, and more addition of charcoal powder resulted in more void filled to cover the aggregate, and less void left in the mixture (VIM).

1. Introduction

Wood Charcoal is residue containing carbon produced from burning wood in high temperature. Charcoal is black-colored, light, and made up of 80% carbon. The charcoal used in this study contained 85% to 98% carbon, and other components such as ash or other chemical substances. Due to the active carbon content in the wood charcoal, mixing charcoals into asphalt was expected to increase ductility, the ability to hold vapor pressure under high temperature (flash point), and other physical properties of the asphalt. Road, where the asphalt is applied, is one of land transportation infrastructures with crucial role for the growth of economy, social, culture, tourism development, and defense and security; all of which are important for national development [1].

The method used in this study was Marshall method. The sequences performed in Marshall method were: testing specific gravity, designing aggregate gradation, designing aggregate composite materials, measuring bulk specific gravity of aggregate, measuring maximum specific gravity of the mix, and measuring Marshall parameter values.

Therefore, this study on the use of charcoal powder and stone dust as filler in asphalt mixture was conducted. The ratio of charcoal powder in the mixture were 0%, 10%, 20%, and 30% of the weight of the stone dust used.

2. Literature Review

2.1. Previous Studies
This study was conducted on asphalt concrete - wearing course (AC-WC) using pen 60 asphalt mixed with 2.5% and 5% carbide waste. The objective of this study was to find out the Marshal perimeters of the asphalt concrete physical properties from different types of asphalt. The Marshall perimeters showed that the optimum asphalt content was 6% for 0% asphalt, 6.7% for asphalt mixed with 2.5% carbide waste, and 6.8% for asphalt mixed with 5% carbide waste [2].

2.2 Percentage Refusal Density (PRD)
Percentage refusal density is the condition in which a mixture reaches the point of maximum density and therefore, cannot become any denser even if the mixture is further compacted. Percentage refusal density (PRD) is the ratio of inter densities to refusal density, expressed as a percentage. That density is an approach to the real condition in the field after asphalt containing mixture is secondarily compressed by pressures from vehicles over several years of the planned service life without getting plastic deformation.

2.3 Marshall Method
Marshall method concept was originally developed by Bruce Marshall in Mississippi State Highway Department around 1939. The method was further refined by U.S. Army Corps of Engineering. Currently, it has become a standard test method in American Society for Testing and Materials (ASTM): ASTM DS1559 Standard Test Method for Resistance to Plastic Flow of Bituminous Mixtures Using Marshall Apparatus. This method is only used for testing hot asphalt concrete mixture containing hard asphalt with certain penetration grade and 1-inch maximum sized aggregate [3]

Voids in the mixture at percentage refusal density, stability, and flow as well as the result of Marshall / Marshall Quotient (MQ) are the result of dividing stability to flow. The formula is as follows:

$$MQ = \frac{MS}{MF} \tag{1}$$

Explanation:
MQ = Marshall Quotient (kg/mm)
MS = Marshal stability (kg)
MF = Flow Marshall (mm)

3. Research Methodology

3.1 The Testing of Pen 60/70 Asphalt
Testing the physical properties of pen 60/70 asphalt made of AC-WC mixture was performed by testing the following properties:
1. Penetration test
2. Softening point
3. Specific gravity
4. Flash point and Fire point
5. Adhesion to aggregate

3.2 The Testing of Aggregate
Choosing the right and qualified aggregates will affect the supportive hardening ability and the success of both road construction and road maintenance. In asphalt containing mixture, aggregate

gives 90% - 95% contribution to the mixture weight. Therefore, the aggregate physical properties are some of the decisive factors of the mixture quality.

The tested physical properties of the aggregate were:

1. Particle size
2. Gradation
3. Cleanliness
4. Hardness
5. Particle shape
6. Texture of surface
7. Absorption
8. Adhesion to asphalt

3.3 Marshall Testing

This experiment used standard test object that was a mold with 101.6 mm diameter and 75 mm (3 inch) height. The test object was obtained using Marshall compaction hammer with the weight of 4.54 kg (10 lbs), diameter of 3.7/8 inch and free fall of 457 mm (18 inch). The test result showed Marshal properties which were affected by the mixture physical properties such as density, voids in the mineral aggregate (VMA), voids filled with asphalt (VFA), voids in the mixture (VIM), stability value and flow value.

4. Result and Discussion

4.1 The Testing of Asphalt Properties

The asphalt used in this research was Pertamina asphalt with 60/70 penetration grade.

Table 1. Test Result Of Pen 60/70 Asphalt Physical Properties

No	Asphalt Properties	Units	Test methods	Specifications		Results
				Min	Max	
1	Penetration 25°c,100 gr, 5 seconds	mm	SNI 06-2456-1991	60	70	70
2	Softening point 5°c	°c	SNI 2434-2011	48	-	49
3	Flash point	°c	SNI 2433-2011	232	-	312
4	Mass lost (using TFOT)	%	SNI 06-2441-1991	-	0.8	0.2073
5	Asphalt solubility in C2HCL3	%	AASHTO T44-03	99	-	99.713
6	Specific gravity	g/cc	SNI 2441-2011	1	-	1.032

4.2 The Testing of Aggregate Properties

The testing was conducted to determine properties of coarse aggregate, medium aggregate, fine aggregate and stone dust. The test was conducted under Indonesian National Standardization (*Standard Nasional Indonesia*, SNI). Gradation observation was based on AC-WC gradation of Department of Public Works Specification (*Spesifikasi Departemen PU*) 2010.

Table 2. Specific Gravity Testing

NO.	AGGREGATE	BULK	S.S.D	APPARENT	ABSORSED
1	Fine Aggregate	2.565	2.611	2.691	1.829
2	Medium Aggregate	2.575	2.625	2.711	1.943
3	Coarse Aggregate	2.694	0.915	2.823	1.002
4	Filler	3.150		3.150	

4.3 Composition and Proportion of Marshall Test Object

Composition and proportion of aggregate were determined using particle analysis result of every aggregate caught in the strainer. Mixture type used was coarse grade suitable for AC-WC mixture according to Department of Public Works Specification (*Spesifikasi Departemen Pekerjaan Umum*) 2010.

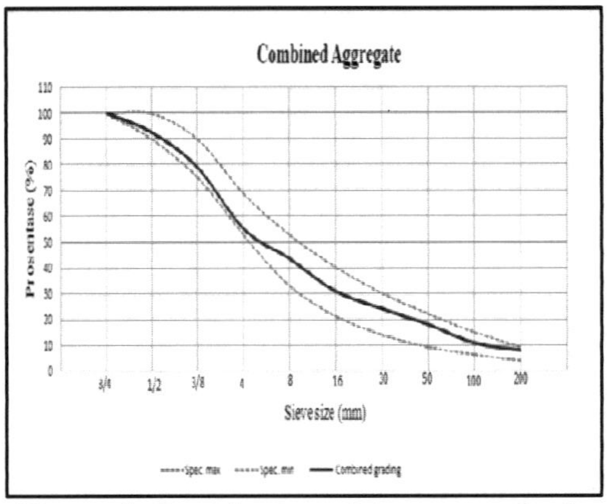

Figure 1. Combination of Aggregate Grade

4.4 Test Result of Marshall Test Object with Charcoal Powder Filler and Test Object without Charcoal Powder

Table 3. Marshall Test

MARSHALL TEST									
Char-coal powder	*No*	*Bulk Sp. Gr*	*V.I.M*	*V.M.A*	*Void Filled*	*Stability*		*Flow*	*Marshall Quotient*
		Combine mix	*%*	*%*	*%*	*Meas*	*Adjust*	*Mm*	*Kg/mm*
0%	1	2.312				80	953	3.2	
	2	2.321				95	1131	3	
	3	2.331				100	1191	3.1	
Average		2.321	4.34	16.97	74.42		1092	3.1	352.2
10%	1	2.269				125	1489	2.9	
	2	2.302				116	1382	3.4	
	3	2.312				100	1191	3.2	
Average		2.298	5.32	17.81	70.18		1354	3.17	427.5

20%	1	2.248				135	1608	4	
	2	2.321				117	1393	3.6	
	3	2.331				110	1310	3.6	
Average		2.3	5.22	17.73	70.54		1437	3.73	384.9
30%	1	2.269				83	989	3.9	
	2	2.344				110	1310	3.8	
	3	2.341				105	1251	3.5	
Average		2.318	4.47	17.08	73.82		1183	3.73	316.9
Spec			**3.0-5.0**	**>15**	**>65**		**>1000**	**>3.0**	**>300**

5. Conclusion

The study concluded that adding charcoal powder to AC-WC mixture give substantial result. The more charcoal powder added to the mixture, the more voids filled to surface aggregate, and less void in the mixture (VIM). In addition, Marshall test data showed that filler containing 30% charcoal powder met every requirement of Department of Public Works Specification (*Spesifikasi Departemen Pekerjaan Umum*) 2010. Filler containing 30% charcoal powder had 1,183 kg stability value, 3.73 mm flow value, 316.9 kg/mm Marshal Quotient, 4.47% VIM, 17.08 % VMA, and 73.82 % void filled.

6. References

[1] Sukirman, Silvia;. (2010). *Perencana Tebal Struktur Perkerasan Lentur.* Bandung: Nova.

[2] Permana, Shezy Nurhayati; Prasetyanto, Dwi; Zurni, Rahmi;. (n.d.). *Studi Penggunaan Limbah Las Karbit Untuk Bahan Tambah Pada Perkerasan Laston AC-WC .*

[3] RSNI M-01-2003;. *Metode Pengujian Campuran Beraspal Panas Dengan Alat Marshall.* Badan Standardisasi Nasional.

[4] Kementerian Pekerjaan Umum Dan Perumahan Rakyat;(2017). *Manual Perkerasan Jalan.* Jakarta: Direktorat Jenderal Bina Marga.

[5] Departemen Pekerjaan Umum: 2002. *Pedoman Perkerasan Lentur Metode Bina Marga (Pt T-01-2002-B)*, Jakarta.

[6] Hendarsin, L Shirley (2000), *Perencanaan Teknik Jalan Raya*. Politeknik Negeri Bandung

[7] Badan Standarisasi Nasional, 1991, SNI 06-2433-1991, *Cara Uji Titik Nyala Dan Titik Bakar Aspal Dengan Alat Cleveland Open Cup*. Departemen Pekerjaan Umum, Badan Penelitian dan Pengembangan Departemen Pekerjaan Umum.

[8] Badan Standarisasi Nasional, 2008, SNI 2417:2008, *Cara Uji Keausan Agregat Dengan Mesin Abrasi Los Angeles*, Departemen Pekerjaan Umum, Badan Penelitian dan Pengembangan Departemen Pekerjaan Umum.

[9] Badan Standarisasi Nasional, 1991, SNI 06-2441-1991, *Cara Uji Berat Jenis Aspal Keras,* Departemen Pekerjaan Umum, Badan Penelitian dan Pengembangan Departemen Pekerjaan Umum.

[10] Badan Standarisasi Nasional, 1991, SNI 06-2456-1991, *Cara Uji Penetrasi Aspal*, Departemen Pekerjaan Umum, Badan Penelitian dan Pengembangan Departemen Pekerjaan Umum.

IOP Conf. Series: Materials Science and Engineering **462** (2019) 012027 doi:10.1088/1757-899X/462/1/012027

Experimental Study of The Performance Characteristic an Induced Draft Cooling Tower with Variates Fillings

E Novianarenti[1,2], G Setyono[1,3] and A G Safitra[4]

[1]Mechanical Engineering Department, Adhi Tama Institute of Technology, Surabaya, Indonesia
[2]Fluid Mechanics and Thermal Laboratory, Adhi Tama Institute of Technology, Surabaya, Indonesia
[3]Internal Combustion Engine Laboratory, Adhi Tama Institute of Technology, Surabaya, Indonesia
[4]Engineering Mechanics and Energy, Electronics Engineering Polytechnic Institute of Surabaya, Indonesia

ekynr@itats.ac.id

Abstract. The aim of this research is designing and testing an Induced Draft Cooling Tower (IDCT). The Cooling tower is the main instrument which used to cool the hot water from the condenser and release heat into the atmosphere at a power plant. Temperature is maintaining as cool as original condition. One of the components of the cooling tower is filler. It is very influential on IDCT performance, where the contact surface between water and air is expanded and the contact time is extended. An experimental study to evaluated the heat and mass transfer coefficient is conducted. A heat transfer correlation is developed; a new variable is defined. The effect of controlling parameters such as range of temperature, heat rejected rates, effectiveness has been analysed. IDCT was filled with a straight, zigzag and wave filler. The results indicate that the tower characteristics increase with increase cold air velocity, with the variation of filler, the highest tower characteristics get from wave filler arrangement.

1. Introduction

Cooling tower, with preceding advantages, is equipment most widely used to release excess heat loads from these processes, such as electric power generation units, thermal and nuclear power plants, chemical, petroleum industries, refrigeration, and air-conditioning systems, into the atmosphere [1]. This device operating based on mass and thermal energy transfer from high-temperature water to coolant air. In the wet cooling tower, water flows over the fillings and thus a direct interface between warm water and coolant air flow occurs. When the bed is in a fully fluidize state, vigorous movement of the wet particle gives rise to intimate mixing of the two flowing fluids. In addition, the residence time of contact for the two phases is high owing to the liquid hold up in the column. These two factors are the reasons for the very high heat transfer coefficient of this tower.

The type of filling material used in the cooling tower has an important point as it provides a very large surface area for evaporative heat and mass transfer to take place from hot water to ambient air and increase contact time between the two fluids. Based on the prior studies, the influence factor of efficiency is mainly the thermal performance of the filling zone because 70% of the heat dissipating capacity depends on the filling zone [2].

Many studies are dealing with the mathematical modelling system which has many advantages such as saving money, but an experimental study is also necessary to know actual performance. This method with different fillings such as splash, honeycomb, ceramic tile, corrugated, and film are reported in the literature [3-5]. For example, Johannes, S [6] established an experimental investigation of the characteristic cooling tower by utilizing the power of capillarity fibres as splash filling. Niken and Arrad [7] developed the variation of splash filling density with a performance of forced draft counter flow cooling tower.

Many researchers have been pursued in this field in order to improve the transport phenomena in this equipment and therefore enhance their performances since the early sixties. Among the existing filling, there is the "VGA" (Vertical Grid Apparatus) type filling which has been initially developed [8] and used in mass transfer equipment. Lemouari performed an experimental investigation of the thermal performance for wet cooling tower filled with a VGA type filling which is 0.42 m high and consist of four galvanized sheets having a zigzag form. He carried out experiments on heat transfer performance of a cooling tower with different parametric studies and yielded a similar model to Gharagheizi [9] which presented an experimental and comparative study on the performance of mechanical cooling tower with two types of film packing. They used vertical corrugated packing (VCP) and horizontal corrugated packing (HCP) having 0.64 meter in high and 0.25 m^2 cross section area. From this research, He reported that the performance of the cooling tower is affected by the water/air mass flow ratio, the type and the arrangement of the packing.

The main purpose of this paper is to carry out an experimental investigation of the performance characteristics of a direct-contact counter flow wet cooling tower filled with a straight, zigzag and wave filler with material acrylic and some variates form in order to increase thermal characteristic with different arrangement than others and to obtain the optimal position pattern.

2. Experimental Setup and Procedure

This research begins with problems in learning. It is a cooling tower performance of energy transfer from hot water and coolant air. This object is, hot water from the condenser assumed that water which is heated by an electric heater. Furthermore, study about the performance of the cooling tower and the developing parameter before. Identification of the influence variable and design of experimental apparatus is to get performance and characteristic.

2.1. Experimental setup

Experimental setup with the identification of boundary variable in order to investigate the characteristic performance of induced draft counter flow wet cooling towers with variated filler conditions is observed. The data reach in the condition equipment and instrument steady.

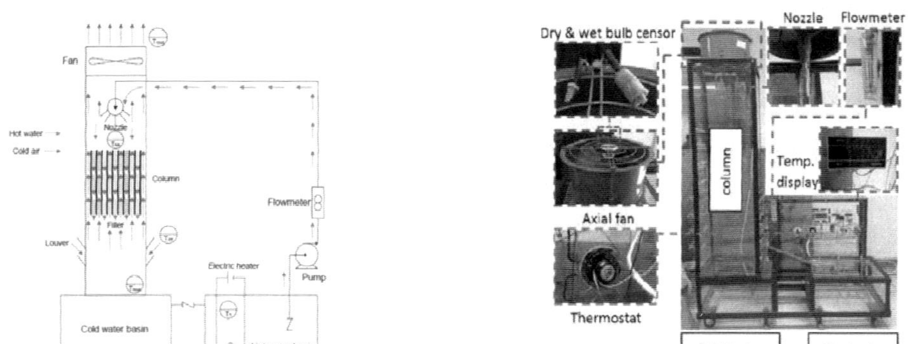

(a) (b)

Figure 1. (a) Schematic diagram (b) Experimental apparatus in real

The purpose of experiments is to consider the optimal position pattern and also find the effect of adding straight, zig zag and wave filler, outlet water temperature and water flow rate on the water temperature difference and cooling tower effectiveness. In what follows, a schematic diagram of the experimental set up is shown in Fig. 1 (a). It consists mainly of a cooling tower which represents the main device used in this test, a cold water basin, a hot water basin which contains electric heaters, a water pump, a flow meter device, a nozzle, a fan, and fillers. The tower's column has a parallel form of dimensions 400 mm x 400 mm x 1400 mm, and is made of Acrylic with frame.

Table 1. Monitor parameters and measurement instruments

Items	Measuring Instruments	Accuracy
Inlet dry and wet bulb temperature	Thermocouple type-K	±1 °C
Outlet air temperature	Thermocouple type-K	±1 °C
Inlet and outlet water temperature	Thermocouple type-K	±1 °C
Circulating water flow rate	Thermocouple type-K	±1 °C

Table 1. explained the instrument and accuracy from the instrument which used for selected item. The considered measurements which were taken consist of the temperatures increase (dry and wet) of the air at the entry and exit of the tower, as well as the inlet and outlet water temperatures. The testing procedure is described in Fig. 1

2.2. Geometry and initial variation

Acrylic is the material used to build the cooling tower column, then the casing IDCT is steel. The Spray *nozzle* is used to spray water throughout the column. On the centre side of the column there is a filler that varies in shape using acrylic. There are three variations of filler geometry used, as shown in fig. 2 (a), (b), and (c).

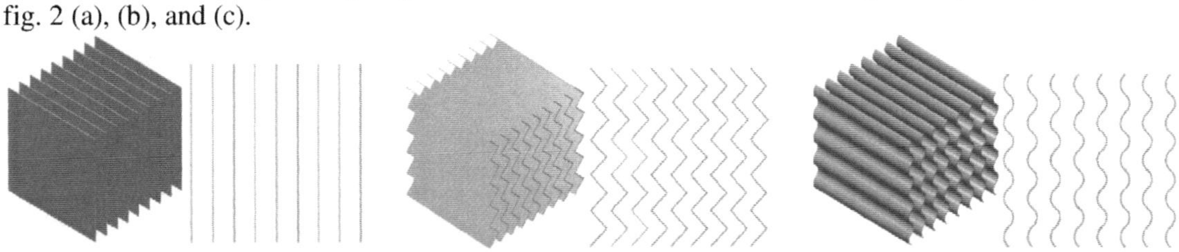

(a) Straight filler (b) Zig zag filler (c) Wave filler

Figure 2. Filler geometry (a) straight, (b) zig zag, and (c) wave

Hot water is sprayed from a spray nozzle then passes through the filler in the centre of the column. The water that leads to the bottom of the column is directly contacted with the cooling air that a sucked using counter flow induced fan. Coldwater (in the reservoir) enter the hot water reservoir again to distributed to the column.

3. Calculation Models of Thermal Parameter

3.1. Range actual and maximum heat transfer

Heat rejected is defined as the difference between hot water entering the cooling tower with hot water coming out of the cooling tower as formulated in equation (1).

$$q_{max} = \left(\dot{m}_{w,in} \cdot h_{w,in} - \dot{m}_{w,in} \cdot h_{w,out}\right) \quad (1)$$

Then, maximum heat transfer is defined as the difference between the heats from hot water entering the cooling tower with cold air coming into the cooling tower as formulated in equation (2).

$$q_{max} = \left(\dot{m}_{w,in}.\, h_{w,in} - \dot{m}_{a,in}.\, h_{a,out}\right) \quad (2)$$

3.2. Effectiveness

The effectiveness of a cooling tower is defined as the ratio of heat rejected to ideal maximum heat, and as a method which used to calculate a capability of heat transfer in the heat exchanger with compared to the calorific value transferred with the possibility of maximum heating value, i.e.

$$\varepsilon = \frac{q_{out}}{q_{in}} \quad (3)$$

3.3. Number of Transfer Unit (NTU)

In the book of Stockers and Jones, as the heat exchanger, a cooling tower performance is also stated in a function which describes characteristics of the cooling tower and basic to know a performance at inlet water temperature and any other wet bulb temperature. The value of $h_c A/c_{pm}$ as known as NTU, commonly known as

$$NTU = \frac{U.A}{m.c_p} \quad (4)$$

Then, by Lu Lu Wenjian Cai, NTU of a counter flow cooling tower i.e.

$$NTU = \frac{ln\left(\frac{1-\varepsilon}{1-\varepsilon.m^*}\right)}{1-m^*} \quad (5)$$

$$m^* = \frac{\dot{m}_a}{\dot{m}_w}.\frac{C_s}{C_{pw}} \quad (6)$$

$$C_s = \frac{h_{a,in}-h_{a,out}}{T_{w,in}-T_{w,out}} \quad (7)$$

4. Result and Discussion

Coordinate x and y-axis in Fig. 3 (a) describe the inlet air velocity and heat rejected. From this graph, it is seen that the Chart show the influence of the whole variation form filler arrangement on the heat rejected by using control velocity of the cooling tower. Trend graphics is linear show that increasing velocity air along with increasing heat rejected.

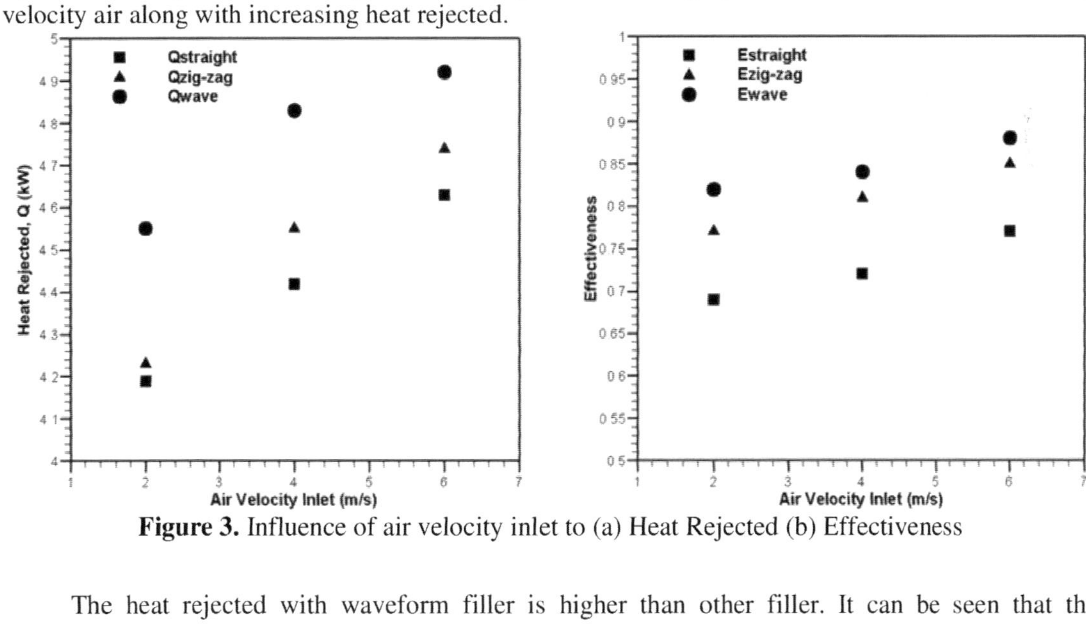

Figure 3. Influence of air velocity inlet to (a) Heat Rejected (b) Effectiveness

The heat rejected with waveform filler is higher than other filler. It can be seen that the influence of heat rejected of the arrangement wave filling is the high intake air velocity because the condition of inlet temperature and outlet temperature is the same. Increasing intake air speed influence moist air mass rate which enters cooling tower ($\dot{m}_{moist}$). When airspeed terrace step by step from 2 m/s; 4 m/s; and 6 m/s, ṁ moist is increased too for all variations. This is in accordance with the equation that has been described that the value of intake air velocity is proportional to the rate of moist air masses. The rate of humid air will affect the rate of the mass of water comes out ($\dot{m}_{w, out}$) of the cooling tower. However, beside all above, heat rejected is also influenced by enthalpy water out ($h_{w, out}$). The temperature of the water coming out ($T_{w, out}$) is decrease when airspeed increased.

Influence of intake air velocity to effectiveness in three variates shows in Fig. 3 (b). By adding intake air velocity of cooling tower will affect the effectiveness. In the intake air velocity, 6 m/s is higher than others velocity variation. Then, it was followed by the effectiveness at the velocity of 4 m/s and 2 m/s. Minimum effectiveness showed at the velocity of 2 m/s on the 0.69 compositions straight filler. Furthermore, the highest effectiveness is obtaining when the air velocity of 6 m/s in 0.88 wave filler arrangement. Amount of filler influenced the characteristics it will be increasing trend graphic of effectiveness.

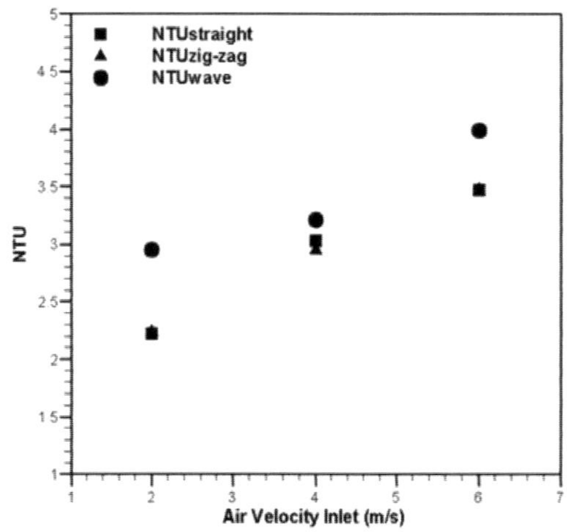

Figure 4. Influence of air velocity inlet to NTU

The effectiveness value increases with increasing air velocity because heat rejected which generate is greater by nearly the same maximum heat. It finds that linearity both of them. Material acrylic with wave filler arrangement can reach the highest value of the effectiveness because of this structure and mesh can accommodate more water from the spray produced by the nozzle, so that the contact between the water with inlet air will be extended with filler media which full of the column. Fig. 4 show the Influence of intake air velocity to NTU. Variation of intake air speed and filler arrangement is affecting Number of Transfer Unit (NTU). The value of NTU in the intake air velocity 6 m/s is 3.99 and it is higher than others velocity variation and arrangement. Between the effectiveness and ratio air to water effective capacitance rate or m*. When intake air velocity increases the value of m* is increase. It is obtained an increase an NTU in all condition of the same speed when the ratio density of the packing material expanded. Since the composition of filler diminishing in each layer can potentially create a hamper heat which will be transferred to the system. NTU is indicating that heat transfer in that process has the better performance. Similarly, when compared with the conditions of the cooling tower with wave filler arrangement compared to others.

5. Conclusion

Several field tests using different variables were performed for an induced draft cooling tower by addition of straight, zigzag and waveforms as the fill material. It was found that the factors affecting the performance of the cooling tower are the air velocity, outlet water temperature, effectiveness, and NTU. The highest heat rejected that happened on the waveforms filler is 4.92 kW, effectiveness of 0.88 and great value of NTU 3.99 on the cooling air velocity 6 m/s with a variation using a wave arrangement.

6. References

[1] M. M. El Wakil, Chapter 7, 1985 *Power Plant Technology,* (New York, St. Louis, San Francisco: Mc Graw-Hill Book Company)
[2] N. Williamson, M. Behnia, S. Armfield 2008 Comparison of a 2D asymmetric CFD model of a natural draft wet cooling tower and 1D model *J. Heat and Mass Transfer.* **51** 2227-2236
[3] J.C. Kloppers, D.G. Kröger 2005 A Critical Investigation into the Heat and Mass Transfer Analysis of Counter Flow Wet-Cooling Towers *Int. Journal of Heat and Mass Transfer.* **48** 765-777
[4] H. Inazumi and S. Kageyama 1975 A Successive Graphical Method of Design of a Cross-flow Cooling Tower *Chem. Eng. Sci.* **30** 717-721
[5] H. R. Goshayshi, J.F. Missenden 2000 Investigation of Cooling Tower Packing in Various Arrangement *Appl. Therm. Eng.* **20** 69–80
[6] J. Susanto 2010 Characteristic of Cooling Tower with "Ijuk" Filling Vocational Diploma Mechanical Eng. (Yogyakarta: UGM)
[7] Elfira Niken, Arrad Ghani 2015 Experimental Study of spacing ratio filling material through Forced Draft Counter Flow Cooling Tower Performance *Electronic Eng. Polytechnic Institute of Surabaya*
[8] M. Lemouari, M. Boumaza, A. Kaabi 2009 Experimental analysis of heat and mass transfer phenomena in a direct contact evaporative cooling tower *Energy Convers. Management* **50** 1610–1617
[9] F. Gharagheizi, R. Hayati, S. Fatemi 2007 Experimental study on the performance of mechanical cooling tower with two types of film packing *Energy Conversion Management* **48** 277–280.

Acknowledgments
Authors wishing to acknowledge the financial support from The Directorate of Research and Community Service, Directorate General of Research and Development, Ministry of Research, Technology and Higher Education, through Beginner Lecturer Research in 2018. The author hopes this research be useful for Institution and added information for the renewable and conversion energy technologies.

Prototype of the Monitoring System and Prevention of River Water Pollution Based on Android

R Sulistyowati[1], A Suryowinoto[1], A Fahruzi[1] and M Faisal[1]

[1]Electrical Engineering Department, Institut Teknologi Adhi Tama Surabaya, Indonesia

971073@itats.ac.id

Abstract. This paper aims at a solution to overcome and reduce river water pollution; factories waste and residential waste that is discharged into river areas that are still used by pond farmers as a source of fresh water in their ponds. If the river water is polluted, it will adversely affect the productivity of the ponds. To overcome the problem of pollution in ponds, it is necessary to research the parameters of pH, temperature and oxygen content in river water. Therefore, flexibility is needed from a system for monitoring the quality and prevention of river water pollution, based on the Android operating system that helps the performance of farmers to find out more about river water pollution wherever they position is. This system design uses Atlas Scientific's DO (Dissolved Oxygen) sensor, v1.1 module pH sensor and DS18b20 temperature sensor. Oxygen content data by DO sensors, pH and water temperature of the river, which have been processed by the system will be displayed on the LCD monitor, then the data will be sent in the form of SMS (Short Message Service) to the user's Smartphone using a modem. If river water pollution is detected, the accuracy rate of this monitoring system is 90% with an average error rate of 3.9% on the measurement of oxygen levels, 2.8% on pH measurements and 1.01% for temperature measurement on river water.

1. Introduction

Ponds are artificial water ponds, which are a growing cultivation medium for shrimp or fish that are kept. In order to obtain optimum results, it is necessary to prepare certain environmental conditions that are in accordance with the cultivation life. The main factor that determines the productivity of ponds is water in the pond, which is a growing medium for the shrimp or fish that are raised. Water quality in accordance with the needs of cultivation commodities needs to be balanced with the adequate quantity of water based on the Android operating system [1]. For traditional ponds, the most important effort to increase pond productivity is to provide water in ponds with good water quality. Ponds generally require brackish water which can be found at tide meetings and water flow from the river. So far, pond farmers have indicated pollution only by looking at the color of the water surface, while according to [2] the indicator or watermark has been contaminated by changes or signs that can be observed through changes in pH or Hydrogen Ion Concentration, colour Change, odour and Sense of Water, changes in Water Temperature, occurrence of sediments, colloids and dissolved materials, microorganisms. Of the many polluted watermark indicators attached above that will be used in the study only three indicators are indicators of changes in pH, changes in water temperature, dissolved oxygen. Changes in each indicator by sensor detection When pollution occurs, farmers take precautionary measures by closing the sluice gate that connects the river and pond water manually. In addition, when they want to put water in the pond, the farmers do not check the quality of the water that will be put into the pond. When the water is put into the pond turned out to be polluted can endanger the growth and survival of the shrimp or fish that are cultivated.

Oxygen plays an important role as an indicator of water quality, because dissolved oxygen plays a role in the oxidation and reduction of organic and inorganic materials. If oxygen is dissolved in water <3 mg/l, it will inhibit the growth of shrimp and fish, and can even lead to death. The causes of the decreasing quality of oxygen dissolved in water include waste that causes a decrease in the quality of pH and temperature in water [3]. Looking at these problems, there needs to be research on water quality monitoring and prevention of water pollution to minimize the risk of contaminated water entering the pond. In this study the author will design a tool that can monitor water pollution by using a pH sensor, temperature sensor, Dissolved Oxygen and prevent contamination with an automatic door opening system.

2. Literature Review

2.1 Sensor Thermometers Digital DS18B20

The Sensor DS18B20 was digital thermometer made by Dallas Semiconductor. DS18B20 has a digital output, so there is no need for an ADC circuit anymore, and has an accuracy of temperature reading and measurement speed has a much better stability than LM35DZ sensor. For temperature reading, the sensor uses a protocol with the one wire bus communication method. DS18B20 is a very practical sensor because it only requires 1 pin of I / O data to be processed by a microcontroller. The physical image of DS18B20 is shown in Figure 1. The DS18B20 digital temperature transducer is used as multi-angle testing method in the process of testing the pitch transport vehicle's temperature. [4]

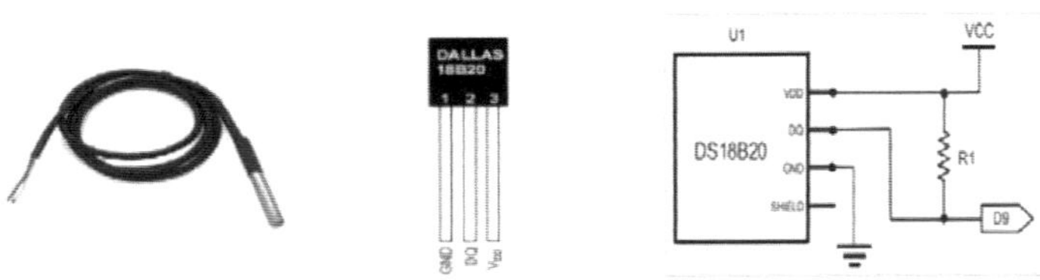

Figure 1. Pin configuration and DS18B20 sensor physical form (source*: Datasheet DS18B20 Dallas semiconductor, Inc.,*2015)

2.2 Sensor PH Probe And Module

In principle, the measurement of the pH sensor is based on the electro-chemical potential that occurs between the solution contained in the glass membrane (electrode) which is known with the solution found outside the glass electrode which is unknown. This is because a thin layer of glass bubbles will interact with hydrogen ions whose size is relatively small and active; the glass electrode will measure the electrochemical (EC) potential of hydrogen ions or termed the potential of hydrogen. Most modern pH meters are equipped with thermistor temperature. Thermistor temperature is a tool to correct the influence of temperature. Between the comparison electrodes and the glass electrodes are arranged in one unit. Sensors commonly used to measure pH are ion-sensitive electrodes or also called glass electrodes. This electrode consists of an electrode rod (made of well-insulated glass) and a glass membrane (thin-walled and sensitive to H ions). A reference there is also a reference electrode. Both of these electrodes exist independently and some are incorporated into one unit, commonly called a combination electrode. A pH measuring sensor element is in the middle, covered by a silver-silver chloride solution (Ag-AgCl). The lower part of the sensor element is related to the glass membrane and contains a silver-silver chloride solution. The ionic contact of the silver-silver chloride solution to the sample occurs through a ceramic connector. The ceramic connector acts as a selective membrane that passes certain ionic currents. Naturally, the glass electrode output impedance is very large (because of the chemical processes that occur on the surface of the electrode), the size is between 50-500 MΩ so that the input impedance is very big. In figure 2.6 is a ph probe sensor which is equipped with a modulee. [5]

IOP Conf. Series: Materials Science and Engineering **462** (2019) 012028 doi:10.1088/1757-899X/462/1/012028

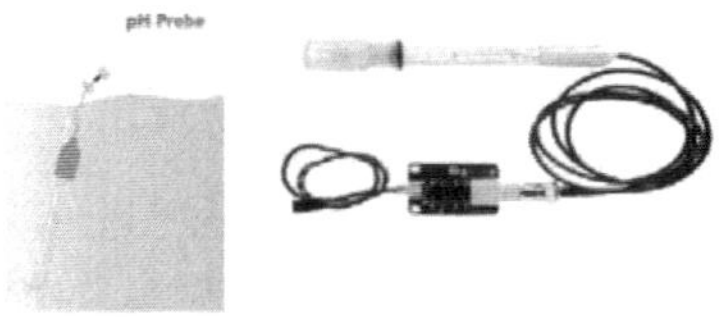

Figure 2. pH Probe Sensor with the module (*source : pH probe module datasheet*)

2.3 Oxygen Sensor Atlas Scientific Kit's Dissolved

The working principle of DO meter is based on polar graphic phenomena that occur between two cathode and anode electrodes. Negative electric voltage is given to the cathode electrode. As a result of this negative voltage, there will be a rapid chemical reaction between water and dissolved oxygen on the cathode surface. Electricity voltage will continue to rise until it reaches saturation value, which is equivalent to the reaction of all dissolved oxygen on the surface of the cathode electrode. The occurrence of a saturated electric voltage is indicated by the almost increase in the reading of the electric current, after a while of being silent at one value even though the voltage value is increased. After passing this saturated voltage value, the electric current continues to rise if the voltage continues to increase. Rising current values are caused by other chemical reactions that have occurred, such as the reaction of breaking H2O water molecules into H and OH ions. [6]

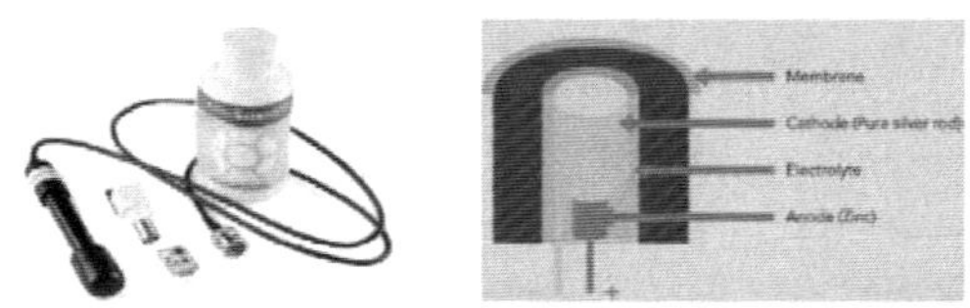

Figure 3. Sensor DO probe with module (*source : Dissolved Oxygen Atlas Scientific datasheet*)

3. Method

3.1 Module The System Mechanic To Mechanical

In this study, the hardware monitoring and prevention system for water pollution consists of servo that prevents the entry of water pollution, accompanied by warnings from alarms (LED and Buzzer circuits), temperature sensors, pH sensors, DO (Dissolved Oxygen) sensors and Arduino Uno. Sending data from hardware to the user is sent via SIM800l communication.

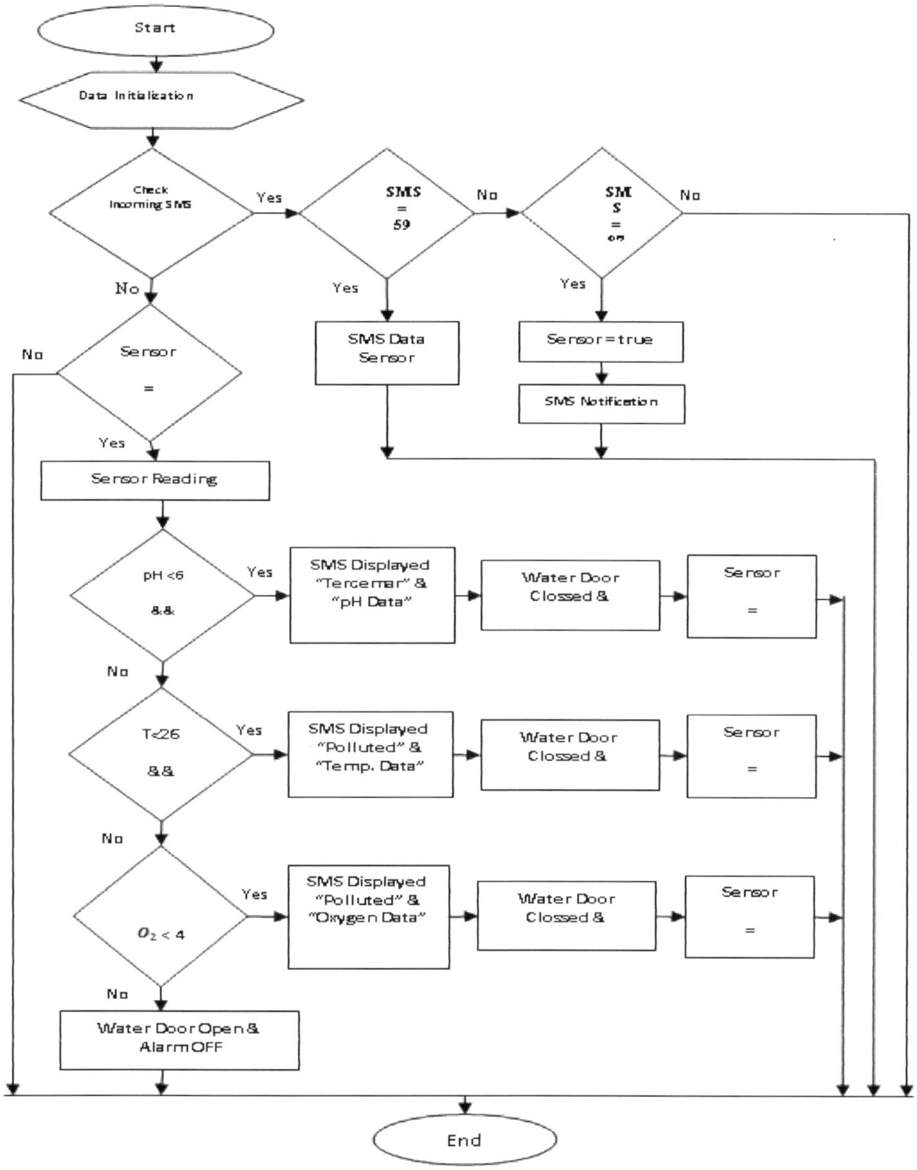

Figure 4. Flowchart of Android-based River Water Pollution Monitoring and Prevention System

A Prototype of monitoring system and prevention of Android-based River water pollution with input of three types of sensors. The sensor used is a pH probe sensor with module V1.1 where the output voltage is affected by the pH concentration in the water. The temperature sensor DS18B20 where the voltage value is linear with readable temperature. Atlas Scientific DO (Dissolved Oxygen) sensor kit module in which the output voltage value is affected by the amount of oxygen dissolved in water. Changes in PH, dissolved oxygen and water temperature are one indication of pollution. If the water is said to be polluted, the servo motor will close the sluice gate as an early precaution against pollution and the SIM800l module will send a notification in the form of an SMS about the water condition.

4. Result and Discussion

4.1 Testing 1 System Effect of Acid pH and Temperature ± 40 °C on Oxygen

Testing 1. System Effect of Acid pH and Temperature ± 40 °C on Oxygen. Testing the effect of temperature and pH on oxygen is intended to determine the results of changes in water conditions after adding pollutants as much as 10 times. The test was carried out randomly within one minute to determine the effect of temperature on the increase or decrease in pH levels of river water. The first step is to take 500ml of random river water samples. Then the water added with acid pollutant in the form of lime juice for 10%, then the oxygen sensor readings were observed. After observing the oxygen sensor readings, the water sample mixture in the acid condition is heated to a temperature of ± 40 ° C.

The equipment used in the test are: Atlas Scientific's Dissolved Oxygen Sensor, V1.1 module pH sensor, DS18B20 Dallas Sensor, LCD circuit, indicator circuit with buzzer, 12 V. power supply.

Table 1. Testing the Effect of Acid Temperature and pH on Oxygen

No	Initial conditions of O_2 River water (ppm)	Initial conditions of water pH	Initial conditions of water Temp.(°C)	Water O_2 after mixed with acid<4 pH (ppm)	Water O_2 acid <4 pH on Temp. ±40°C
1	5,61	7,11	28,86	4,08	3,03
2	5,63	7,13	28,35	4,18	3,08
3	5,60	7,14	28,59	4,04	3,11
4	5,59	7,23	28,01	3,96	3,21
5	5,60	6,87	28,86	4,01	3,15
6	5,62	6,93	28,35	4,03	3,32
7	5,62	7,17	28,86	4,03	3,25
8	5,62	7,08	28,37	4,09	3,22
9	5,61	6,90	28,35	3,99	3,03
10	5,59	6,99	28,86	3,96	3,28
Average Oxygen				4,037	3,168

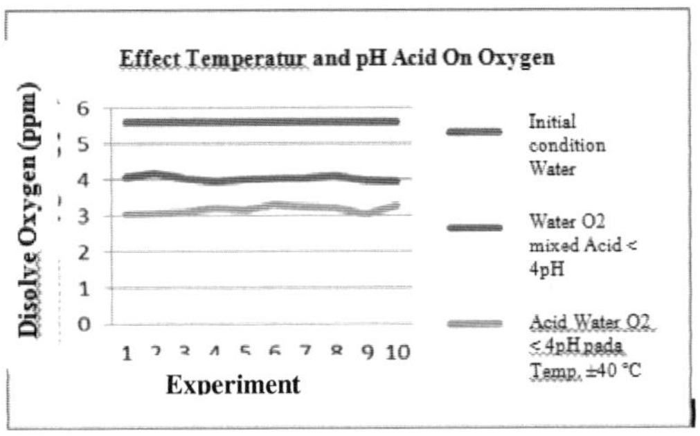

Figure 5. Graph Effect of Temperature Changes and Acid PH on Oxygen

4.2 Testing 2 System of Effect of Base pH and Temperature ± 40 ° C on Oxygen

Table 2. Testing the Effect of Base Temperature and pH on Oxygen

No	Initial conditions of O_2 river water (mg/L)	Initial conditions of pH water	Initial conditions of temp. (°C)	Water O2 Base Basa >9 pH (mg/L)	Water O_2 Base >9 pH on Temp. ±40°C (mg/L)
1	5,61	7,11	28,86	5.00	3.86
2	5,63	7,13	28,35	4.99	3.67
3	5,60	7,14	28,59	5.03	3.86
4	5,59	7,23	29,01	5.15	3.93
5	5,60	6,87	28,86	5.01	3.62
6	5,62	6,93	28,35	4.98	3.47
7	5,62	7,17	28,86	4.99	3.52
8	5,62	7,08	28,37	5.09	3.54
9	5,61	6,90	28,35	4.93	3.61
10	5,59	6,99	28,86	5.01	3.67
Average Oxygen				5,018	3,675

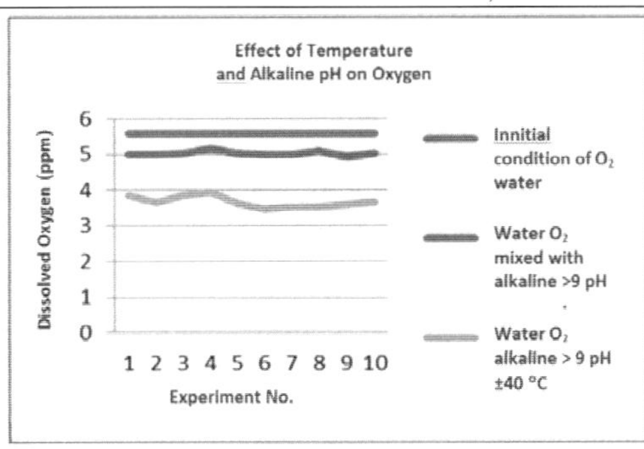

Figure 6. Graph Effect of Changes in Temperature and PH Base on Oxygen

4.3 Testing 3 Sluice Closure Response to Polluted Conditions

Testing 3. Sluice Closure Response to Polluted Conditions Testing of the sluice closure response to polluted conditions is intended to determine the length of time the closure of the sluice door when water pollution is detected as an early preventive measure against pollution.

Table 3. The response of closure of sluice gates when polluted conditions

Experiment no.	**Contitions**	**Motor Servo Response time**	
		Time (s)	**Door status**
1	Polluted	0,66	clossed
2	Polluted	1,1	clossed
3	Polluted	1	clossed
4	Polluted	0,93	clossed
5	Polluted	1,21	clossed
6	Polluted	1,48	clossed
Average time		**1,03**	

IOP Conf. Series: Materials Science and Engineering **462** (2019) 012028 doi:10.1088/1757-899X/462/1/012028

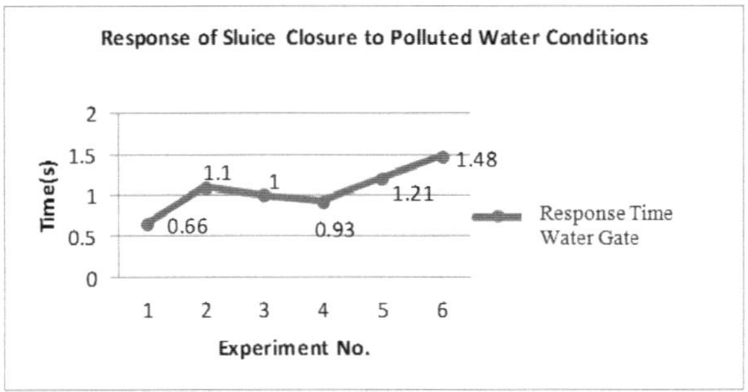

Figure 7. Results of Graph Response of Closing Time of Water Gate to Polluted Water Conditions

5. Conclucions

From the overall test, the average error results from the testing of 3.9% oxygen levels were measured at a pH measurement of 2.8% and 1.01% for temperature measurement. When the acid and alkaline solution is added the sluice closes with an average response time of 1.03 seconds, then sms notification that the contamination can be sent. Tool success rate is 90%. The sluice gate can function as an early prevention of pollution because water pollution leaks are not detected at the time of testing but often experience obstacles to the slow process of sending SMS notifications.

6. References

[1] Djoel Simata, (2011). *"Chemical Engineering, Dissolved Oxygen(DO)".*

[2] Deng Yi, Liu Saojun, ,Liu Zhongwei, and Zhou Yuca, "Research on Temperature Testing System of Pitch Transport Vehicle Basing on DS18B20 Mode". 2007 Chinese Control Conference IEEE, ISSN: 1934-1768, Pages: 456 – 458.

[3] Febriani Rinanda, (2013) *"Temperature Sensor DS18B20 Arduino"*

[4] Ariansyah Yudi. (2012). *"Design and Development of Detecting pH in Water with Microcontroller Based LCD Output"*, Politeknik Negeri Sriwijaya

[5] WS Davidson, "Water quality monitoring", 1994 Conference Record Southcon IEEE, ISBN: 0-7803-9988-9.

[6] Riny Sulistyowati Dedet Candra Riawan Mochamad Ashari 2016 PV Farm Plaecement and Sizing using GA for Area Development Plan of Distribution Network Proceed ISITIA IEEE pp. 513-519.

[7] Jacob Fraden, (1993) "Handbook of Modern Sensors" 4th Editition. Springer, New York.

[8] Andy Suryowinoto, Abdul Hamid, (2017) "Remote Sensing As Preliminary Hazard using Android Operating system" Journal INFORM Vol.2 No.1, July 2017, E-ISSN : 2581-0367.

[9] Enrique Lopez Manas, Diego Grancini (2106) "Android High Performance Programming" , PACKT Pub.

[10] Xuying Yuan, Yijin Wu, et al, (2009) "Early Warning and Emergency Response system for Water Pollution Accidents in Wuhan City, China" Environmental Science and Information Application Technology, International Conference on Wuhan, China July 4, 2009. IEEE ISBN: 978-0-7695-3682-8. pp: 171-174

IOP Conf. Series: Materials Science and Engineering **462** (2019) 012029 doi:10.1088/1757-899X/462/1/012029

A Comparative Study of Single-Tuned Filter and Detuned Reactor for Improve Power Quality in Microgrid

F A Santoso, M Syai'in, A S Setiyoko

Marine Electrical Engineering Department, Shipbuilding Institute of Polytechnic Surabaya

Abstract. Propose of this study is to analyze the performance of the single-tuned filter and detuned reactor to improve power quality (voltage profile, power factor, and harmonic distortion) in a microgrid. Nowadays power quality becomes the focus of the electrical company as well as commercial customer and industry. The load consists of two types, linear load, and nonlinear load. Both of them cause a reduction of power quality as voltage drop, low power factor, and harmonic distortion. Those problems can make losses of electrical usage, broken on electrical devices and can stop the production process in the industry. Single-tuned filter and the detuned reactor is the type of harmonic filter that commonly used to mitigate harmonic distortion. The harmonic filter also has benefit to improve power factor and voltage profile on the system. All filters will be simulated on ETAP to get the performance of each filter on different condition. The most dominant load in this study is a 6-pulse battery charger.

Keywords: Harmonic distortion, Power factor, Voltage drop, Single-tuned filter, Detuned reactor

1. Introduction

A good assumption for most utilities in the world is that the sine-wave voltage generated in a central power station is very good. Because of that nowadays power quality became quite concerned with utility engineers. Power quality is about compatibility between the quality of the voltage supplied from the electric power system and the proper operation of end-use equipment. There are two categories of power quality that need to be considered steady state (continuous) power quality and disturbances. Steady-state power quality characteristics include voltage regulation, harmonic distortion, unbalance and flicker.

Drop voltage is phenomena defined as the amount of voltage loss that occurs through all or part of a circuit due to impedance. Drop voltage can cause low intensity in a lightning system, miss-operation in control system and decrease motor starting torque. Indonesian Electric Company recommends limiting the voltage drop for low-voltage 4 percent of the circuit voltage.

Power factor is important in the AC system like 3-phase or single-phase. Power factor is the ratio of the real power that is used to do work and the apparent power that is supplied to the circuit. The power factor can get values in the range from 0 to 1.

Harmonic distortion is caused by nonlinear devices in the power system. A nonlinear load device is one in which the current is not proportional to the applied voltage. The most commonly used standard of harmonic distortion is IEEE 519-2014.

The single-tuned filter is a passive filter which consists of inductance, capacitance and resistance element configured and tuned for a harmonic orde. They are commonly used and are relatively inexpensive compared with other means for eliminating harmonic distortion.

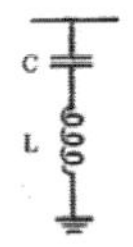

Figure 1. Single-tuned filter circuit.

The use of detuned reactors thus prevents harmonic resonance problems, avoid the risk of overloading capacitors and contributes to reducing harmonic distortion in the network. The tuning frequency can be expressed by the relative impedance of the reactor (in %), or by tuning order, or directly in Hz.

Table 1. Characteristic of detuned reactor.

Impedance (p)	Function
5.67%	Is used where the dominant in 5th harmonic
7.00%	Is used where optimal protection to capacitor and harmonic reduction is achieved.
14.00%	Is to be used where system is rich in 3rd harmonics (3, 9, 15)

2. Methodology

2.1 Study Flowchart

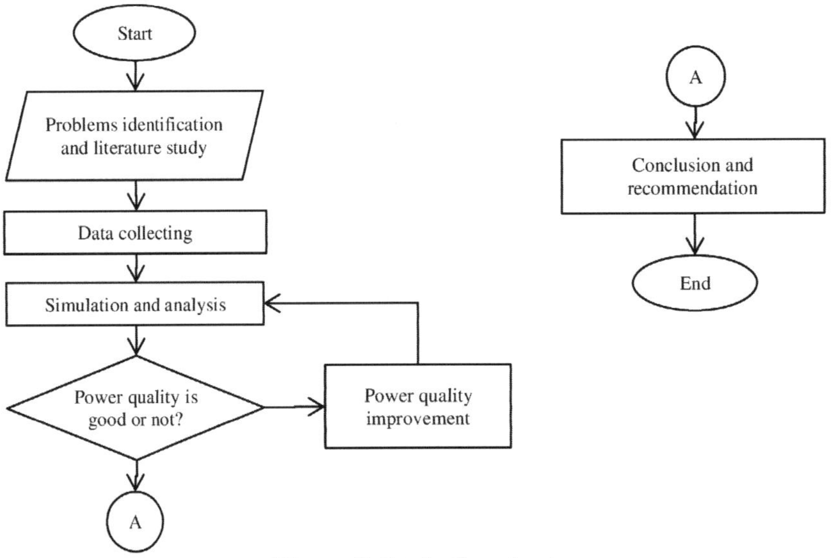

Figure 2. Study flowchart.

The methodologies conducted in this are:

1) Problems identification and literature study: site survey and collect literature.
2) Data collecting: One-line diagram, Plant layout, harmonic spectrum and product catalogue.
3) Simulation and analysis: simulating system and analysis power quality from the simulation.
4) Power quality improvement: if power quality in a bus is low we should install harmonic filter on that bus to improve the power quality.
5) Conclusion and recommendation.

2.2 Study Case

Simulating the system on different study case to get filter design that can work in different condition. For this research selected 5 different condition.

Table 2. Study case.

Condition	Io/Charger (A)	Vo/Charger (V)	P/Rectifier (kW)
1	10.0	288	13
2	15.0	240	17
3	23.0	320	35
4	33.5	256	41
5	46.5	192	41

Figure 3. 6-pulse battery charger on the system.

3. Power System Overview

This microgrid is a automotive battery manufacturer. Supplied by national electrical company with 3.465 kVA rated power as main power supply. Main distribution voltage from main transformers are 20kV step-down to 0.38 kV. The LV 0.38 kV distributed to others bus (LVMDP 1, LVMDP 2 and LVMDP 3). Existing capacitor bank capacity is 1,005 kVAR.

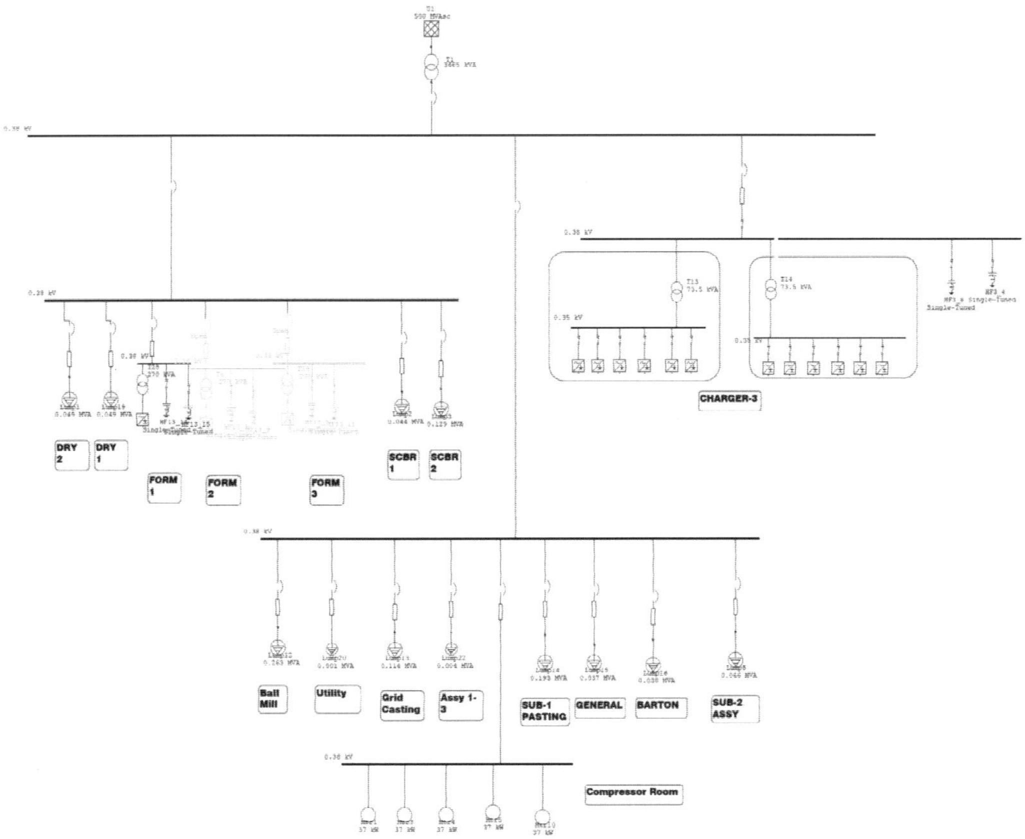

Figure 4. Overall single line diagram.

IOP Conf. Series: Materials Science and Engineering **462** (2019) 012029 doi:10.1088/1757-899X/462/1/012029

4. Result and Discussion

4.1 Parameter Analysis

Software simulation using ETAP conducted in this study to find performance of single-tuned filter and detuned reactor to find the performance of both filter in 5 different load conditions.

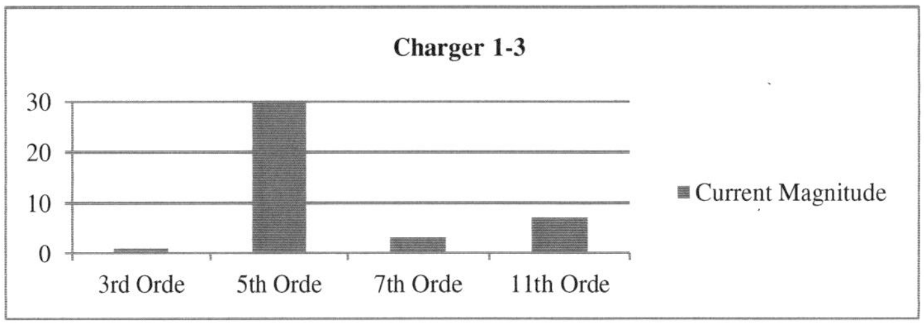

Figure 5. Harmonic spectrum on charger.

Figure 5 shows the harmonic spectrum based on measurement. The most dominant orde is 5th.

Table 3. Current harmonic analysis.

Bus	CM Limit <11th (%)	CM 3rd (%)	CM 5th (%)	CM 7th (%)	CM Limit >11th (%)	CM 11th (%)
Charger 1-3 $Isc/I_L = 25$	7	1	30	3	3.5	7

Yellow = exceeding limit, CM = Current Magnitude

From table 3 and with IEEE 519-2014 standard about harmonics, harmonic on bus charger 1–3 which exceeding the limit are 5th harmonic and 11th harmonic.

Table 4. Bus with undervoltage condition.

Bus	% Voltage on Condition			Standard
	3	4	5	(%)
Ball Mill	95.93	95.78	95.72	96–104
Grid Casting	96.00	95.85	95.79	96–104

Yellow = exceeding limit

From table 4 bus where happens undervoltage are Ball Mill and Grid Casting in condition 3, 4 and 5.

4.2 Filter Specification

Table 5. Single-tuned filter specification.

Bus	Condition	Orde	ΔQ (kVAR)	C (μf)	L (mH)
Charger 1-3	1	5	2.023	44.623	9.091
		11	1.759	38.789	2.160
	2	5	2.646	58.354	6.952
		11	2.138	47.101	1.779
	3	5	5.447	120.140	1.060
		11	4.397	96.973	0.271
	4 and 5	5	7.529	166.050	2.440
		11	5.151	113.600	0.737

Table 6. Detuned reactor specification.

Qs Detuned Reactor (kVAR)	L (mH)	P (%)	Q_N Kapasitor (kVAR)	Capacitor Rating Voltage (Volt)	Condition
50	0.61	5.57	60	415	
75	0.41	5.57	85		
100	0.31	5.57	115		

4.3 Case Studies

4.3.1 Case 1. The experiment, in this case, aims to identify the level of harmonics, voltage profile and power factor if charger set to condition 1.

Table 7. Current harmonic in condition 1.

Bus	Orde	Limit (%)	CM Without Filter (%)	CM Installed Single-Tuned Filter (%)	CM Installed Detuned Reactor (%)
Charger 1-3	1	7.0	0	0	0
	3	7.0	2	2.15207	0.304147
	5	7.0	30	5.92453	2.986630
	7	7.0	3	3.21645	0.338990
	11	3.5	7	1.90088	0.840195

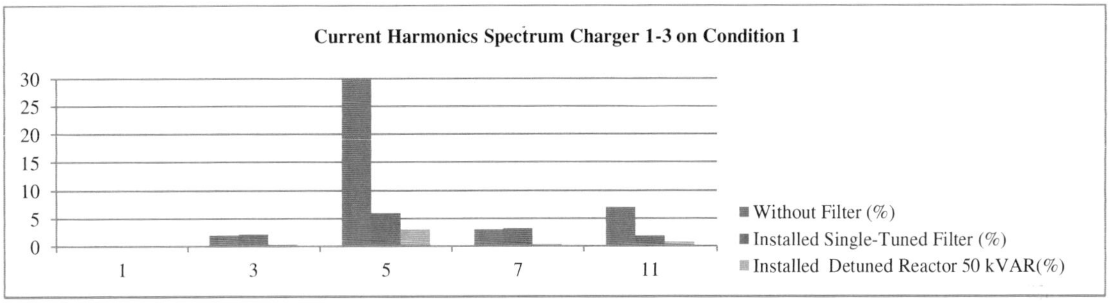

Figure 6. Current harmonics spectrum on condition 1.

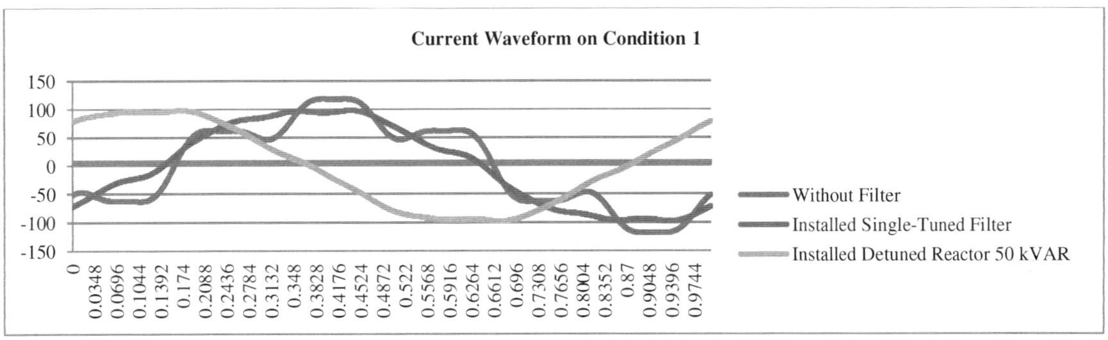

Figure 7. Current waveform on condition 1.

Figure 6 shows the spectrum on each different filter. From the figure can be seen that the current harmonic decreased and all harmonic can meet IEEE 519-2014 standard (written on table 7). Figure 7 shows the waveform on each different filter. Both of harmonic filter can maintain the current waveform to be pure sine wave. But on detuned reactor happen phase shifting 90°.

Table 8. Capacitor needed on condition 1.

Filter	Power (kW)	Early PF (%)	Target PF (%)	Capacitor Needed (kVAR)
Without filter	1381	74.0	95	816.347
Single-Tuned Filter	1381	74.0	95	816.347
Detuned Reactor 50 kVAR	1408	93.4	95	75.800

Table 8 shows the capacitor capacity needed for supplying reactive power on the system. If without the filter and using filter need 816.347 kVAR, but if using Detuned reactor only need 75.8 kVAR to gain 95% PF on the kVARH meter. Voltage profile and THDv still meet standard on condition 1.

IOP Conf. Series: Materials Science and Engineering **462** (2019) 012029 doi:10.1088/1757-899X/462/1/012029

4.3.2 Case 2. The experiment, in this case, aims to identify the level of harmonics, voltage profile and power factor if Charger set to condition 2.

Table 9. Current harmonic in condition 2.

Bus	Orde	Limit (%)	CM Without Filter (%)	CM Installed Single-Tuned Filter (%)	CM Installed Detuned Reactor (%)
Charger 1-3	1	7.0	0	0	0
	3	7.0	2	2.15340	0.531834
	5	7.0	30	4.70940	5.608200
	7	7.0	3	3.21850	0.626469
	11	3.5	7	1.46996	1.538540

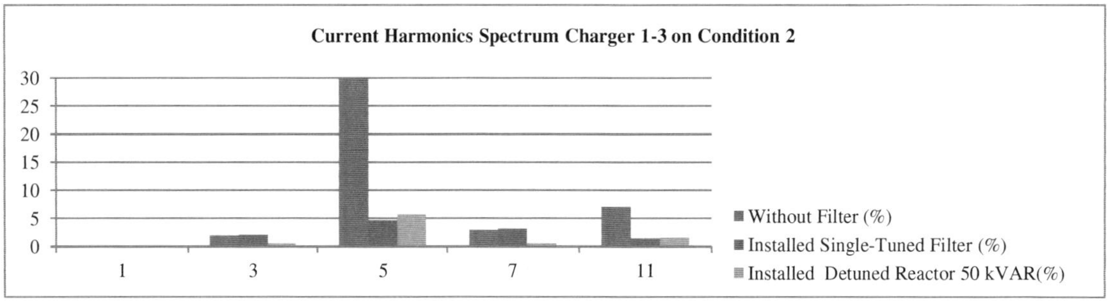

Figure 8. Current harmonics spectrum on condition 2.

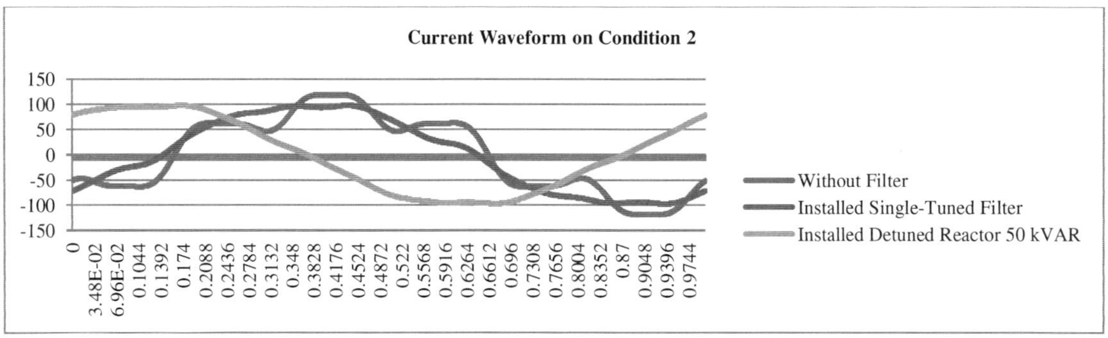

Figure 9. Current waveform on condition 2.

Figure 8 shown the spectrum on each different filter. From the figure can be seen that the current harmonic decreased and all harmonic can meet IEEE 519-2014 standard (written on Table 9). Figure 9 shown the waveform on each different filter. Both of harmonic filter can maintain the current waveform to be pure sine wave. But on detuned reactor happen phase shifting 90°.

Table 10. Capacitor needed on condition 2.

Filter	Power (kW)	Early PF (%)	Target PF (%)	Capacitor Needed (kVAR)
Without filter	1402	73.3	95	840.25
Single-Tuned Filter	1417	85.2	95	405.00
Detuned Reactor 50 kVAR	1530	88.5	95	302.30

Table 10 shown capacitor capacity needed for supplying reactive power on the system. If without filter 840 kVAR, if using Single-Tuned Reactor need 405 kVAR and if using Detuned reactor only need 302.3 kVAR to gain 95% PF on the kVARH meter. Voltage profile and THDv still meet standard on condition 2.

4.3.3 Case 3. The experiment, in this case, aims to identify the level of harmonics, voltage profile and power factor if charger set to condition 3.

Table 11. Current harmonic in condition 3.

Bus	Orde	Limit (%)	CM Without Filter (%)	CM Installed Single-Tuned Filter (%)	CM Installed Detuned Reactor (%)
Charger 1-3	1	7.0	0	0	0
	3	7.0	2	2.15	0.700
	5	7.0	30	4.14	6.900
	7	7.0	3	3.30	0.810
	11	3.5	7	0.17	2.023

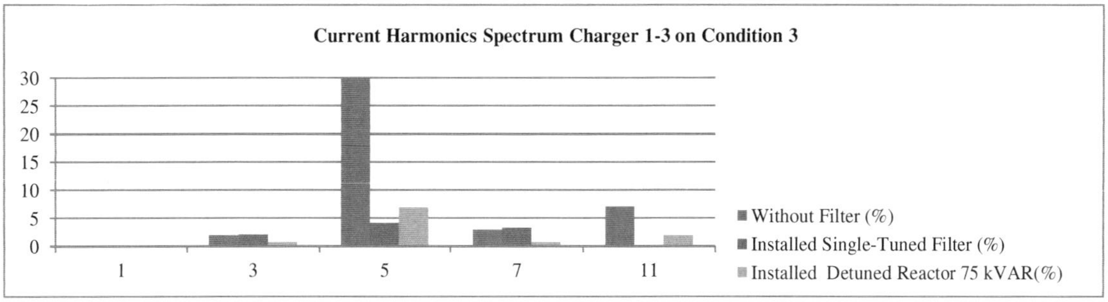

Figure 10. Current harmonics spectrum on condition 3.

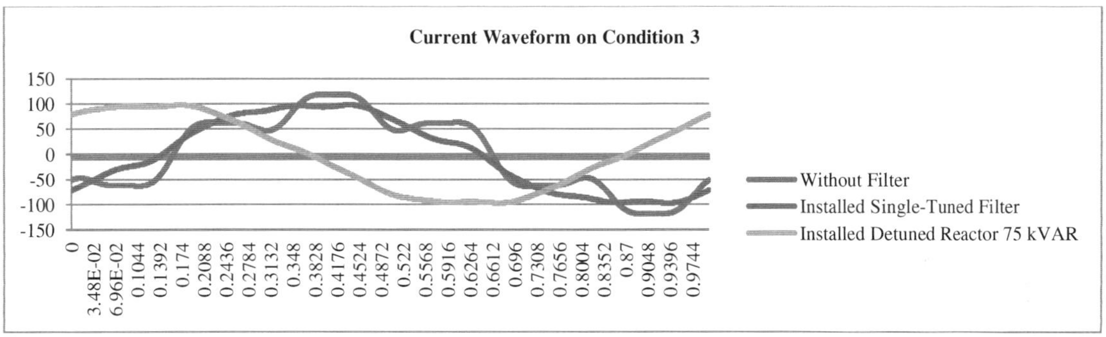

Figure 11. Current waveform on condition 3.

Figure 10 shown the spectrum on each different filter. From the figure can be seen that the current harmonic decreased and all harmonic can meet IEEE 519-2014 standard (written on table 11). Figure 11 shown the waveform on each different filter. Both of harmonic filter can maintain the current waveform to be a pure sine wave. But on detuned reactor happen phase shifting 90°.

Table 12. Capacitor needed on condition 3.

Filter	Power (kW)	Early PF (%)	Target PF (%)	Capacitor Needed (kVAR)
Without filter	1505	74.4	95	857.0
Single-Tuned Filter	1506	75.8	95	800.0
Detuned Reactor 50 kVAR	1531	93.9	95	57.5

Table 12 shown capacitor capacity needed for supplying reactive power on the system. If without filter 857 kVAR, if using Single-Tuned Reactor need 800 kVAR and if using Detuned reactor only need 57.5 kVAR to gain 95% PF on the kVARH meter.

Table 13. Voltage profile on condition 3.

Bus	%V Without Filter	%V Installed Single-Tuned Filter	%V Installed Detuned Reactor 75kVAR
Ball Mill	95.93	95.96	96.01
G Casting	96.00	96.03	96.08

From table 13 shown voltage profile increased after installing harmonic filters. But detuned only detuned reactor can increase voltage profile to meet the standard. THDv still meets the standard IEEE 519-2014 on condition 3.

4.3.4 Case 4. The experiment, in this case, aims to identify the level of harmonics, voltage profile and power factor if charger set to condition 4.

Table 14. Current harmonic in condition 4.

Bus	Orde	Limit (%)	CM Without Filter (%)	CM Installed Single-Tuned Filter (%)	CM Installed Detuned Reactor (%)
Charger 1-3	1	7.0	0	0	0
	3	7.0	2	2.160	0.6500
	5	7.0	30	2.020	5.2300
	7	7.0	3	3.266	0.6234
	11	3.5	7	0.420	1.5700

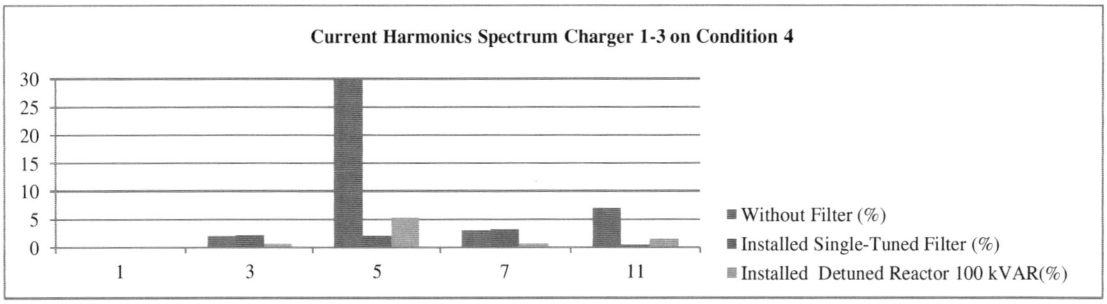

Figure 12. Current harmonics spectrum on condition 4.

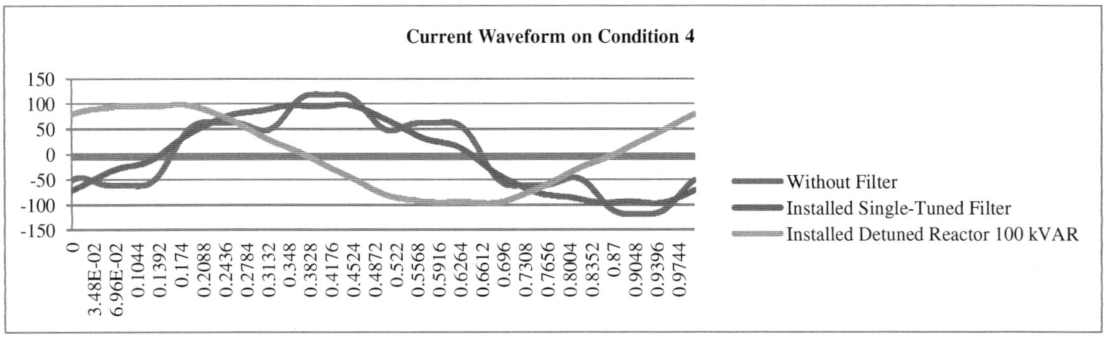

Figure 13. Current waveform on condition 4.

Figure 12 shown the spectrum on each different filter. From the figure can be seen that the current harmonic decreased and all harmonic can meet IEEE 519-2014 standard (written on Table 14). Figure 13 shown the waveform on each different filter. Both of harmonic filter can maintain the current waveform to be pure sine wave. But on detuned reactor happen phase shifting 90°.

Table 15. Capacitor needed on condition 4.

Filter	Power (kW)	Early PF (%)	Target PF (%)	Capacitor Needed (kVAR)
Without filter	1631	75.5%	95%	880.5
Single-Tuned Filter	1632	77.2 %	95%	807.3
Detuned Reactor 50 kVAR	1662	93.4%	95%	89.5

Table 15 shown capacitor capacity needed for supplying reactive power on the system. If without filter 880.5 kVAR, if using Single-Tuned Reactor need 807 kVAR and if using Detuned reactor only need 89.5 kVAR to gain 95% PF on the kVARH meter.

Table 16. Voltage profile on condition 4.

Bus	%V Without Filter	%V Installed Single-Tuned Filter	%V Installed Detuned Reactor 75kVAR
Ball Mill	95.78	95.79	96.01
G Casting	95.85	95.86	96.08

From table 16 shown voltage profile increased after installing harmonic filters. But detuned only detuned reactor can increase voltage profile to meet the standard. THDv still meets the standard IEEE 519-2014 on condition 4.

4.3.5 Case 5. The experiment, in this case, aims to identify the level of harmonics, voltage profile and power factor if charger set to condition 5.

Table 17. Current harmonic in condition 5.

Bus	Orde	Limit (%)	CM Without Filter (%)	CM Installed Single-Tuned Filter (%)	CM Installed Detuned Reactor (%)
Charger 1-3	1	7.0	0	0	0
	3	7.0	2	2.160	0.680
	5	7.0	30	2.015	5.531
	7	7.0	3	3.260	0.660
	11	3.5	7	0.400	1.680

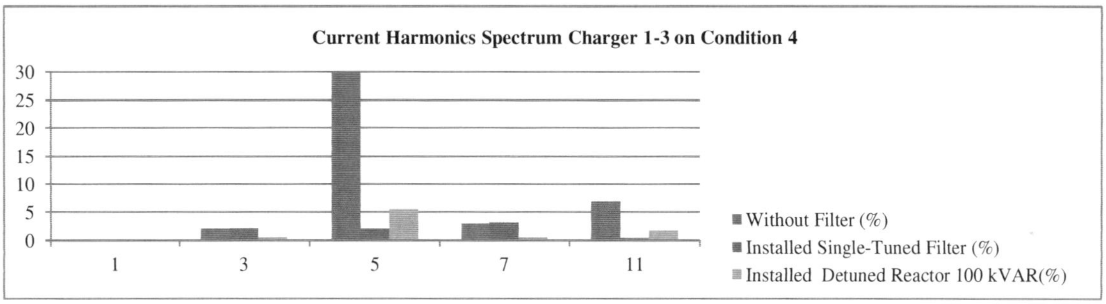

Figure 14. Current harmonics spectrum on condition 5.

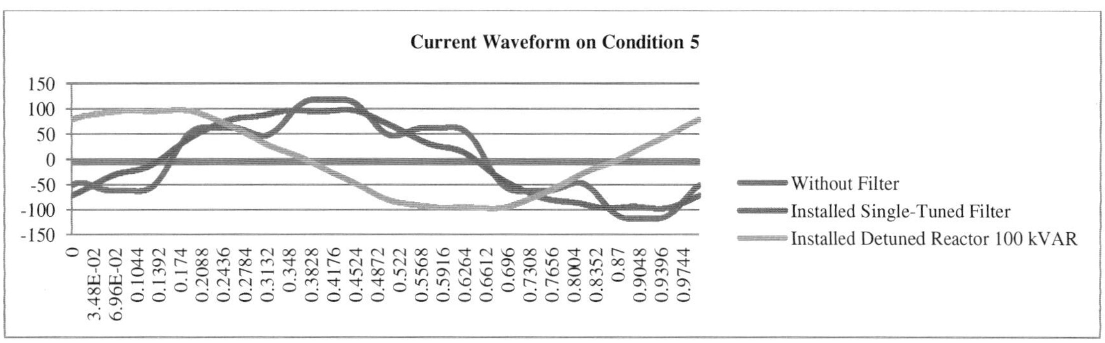

Figure 15. Current waveform on condition 5.

Figure 14 shown the spectrum on each different filter. From the figure can be seen that the current harmonic decreased and all harmonic can meet IEEE 519-2014 standard (written on Table 17). Figure 15 shown the waveform on each different filter. Both of harmonic filter can maintain the current waveform to be pure sine wave. But on detuned reactor happen phase shifting 90°.

Table 18. Capacitor needed on condition 5.

Filter	Power (kW)	Early PF (%)	Target PF (%)	Capacitor Needed (kVAR) (%)
Without filter	1546	74.8%	95%	863.6
Single-Tuned Filter	1548	76.6%	95%	790.3
Detuned Reactor 50 kVAR	1585	98.4%	-	-

Table 18 shown capacitor capacity needed for supplying reactive power on the system. If without filter 863.6 kVAR, if using Single-Tuned Reactor need 790.3 kVAR and if using Detuned reactor only do not need the capacitor to gain 95% PF on the kVARH meter. Because power factor on the system is up to > 95%.

Table 19. Voltage profile on condition 5.

Bus	%V Without Filter	%V Installed Single-Tuned Filter	%V Installed Detuned Reactor 75kVAR
Ball Mill	95.72	95.90	96.49
G Casting	95.79	95.97	96.55

From table 19 shown voltage profile increased after installing harmonic filters. But detuned only detuned reactor can increase voltage profile to meet the standard. THDv still meets the standard IEEE 519-2014 on condition 5.

Conclusion

Based on the five study cases that have been done can be concluded:

1) Installation single-tuned filter and the detuned reactor can improve power quality on the system (voltage profile, capacitor load [PF] and harmonic distortion).
2) Electrical disturbances those happen on this microgrid is under voltage and current harmonic distortion. Voltage harmonic still meet IEEE 519 standard.
3) Installation detuned reactor needs autotransformer to maintain 90° leading condition effected by the detuned reactor.

Reference

[1] Syai'in M, Rohiem N H, Tobing R K, Atmoko M A, Adiatmoko M F, Soeprijanto A and Hatta A M 2017 Harmonics monitoring of car's inverter using discrete fourier transformation *Information Technology, Computer, and Electrical Engineering* (ICITACEE), 4th International Conference on 2017 Oct 18 (pp. 102-107). IEEE.

[2] Husnayain F, Purnomo N D, Anwar R and Garniwa I 2014 Harmonics mitigation for offshore platform using active filter and line reactor methods *Electrical Engineering and Computer Science* (ICEECS), 2014 International Conference on 2014 Nov 24 (pp. 331-336). IEEE.

[3] Simpson J I, et al. 2008 *Requirement and Evaluation of an active filter in an catual industrial installation* IEEE.

[4] Akagi H 2006 Modern active filters and traditional passive filters. *Bulletin of the polish academy of sciences technical sciences* **54**.

IOP Conf. Series: Materials Science and Engineering **462** (2019) 012030 doi:10.1088/1757-899X/462/1/012030

Atrium Form and Thermal Performance of Middle-Rise Wide Span Building in Tropics

D P E Laksmiyanti[1] and R P Salisnanda[1]

[1]Architecture Departement, Institut Teknologi Adhi Tama Surabaya, Indonesia

dianpramita@itats.ac.id

Abstract. Energy use of a building usually increases as long as an increase of the surface area of the envelope. Heat transfer of the building through the envelope can be reduced by perfect design. Right building form and materials can reduce the heat transfer. Middle rise building has a quite high surface area of the roof, and in a tropical country, roof accepts the highest irradiance thus bring much heat to the building. In the same volume, compact building form is better in thermal performance, but large building usually need an atrium to help the natural lighting in the room. The atrium uses transparent roof or skylight. This transparent roof definitely can increase the heat gain to the building significantly. So this research aims to find the best form of atrium roof in middle-rise and large span building in the tropical region. This research aims to analyse the thermal performance of middle-rise and large span buildings and its relation to the form of the atrium. A sample of middle-rise office buildings in Surabaya was taken randomly. The simulation was conducted to predict thermal and cooling energy performance. Energy performance of the building was found to bear some relations to form of roof.

1. Introduction

One indicator of success in designing energy-efficient buildings is the achievement of optimisation of energy-saving lighting and ventilation in buildings. Moreover, the primary parameter that is crucial to the optimisation of energy is the total energy use of the building's operations for one year.

The most significant energy use in the building sector in Indonesia is in energy cooling and the energy for lighting. The biggest cooling energy is due to the design of the building envelope which is not suitable for the tropics. Previous research shows that buildings with octagon form consume the least cooling energy, this also supports the Crawford [1] statement which states that the more compact the design of a mass of the building, the smaller the energy consumption of cooling. Another problem if using a large model is natural lighting that cannot reach the interior of the room. For this reason, wide-span buildings usually add skylights to the atrium so that natural light can enter the building. For the tropics, building envelopes that affect thermal comfort in a building consist of walls in each orientation and roof, therefore in addition to the composition of materials and openings on the walls, it is also necessary to pay attention to the roof design of the building.

The problem that often arises for building designs with lots of openings is the solar pass that comes in to make the heat gain in buildings increase, and the operational energy for cooling eventually increases as well. Especially for the roof, this horizontal surface side receives the largest radiation all year round compared to the other side of the building envelope, so it needs to be set the standard percentage of skylights in the atrium.

2. Literature Study

2.1 Energy Efficient Building

There are five basic requirements of a building. First, buildings must be made based on local development culture; second, buildings must be able to control the internal environment for the convenience of building users. Third, buildings must support users' social activities, fourth, they must be able to react to external forces, such as traditional building structures that can withstand earthquakes or land shifts or designs that are able to address climate problems. Fifth, buildings meet the first to fourth requirements with reasonable and economic costs in the use of energy, both embodied the energy and operational energy [2]. Energy-efficient buildings are buildings that have an energy balance [3-4]. The factors include:

- Optimising the use of high internal heating load
- Avoid overheating
- Protection from solar irradiance and glare
- Optimisation of natural lighting
- Flexibility in replacement and maintenance of equipment in buildings without excessive energy consumption

2.2 Building Form and Energy

Yeang [5] say that the shape of the building will affect the energy used in buildings. Rectangular and north-south oriented structures will be better because the east and west sides get substantial radiation, especially in the morning and evening so this side is attempted to have a small surface area. Szokolay [6] explains that for buildings in Jakarta, the broad north side will receive far greater radiation because this year the side will receive greater radiation than the south side. The best form of the building is a form that has a low surface area so that the field that receives radiation and propagates heat by conduction will be smaller. Depecker [2] says that the shape will be very influential with the cooling load, the more fat the building and the more compact, the lower the cooling capacity will be. According to Heerwagen and Ling [7, 8], changing the shape by minimizing incoming radiation will have a more significant effect on reducing cooling load compared to adding wall insulation, as in Noerwasito's study which shows that buildings with a higher ratio of floor area to circumference small will have a lower cooling load so that jagged structures will be better than square buildings with straight walls, but Crawford [1] states that the orientation and floor area of the building does not significantly influence the cooling load. The use of material has more influence on the energy used in buildings.

Olgyay [9] recommends a rectangular shape with a width and length ratio of 1: 1.7 is the most effective form for a tropical climate. This form is considered to be the most effective way to overcome cooling and heating problems in high concentrations. Juniwati [10] suggested a cylindrical shape because this form is best for minimising solar heat gain and solar heat loss in tall buildings. Ling [8] links the way of a high-rise building in a tropical climate with insolation and gets the results of a cylindrical and square building shape with a ratio of width and length of 1: 1 to minimise the influence of direct and diffuse solar insolation. Juniwati [10] produced the same rate as Olgyay; he stated that the building with a north-south orientation and W / L ratio was 1: 1.7. in the same volume, the Octagonal shape has the best thermal performance than a rectangle, L shape and H shape [11].

2.3 Building Envelope and Cooling Energy

Buildings in different locations should also have different sheath designs. The building envelope design concept must be able to respond to climate problems and provide benefits concerning investment costs and building operations. Outside air temperature and wind speed affect site conditions. For sites with lousy weather, pure fabrication and installation materials are recommended. Noerwasito [12] classifies buildings based on their building skin, thin sin, prone skin, thick skin, buffering, and valve effect. Whereas Olgyay [9] classifies facade types based on the structure and how to install building elements.

2.4 Roof Design
Horizontal openings have two advantages:

a. Allowing illumination is not equally fair in a vast interior area, while natural light from openings is limited (Figure 1).
b. Horizontal openings also receive more light than vertical openings.

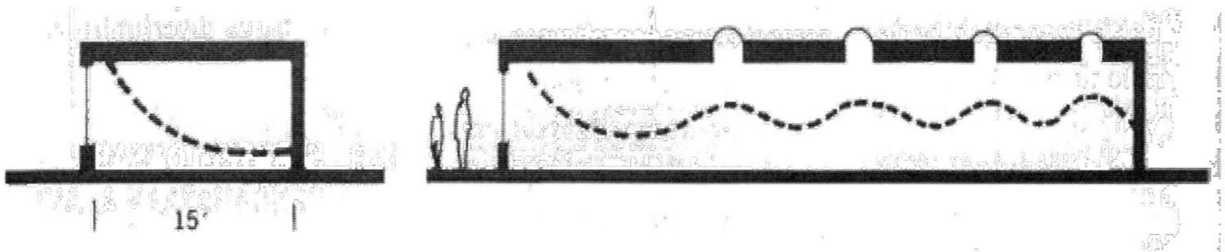

Figure 1. When natural lighting from openings is limited to the outer wall, openings on the roof will be able to flatten the lighting along a limited area of space

3. Method

This research aims to formulate the design of the atrium roof, in this case, the atrial shape of a middle-rise wide-span building to achieve cooling energy efficiency. The experimental method with simulation tactics was used to examine the relationship between the independent variable (atrium form) and the dependent variable (cooling energy). The software used in this simulation is Ecotect 2011.

3.1 Base Case Experiment
According to previous research, the best ratio of fenestration roof and the opaque roof is less than 10% (Laksmiyanti 2017). The object study of this research has been chosen randomly. Most of the multi-storey building has a rectangular. Laksmiyanti [11] found in Surabaya 63.3% middle rise building in Surabaya has a rectangular form, so the base case for this research is building in Surabaya which has a rectangular shape and having an atrium.

G Building in ITATS has been selected for the base case, and it has skylight type for the atrium (fig 3.1). In this base case, material for the opaque roof is galvalume, which is usually used in large span building. Material for fenestration roof is polycarbonate, as the real condition of the building.

3.2 Scenario Experiment
There is four common design in the atrium. So the model of an experiment in this research as his theory. Each model has 10% of fenestration roof (fig.2)

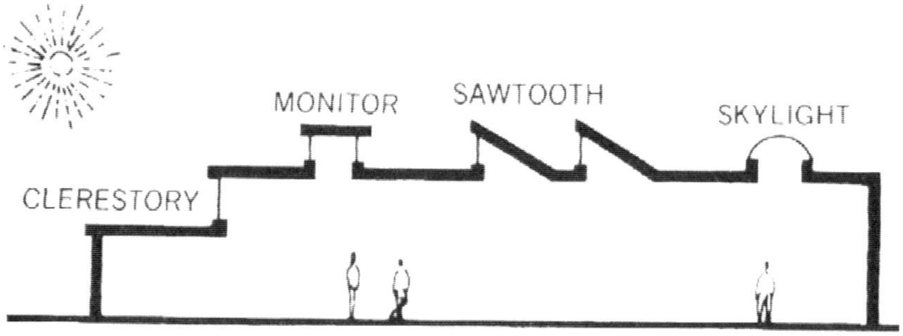

Figure 2. Illustration of atrium form in G building ITATS from right: Skylight for Base Case, monitor for model 1, sawtooth for model 2, clerestory for model 3.

4. Result and Discussion

4.1 Heat Flow to The Building

There is five type of heat gain in this building such as:

a. Conduction heat flow (sQc)
This is a process of heat transfer from outdoor to the indoor environment through the building envelope. This heat transfer happens because of the temperature difference between indoor and outdoor. If the outdoor having higher temperature, the heat will enter the building and vice versa.

b. Indirect solar irradiance heat flow (sQs)
This heat comes from sun radiation which is through the opaque envelope of the building. This heat flow happens on the day, but because it happens on opaque material, there is delay depends on the thermal character (time lag) of the material.

c. Direct solar irradiance heat flow (sQg)
This heat transfer happens because of the radiation of sun through the fenestration envelope of the building. There's no delay of the heat transfer.

d. Ventilation heat flow (sQv)
It's a convection heat flow. Heat transfer by the air flow of the building. All scenario experiment has the same volume of opening, same type of air conditioning and same air change per hour so the conduction heat flow will be same for all model (table 1)

e. Internal heat gain (sQi)
This is heat gain from electrical equipment in the building. Because the floor plan of the building is the same in all scenario experiment, so this kind of heat gain will not be discussed in this paper.

Table 1. Total heat flow in each model

Heat Gain	BC	Model 1	Model 2	Model 3
sQc (kWh/year)	1089.69	1091.43	1089.68	1090.72
sQs (kWh/Year)	43.29	44.01	43.28	43.30
sQg (kWh/Year)	171.28	164.61	170.68	171.71
sQv (kWh/Year)	53.17	53.17	53.17	53.17

There are not a significant difference in conduction heat gain on each model because of the area of the roof materials almost the same. The contents of the roof are precisely the same, so each model has similar thermal properties and conduction heat gain. There's no significant change in radiation heat flow on model 2 and 3 instead of model 1. The fenestration roof in model 1 face to south and north. Horizontal plane will accept the biggest irradiance and highest intensity, so model 1 enters the higher irradiance heat flow to the building than base case. Model 2 has inclined atrium, so the irradiance is lower.

Direct solar gain (sQg) in model 1 is the lowest of all models. sQg is direct irradiance trough transparent building envelope. Base case has a fenestration roof in the horizontal plane so that it may accept a quite high amount of this heat gain. The sawtooth shape and celestory have sQg which is not too much different because it has the same translucent roof area that reaches the north side, while model 1 has only half. For the city of Surabaya, the number of irradiation for the year on the north side is higher than the south side, so that model 1 has the smallest number among all models.

4.2 Cooling Energy of Building

According to the data of the total amount of energy use for one year, model 2 has the worst performance of all models (Figure 3). if compared with the base case, models 1 and 3 have almost the same total energy use because the amount of heat flow into the building is also nearly the same. Referring to table 1, the highest heat propagation comes from propagation by conduction. The distribution of conduction heat is influenced by the surface area of the shell and the thermal properties of the sheath, especially the conductance of the material. Model 2 has a sloping roof so that compared to the roof cover area on model 2 is bigger than the base model. In addition to the large surface area, galvalume has a high conductivity value, reaching 50 W / mK so that the conduction heat gain increases substantially resulting in improved cooling load and automatic cooling energy usage increases

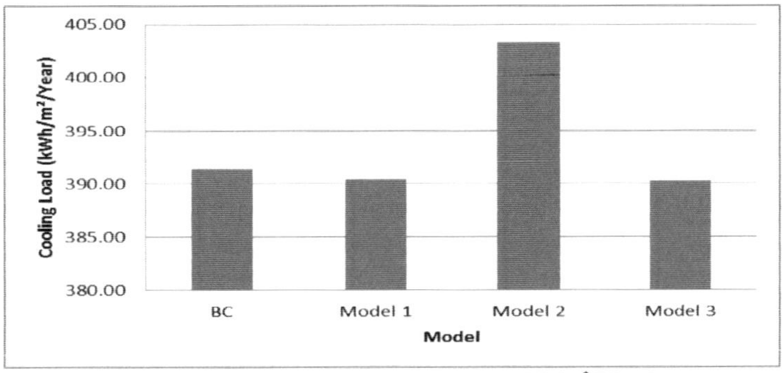

Figure 3. Cooling load on each model (kWh/m^2/Year)

5. Conclusion

From the results and discussion above it can be concluded that atrial roof design that is good for long span buildings in the tropics is in the form of a monitor. This form has a smaller area of sheath than the others so that the incoming conduction heat is also lower. The division of transparent fields on the north and south sides makes this building have a small direct and indirect irradiance heat gain.

6. Refences

[1] Crawford Robert H, Czerniakowski Isabella, Fuller Robert J (2010), "A comprehensive framework for assessing the life cycle energy of building construction assemblies", Science Direct, Architectural Science Review 53 (2010) 288–296.

[2] Depecker P, C. Menezo, J. Virgone, S. Lepers (2001), "Design of buildings shape and energetic consumption", Science Direct, Building and Environment 36 (2001) 627–635

[3] Gonzalo Roberto dan Karl J Hbermann (2006), Energy Efficient Architecture, Birkhauser, Berlin

[4] Gratia Elisabeth dan Herde Andre De (2003), "Design of low energy office building", Science Direct, Energy and Buildings 35 (2003) 473–491

[5] Yeang Ken (1996), Bioclimatic Skyscraper. London: Artemis London Limited

[6] Szokolay, S.V. (1987), Thermal Design of Buildings. RAIA Education Division, Canberra, Australia.

[7] Heerwagen Dean, 2004, Passive and Active Environmental Controls Informing The Schematic Design of Building, Mc Graw Hill, New York

[8] Ling C.S (2007), Minimising Solar Insolation In High Rise Buildings Trough Self Shaded Form, Universiti Teknologi, Malaysia

[9] Olgyay, V (1972). Design with Climate- Bioclimatic approach to architectural regionalism, Princeton University Press, New Jersey

[10] Juniwati Anik dan Asri Dinapradipta (2008), "Optimum Façade Design For Energy Efficient High-Rise Office Building In Hot Humid Tropics", Proceedings 9th SENVAR + 2nd ISESEE 2008, Malaysia

[11] Laksmiyanti Dian P.E (2015), Kinerja Bentuk Bangunan Perkantoran Bertingkat Menengah Di Surabaya Terhadap Efisiensi Energi Pendinginan, Jurnal IPTEK ITATS

[12] Noerwasito Totok (2006), Pengaruh Disain dinding dan Ruang terhadap Temperatur dalam Ruangan yang Berdinding Beton Ringan di Surabaya, untuk Efisiensi Energi. Surabaya: Jurusan Arsitektur ITS.

IOP Conf. Series: Materials Science and Engineering **462** (2019) 012031 doi:10.1088/1757-899X/462/1/012031

The Application of Adaptive Concept Form of Tissue Culture Laboratory Building in Black Orchid Research and Development Center in Samarinda

R Fajarini[1], I Ratniarsih[1] and Sukarnen[1]

[1]Department of Architecture, FTSP, Institut Teknologi Adhi Tama Surabaya, Surabaya, Indonesia

rizkifjrini@gmail.com

Abstract. Black Orchid is an endangered species in East Kalimantan in the last five years. If it's not conserved, this endemic species will extinct. The research and development centre aims to research and develop black orchids so that the population is maintained. In this research and development, the main facilities are tissue culture laboratory and greenhouse with adaptive as the concept form. The adaptive concept was applied to create a form of building that can adapt to the environment, without having to reject the use of modern materials and technology. The shape of the structure based on the shape of regional architecture of East Kalimantan which adapts to the local cultural values to create harmony with the surrounding environment. The characteristic of the building has many openings, and in the front facade, there are typical of Dayak carvings with philosophy as protection of the building.

1. Introduction

1.1 Background

Black orchids' population is currently decreasing, and even their presence in nature threatened with extinction due to excessive harvesting because black orchids are in high demand by the public. Another reason that causes a decline in the black orchid population is the growing habitat that damaged by logging and land conversion and the flowering period is very short (withering), and flowers are relatively difficult to cross. Design of Tissue Culture Laboratory Facilities at the Center for Black Orchid Research and Development provides solutions for orchids vegetative development so that the population maintained and balanced. The research facility was designed to adapt to the surrounding environment and culture so that the design can give a cultural image of East Kalimantan and able to adapt to the environment.

Figure 1. Black Orchid (Coelogyne pandurata)

IOP Conf. Series: Materials Science and Engineering **462** (2019) 012031 doi:10.1088/1757-899X/462/1/012031

1.2 Problems
Based on a review of the existing background it is seen that there is a problem in the application of the form, namely how to design an area consisting of black orchid research buildings that are oriented to the surrounding, nature, and culture.

1.3 Benefits
The design of the Tissue Culture Laboratory expected to become a building that provides research facilities for black orchids' ecosystem balance and to develop endangered species from extinction.

2. Literature Review
Adaptive based on KBBI 2008 is easy to adjust (self) to the situation. Meanwhile adaptive research [1], Ching [2-7], concerned with buildings that designed to adapt to their environments, their inhabitants and objects as well as those buildings that are entirely driven by internal data. The term is trying to merge what people imply when they talk about flexible, interactive, responsive or indeed media architecture, the large variety of recent publications demonstrates the mounting interest in this emerging.
Lamin (long house), is Dayak traditional house, have been known as one of unique architecture in east Borneo, especially in Kutai Kartanegara county. Lamin (long house), as a living houses not only physical formation but also as community system, so lamin as communal living houses [8].

3. Research Methods
The research method used is using descriptive research which aims to see, know and interact directly with the black orchid's living environment. With this research can give information and knowledge that is very important in the fundamental aspects of the Planning and Design of the Black Orchid Research and Development Center in North Samarinda, East Kalimantan [9-10].

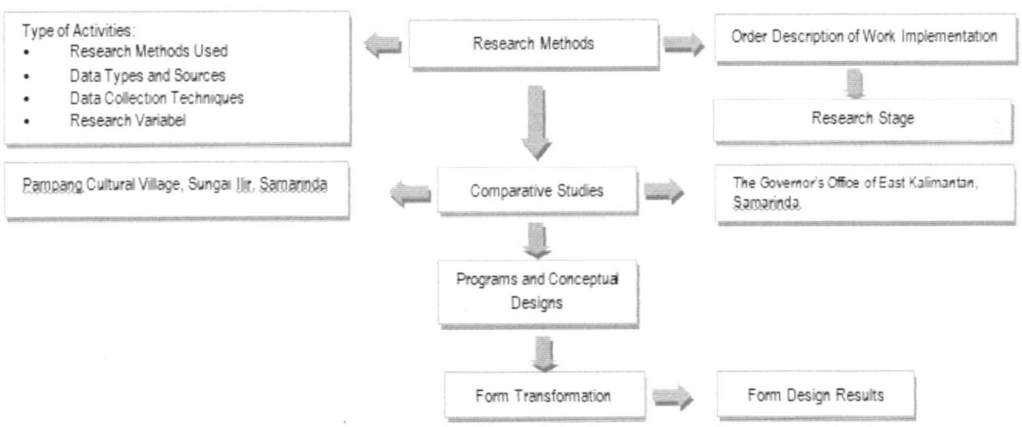

Figure 2. Research Stage Diagram

4. Comparative Study
Case Study Form 1: Pampang Cultural Village, Samarinda

Figure 3. Pampang Cultural Village

Pampang Culture Village is a tourist attraction of Kota Samarinda with traditional East Kalimantan building forms. The main characteristic of this building is the stairs made of tree wood. Surrounding the Lamin House there are many Blontang statues which depict gods as guardians of the house or village. Right at the corner of the roof is usually given a dragon headdress as a symbol of grandeur, nobility, and heroism.

Case Study Form 2: Office of the Governor of East Kalimantan, Samarinda

Figure 4. The Governor's Office of East Kalimantan

The Governor's Office is located in Samarinda City with a building that combines modern architecture with traditional architecture. Traditional forms are found on the gable roof and dragon carvings on the roof corner. While the material on the roof is metal roof with the main material of the building is concrete and steel.

5. Program and Design Concept

5.1 Form Design Program

In the design of the Tissue Culture Laboratory there is a form design program, namely; using colors and building materials that are environmentally friendly, designing roofs, bodies, and building legs that aligned to create unity in the building. Laboratory buildings designed with organic formations so can be blend with the surrounding environment.

5.2 Map of Site Location and Boundary of Site

The building located on Jalan Rimbawan Dalam, the site is a form of terrains with an area of about 1.6 hectares. This site was chosen as the site of the Black Orchid Research and Development Center because of the land was designated as conservation forest for rare plants from the Samarinda Botanical Garden.

Figure 5. Site Location Tanah Merah

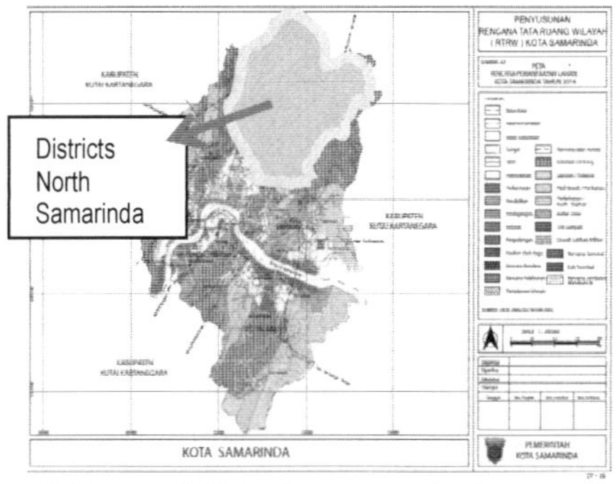

Figure 6. Districts – north Samarinda

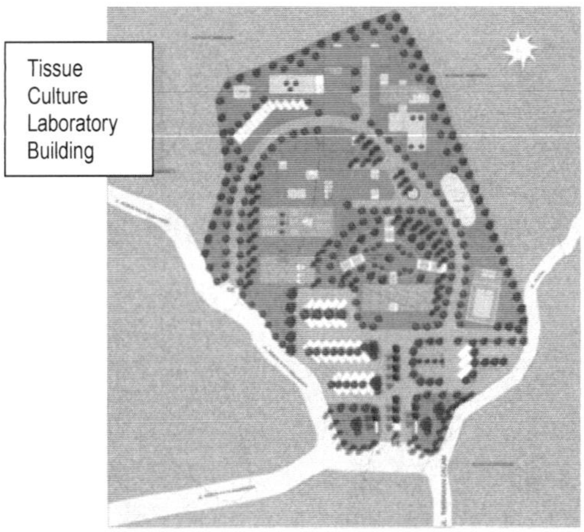

Figure 7. Layout Plan

5.3 Building Boundaries

The boundaries area of the Tissue Culture Laboratory are the North Side is the side entrance on the site, the southern side of the border is a pond and garden. The East Side is a Pest and Disease Laboratory, a greenhouse and a parking area, while for the West side of the building is a greenhouse and an administration and management building.

5.4 Design Concept

In the design of Tissue Culture Laboratory building the concept used is an adaptive concept that aims to create a form of building that can adapt to the environment, without having to reject the use of modern materials and technology. The shape of the building adapts to the local cultural values to create harmony with the surrounding environment.

6. Design Transformation

6.1 Transformation of Building Forms of Tissue Culture Laboratory

Transformation form is adaptive. An adaptive form can adjust to the environment without changing the surrounding's condition. The adaptive form also adapts to the buildings that are around and oriented to the original architecture of East Kalimantan.

6.2 Roof Form

The roof shape is a saddle-shaped roof. The roof used shingles as materials because it is easy to get in the area, besides this roof cover lasts about 25 years. With the shingle roof, the room inside the building can feel cool. The sloping roof is very sloping because it was located in the tropics, and facilitates the flow of rainwater, so it does not absorb on the roof of the shingle.

Figure 8. Saddle-shaped Roof

6.3 Building's Body Form

The building body has many openings so that air circulation can run well. The wall is made of woodplank material to give a natural impression to the building.

Figure 9. Building's Body Form

6.4 Building's Legs Form

Legs of buildings consist of columns and stairs made of wood. The stage-shaped staircase the building serves as the main entrance to the building. The bottom of the building is not walled so that air circulation can run smoothly.

Figure 10. Building's Legs Form

6.5 Whole Form Transformation

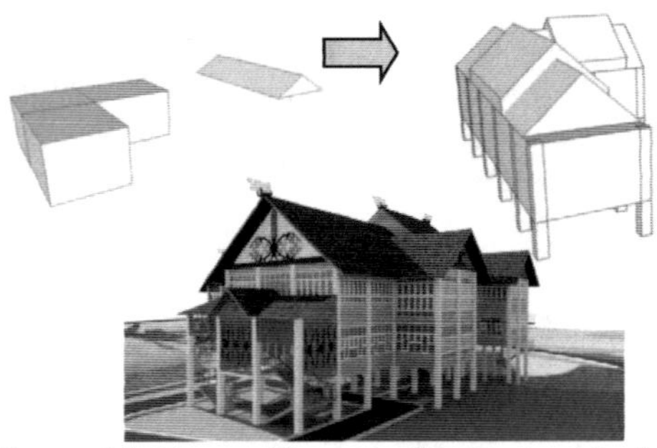

Figure 11. Form a Rectangular Base and triangle

It undergoes a change of shape with the addition of volume. The development of forms from the addition of square and pyramid volumes is then complemented with other forms such as carving and adding windows, doors and stairs so that the shapes become complex.

The last result of the form is a building with a shingle material that can adapt to the surrounding environment and culture. Wood plank was used for building material on the wall so that the building is not too massive and can give a natural impression that can adjust to the architecture of East Kalimantan. At the foot of the building, there is a column and stairs of the building that serves as the main entrance to the building.

7. Design Results

7.1 Form Design

With an adaptive concept, the design of the building itself uses the shape of a tropical building that combined with the traditional building form of East Kalimantan, namely Rumah Lamin Panjang. With distinctive characteristics:

- Forms that can adapt to the surrounding environment.
- On every building has many openings that serve to lower the use of air and artificial lighting.

The shape design results in the building of a tissue culture laboratory with a characteristic building shape that elongated and has many openings.

Figure 12. Front View of Tissue Culture Lab

In the front view, there are typical Dayaks carvings, a shield form with the philosophy of protecting the building as if for the main function as a sunscreen so that the front area of the building will be protected from sunlight.

Figure 13. Right View of Tissue Culture Lab

On the right side of the lab building, there is a carving shield with a meaning as a protector of the good side of the building. The shape of the building in the form of a rectangle provides effective spaces so that it can accommodate all activities inside the building. In addition, rectangular formations also make it easier to place lighting and ventilation on buildings.

Figure 14. Left View of Tissue Culture Lab

Figure 15. Back View of Tissue Culture Lab

On the left side of the building, there is a greenhouse which is a place to grow orchids with attention to the living conditions of orchid plants. The shape of the greenhouse also adapts to the surrounding environment and culture so that it can adjust to the main building.

The shape of the roof at the Research and Development Center is in the form of a saddle that can protect the building and the room inside from heat and rain properly. Because the building is a stilt house, the foot of the building is the columns, stairs, and terraces of the building.

8. Conclusion

After carrying out the discussion, it could be concluded that the concept of adaptive form in the building of the Tissue Culture Laboratory was applied to the material, shape, and appearance of the building. For materials used are materials that are able to adapt to the environment such as shingle roofs that are easily found in East Kalimantan. The form of adaptive concepts is applied to saddle-shaped roofs that are able to adapt to the weather while the building's appearance adapts the shapes and ornaments of East Kalimantan so that it can adapt to the surrounding culture.

9. References

[1] Adi A. P. Ni kade, Astarini I. A, Astiti and A. Ni Kade. 2014. *Aklimatisasi Anggrek Hitam (Coelogyne pandurata Lindl.) Hasil Perbanyakan In Vitro Pada Media Berbeda.* ISSN: 2337-7224. Page 9-10.

[2] Ching, Francis D. K. 1996. *Arsitektur : Bentuk, Ruang dan Susunannya.* (2nd ed.). (Ir. Nurahma Tresani Harwadi, MPM., Trans). Jakarta : Erlangga.

[3] Departemen Pendidikan dan Kebudayaan. 1997. *Kamus Besar Bahasa Indonesia*. Jakarta: Balai Pustaka. Indonesia

[4] Noviana, Mafazah. 2013. *Konsep Arsitektur Berkelanjutan Arsitektur Vernakular Rumah Lamin Suku Dayak Kenyah.* Politenik Negeri Samarinda. Samarinda.

[5] Rachima M. Siti., Denny Sudharnoto., dan Agus S. Sadana. 2006. *Mengenali Arsitektur Rakyat Dari Perubahan Bentuk Lamin Masyarakat di desa Pampang-Samarinda.* Universitas Pancasila. Jakarta

[6] Rapoport, A. 1969. *House, Form and Culture*. Prentice Hall. New York.

[7] Sedyawati, Edi, EKM. Masinambow, Tjahyono, Gunawan. 1995. *Konsep Tata Ruang Suku Bangsa Dayak Kenyah di Kalimantan Timur.* Departemen Pendidikan dan Kebudayaan. Jakarta.

[8] Widayati, Rusfina. 2014. Thesis S2. *Konsep Spasial Lamin Adat Suku Dayak Kenyah di Kabupaten Kutai Kartanegara*. UGM. Yogyakarta.

[9] Winzeler L. Robert. 2004. *The Architecture of Life and Death in Borneo*. University of Hawaii. Honolulu.

[10] Yuuwono B. Abito. 2015. *Peran, Fungsi, dan Makna Arsitektur Rumah Lamin Dalam Budaya Suku Dayak di Kutai Barat Kalimantan Timur.* Fakultas Teknik Universitas Tunas Pembangunan Surakarta. Indonesia.

IOP Conf. Series: Materials Science and Engineering **462** (2019) 012032 doi:10.1088/1757-899X/462/1/012032

The Use of Co-solvent for Insitu Transesterification of Microalgae with Base Catalyst under Microwave Irradiation

U Kalsum[1], A Roesyadi[1] and M Mahfud[1]

[1]Department of Chemical Engineering, Institut Teknologi Sepuluh Nopember, Surabaya, 60111 Indonesia

umkal29@gmail.com

Abstract. The utilization of co-solvent was considered as one of several methods to improve yield of methyl ester on in situ transesterification of microalgae. The research aims to study the effect of co-solvent in the in situ transesterification of *Chlorella Sp* using base catalyst under microwave irradiation. The in situ transesterification reaction carried out in a microwave reactor using 1% of base catalyst with an addition 30 ml of n-hexane as a co-solvent. The microwave was equipped with a reflux condenser and the reactor was a round bottom flask equipped with magnetic stirrer. The reaction study was performed by varying the reaction time (10, 20, 30, 40, 50) min at 450 W of microwave power. Based on the experiment result obtained that the optimal yield of methyl ester was 30.08% at 40 min. The analysis result obtained that the component of Fatty Acid Methyl Ester(FAME) from *Chlorella Sp* was a medium chain methyl ester (C_{12}-C_{20}) with Saturated Fatty Acid (SAFA)s as the dominant component.

1. Introduction

The increasing of energy consumption in the transportation sector has the big impact in the decreasing of fossil fuel energy stock and the increasing of environment pollution. The most effect contributes global warming and the Green House Gas (GHG) emission. The global warming is a serious problem which impacts to the human life and environment and needs a huge solution to solve it.

Microalgae is an autotrophs microorganism has high lipid content which prospective to develop into biodiesel. Biodiesel from microalgae is one of renewable energy sources that has potential to replace fossil fuel energy and decrease the green house gas emission. The economical production of biodiesel from microalgae mainly depends on the energy for process and type of lipid extraction process used. Lipid extraction method of microalgae is divided into mechanical and chemical methods. The mechanical methods involve microwave -assisted extraction, ultrasonic assisted extraction and press; while chemical methods imply soxhlet extraction, ionic liquid extraction and supercritical extraction [1]. In situ transesterification is a method to eliminate the extraction process and improve the conversion performance [2]. It is reported that in the in situ trans esterification oils in the material directly contact with alcohol alkalized or alcohol acidified and the process acts as a pre-extraction reaction with alcohol [3]. The in situ transesterification method has the potential to simplify the conversion process, reducing the number of unit operation and the process cost for biodiesel production [4]. However, an optimization of reaction condition still required on the process extraction and trans esterification which occur simultaneously [5]. The usage of co-solvent is one of several methods to accelerate reaction and improve yield in the in situ trans esterification process. In situ transesterification process from wet microalgae using chloroform-methanol solvent and sulfuric acid has been researched with conversion yields as high as 91% [6]. In order to reduce the reacting methanol volume and increase in situ transesterification process, several co-solvents have been used to enhance yield of lipid to methyl ester [7]. In this study, the effect of microwave on in situ transesterification of microalgae using base catalyst and addition of n-hexane as co-solvent in the reaction were evaluated. This work optimized the experimental parameters on conversion of yields for in situ transesterification of microalgae.

2. Methods

2.1 Materials

Dry microalgae powdes (*Chlorella Sp*) were supplied from Balai Budi Daya Air Payau Takalar, South Sulawesi. Sodium hydroxide, methanol and n-hexane were purchased from Brataco Chem.

2.2 Experimental Procedures

2.2.1 Extraction Microalgae Lipids. Extraction lipid of microalgae aims to determine the fatty acid component contained in microalgae lipids. The extraction process carried out using mixture of methanol - n-hexane solvent and a microwave as a heating source. Ten grams of dried Chlorella vulgaris were mixed with 100 ml of methanol-hexane solution. The mixture was subjected to the microwave for 50 min with 450 W of microwave power for extraction process. After filtration, the filtrate was then destilled to evaporate remains solvent and lipids were weight and analysis by GC MS.

2.2.2 In situ Transesterification and In Situ Transesterification with Co-Solvent. In situ transesterification experiments were carried out in a microwave reactor. The microwave reactor was equipped with a reflux condenser to provide condensation of evaporation from solvent mixture. A magnetic stirrer was used to assure a uniform mixture with a constant mixing speed. Ten grams of dry microalgae powder were mixed with solution of methanol and sodium hydroxide catalyst. The mixture was then expose to microwave irradiation with power of 450 W, reaction time 20-50 min; catalyst concentration 1% and dry algae to methanol (w/v) 1:12 and n-hexane volume 30 ml. The samples were cooled and filtered before the separation steps. The same experiment was carried out with an addition 30 ml of n-hexane in the reaction mixture to observe the effect of co-solvent in the in-situ transesterification reaction. After filtration, the samples were separated in a separating funnel before purification stage and analysis by GC MS.

3. Result and Discusssion

3.1 Characterization of Microalgae Lipids

The analysis of *Chlorella Sp* lipids showed that the fatty acid components produced from the extraction varied from C14-C18 which are saturated fatty acids (SAFA), monounsaturated fatty acids (MUFA) and polyunsaturated fatty acids PUFA). The fatty acids are dominated by palmitic acid C16: 0 and C18: 1 oleic acid. Both components of these fatty acids and some of the various fatty acids in microalgae have the potential to be developed into biodiesel. The raw materials for diesel fuels are hydrocarbons containing 8-10 per molecular carbon atoms. The number of carbon atoms found in vegetable oil is around 16-20 carbon atoms per molecule. Therefore, microalgae oil must be trans esterified before being used as fuel.

3.2 The Effect of N-Hexane Addition on in Situ Transesterification of Microalgae Under Microwave Irradiation

In situ transesterification with co-solvet addition as co-solvent into the reaction mixture were studied at 450 W of microwave power, 1:12 (w/v) microalgae biomass to methanol, 1% of KOH and difference reaction time (10 to 50) min. Co solvent volume was investigated by study the effect of n-hexane use during in situ transesterification with microwave and compare the results with direct transesterification in the absemce of co-solvent.

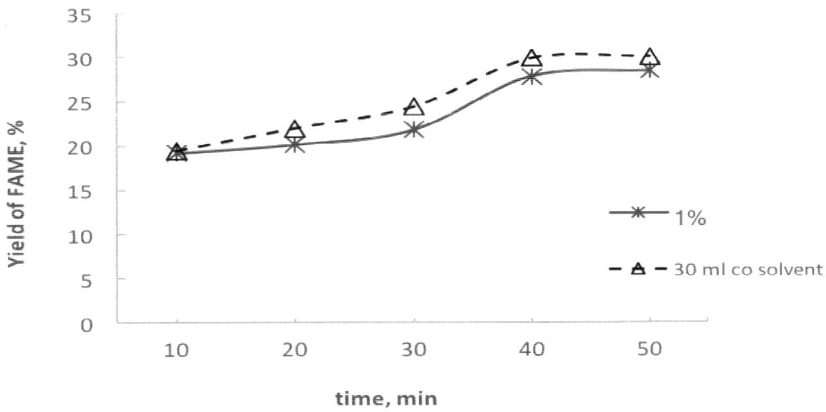

Figure 1. Yield of FAME from in situ transesterification with 1% of KOH, 30 ml of n-hexane and 450 W of power

Figure 1 shows yield of FAME obtained by in situ transesterification with KOH under microwave irradiation and in situ transesterification with co-solvent. As shown in fig 1, Yield of microalgae FAME increased significantly with the increase in reaction time for both reactions. The significant increase in yield starts from 10 min to 50 min of reaction. The influence of reaction time on the increase in yield of methyl ester under microwave irradiation caused by the longer contact time that induces more disrupted cell walls and faster triglycerides are released into the reaction mixture, therefore there is sufficient time for reactant mixture to form biodiesel [8]. The influence of microwave use as a heater in this process was also presented by Guerra et al., 2013 which states that microwaves can accelerate organic synthesis with a short reaction time due to an increase in reaction rate. Microwave energy produces electromagnetic waves, wherein the amount of wave energy per unit of time is reflected as the strength. Heating process with microwaves can take place very quickly because the absorption capacity of microwave energy from polar molecules is very high, and the presence of heat transfer in microwaves which rapidly leads to molecular material cause an increase in yield of product. Based on the nature of polarization, the homogeneity between alcohol and oil occurs rapidly due to the movement of the dipole molecule without changing the molecular structure of the material. Therefore, the addition of supplied power leads to the heating at the molecular level that takes place very quickly and affects the speed of the reaction rate [8, 9, 12]. As stated by Arrhenius law that with the longer reaction time the contact between reactant molecules is also increased and yield of product is likely to increase. But if there has been an equilibrium, the increasing time will not affect the yield of product.

The effect of n hexane addition in the in situ transesterification also displayed in figure 1. From the graph can be seen that yields obtained from reaction with hexane were higher than yield by reaction with the absence of hexane. At the end of the reaction, the highest yield with hexane addition was 30.08%, which was 1.84% higher than yield resulted by in situ transesterification without n-hexane. The effect of n-hexane as co-solvent on the increase in yield FAME caused by the ability of hexane to increase the solubility of alcohol in the lipid and the power of extraction before reacting with alcohol. The role of n-hexane as co-solvent was a mediator during the reaction under microwave which can increase the mass transfer between catalyst and methanol. In addition, n-hexane is a non-polar solvent which transparent and able to pass through the microwave irradiation without receiving a heating process. Meanwhile, the polarity nature of methanol can absorb microwave energy strongly due to the permanent moment dipole. As a result, an extractive transesterification in this study can be provided by the combination of polar and non-polar solvent use.

3.3 Analisys of Fatty Acid Methyl Ester (FAME)

The analysis to *Chlorella Sp* Fatty acid methyl ester showed the composition of methyl ester from Chlorella Sp. The fatty acid compound was a medium chain methyl ester, MCFA with (C_{12}-C_{20}) which consist of esters from lauric acid (C12:0), myristic acid (C14:0), palmitic acid C16:0), palmitoleic acid (C16:1), stearic acid (C18:0), oleic acid (C18:1n9), vaccenic acid (C18:1n7), linoleic acid (C18:2), arachidonic acid(C20:4) and eicosapentanoic acid (C20:5).

Table 1. Fatty Acid Methyl Ester Composition from *Chlorella Sp*

Compounds	Formula	Mass (%)
Palmitic acid ME	$C_{16}H_{32}O_2$	26.93
Myristic acid ME	$C_{14}H_{28}O_2$	0.76
Palmitoleic acid ME	$C_{16}H_{30}O_2$	2.25
Eicosapentanoic ME	$C_{22}H_{32}O_2$	1.66
Linoleic MEcatalyst	$C_{19}H_{34}O_2$	3.95
Linolenic ME	$C_{19}H_{32}O_2$	1.04
Oleic ME	$C_{19}H_{36}O_2$	14.55
7.10 Hexadecadienoic ME	$C_{17}H_{30}O_2$	0.76
10-Heptadecen-8-ynoic acid	$C_{18}H_{30}O_2$	1.08
Octadecanoic acid ME	$C_{18}H_{36}O_2$	3.96
Others		36.47

4. Conclusion

Base in situ transesterification with co-solvent was performed for biodiesel production from microalgae. The evaluated the process parameters required to produce biodiesel using microwave as a heating source. From the parameters evaluated (methanol volumes, reaction time and microwave powers) the highest yield was obtained at 50 min with biomass to methanol ratio, 1:12 and power of 450 W. It was represented that 30.08% of yield can be obtained from base catalyzed in situ transesterification with c-hexane addition. From analysis, FAMEs profile was identified and the majority of FAME produced from palmitic acid, stearic acid and oleic acid.

5. References

[1] Mubarak M, Shaija A and Suchithra T V 2015 *J Alga Research* 7 117
[2] Georgogianni K G, Kontominas, Pomonis P J, Avlonitis D and Gergis V 2008 *J Fuel Process Technol* 89 503- 509
[3] Ma G, Hu W, Pei H, Jiang L, Song M and Mu 2015 J Energy Convers. Manage.2015 90 41-49
[4] Haas MJ and Wagner K 2011 *J Lipid Sci Technol*, 113, 1219
[5] Dianursanti, Religia P and Wijanarko A 2015 *P Procedia Environ.Sci.* 23 412-420
[6] Im H, Lee J W and Kim B 2015 *J Bioresource Technol 193 386-392*
[7] Ehimen E.A, Sun Z.F and Carrington C.G 2010 *J Fuel* 89 627
[8] Patil P D, Gude V G, Mannarswamy A, Deng S, Cooke P and Munson-McGee 2011 *J Bioresource Technol* 102 118 – 22
[9] E.Martinez-Guerra 2013 *J Bioresource Technology* 156 240
[10] Chisti.Y 2007 J *Biotechnol. Adv*, 25 294
[11] Quintan A T, Katoh S and Goto M 2011 *Microwave-Assisted Synthesis of Biofuels*, Biodiesel production- recent development and prospect (Croatia: Intech) chapter 16 pp 415
[12] Martinez-Guerra E, Gude V G, Mondala A, Holmes W and Hernandez R 2014 *J Bioresource* Technol 156 240 – 7

Acknowledgement
This research was funded by the Directorate General of Indonesian Higher Education through The Doctoral Dissertasion Grant. We also thanks for Mr. Rifai help for analysis Algae's FAME.

IOP Conf. Series: Materials Science and Engineering **462** (2019) 012033 doi:10.1088/1757-899X/462/1/012033

Application of Electrocoagulation Methods to Reduce BOD and COD Content in The Soft Drink Industry's Wastewater with Addition Bittern

S Julaika[1], A P Dewi[1] and U H Cintia[1]

[1] Department of Chemical Engineering, Faculty of Industrial Technology, Institut Teknologi Adhi Tama Surabaya, Surabaya, Indonesia

shofiyyaj@itats.ac.id

Abstract. Wastewater from the soft drink industry must be treated before release to reservoir. One of many treatments of wastewater is electrocoagulation method. Five litres of wastewater was agitated using a magnetic stirrer with the addition of bitterns in a tank electrochemical device. It was started by 100 rpm of agitation for 3 minutes and then the agitation was decreased to 50 rpm for the certain time. The filtrate was separated and analysed Chemical Oxygen Demand (COD) and BOD. Percent of COD removal tends to increase with the length of the electrocoagulation process and vice versa for BOD. Different things occur in percent COD and BOD removal, where the best percent removal is obtained at 0.5% bitter addition. The highest percentage of COD removal reached 98.31% at 6 hours and the highest percent of BOD removal reached 95.00% within 2 hours.

1. Introduction

Wastewater is one of the problems in the industry. Wastewater must be treated before released to water bodies, waterways, reservoir, or rivers so that this wastewater does not contaminate the environment. In this case, the methodology of wastewater treatment must be able to minimize contaminant in wastewater, even in the soft drink industry. The wastewater of the soft drink industry with all of the flavour contains organic compound about 500-3000 mg/l [1], [2]. Technologies used to treat it are plasma technology to reduce BOD, COD, TSS, and color contents [3], [4], using chitosan as a coagulant to reduce BOD, COD, and TSS contents [5] or microbiologically using the aerobic system uses activated sludge [1], using the aerobic system an oxidation ditch [2], aerobic and anaerobic systems [6].

Another method that may be able to treat wastewater of the soft drink industry is electrocoagulation. This method is a simple and efficient method for wastewater treatment [7]. The principle used is electrochemistry, mainly electrolysis, that it use electrical energy to degrade and coagulate compounds in wastewater in a present of anode-cathode. The electrocoagulation can remove or reduce organic and inorganic contaminant from wastewater. Implementation of this method were to remove fluoride from aqueous solutions [8], to reduce organics for washing wastewater reclamation [9], to reduce Chemical Oxygen Demand (COD) levels from baker's yeast wastewater [10], to remove heavy metal in industrial water [11], to remove sulphate and other anionic materials from brackish well water [12], to treat wastewater from textile industry [13]–[15], to treat domestic wastewater [16], to treat wastewater for slaughterhouses [17], and to treat wastewater for skin tanning industry [18].

Besides electrocoagulation method, there are materials as a coagulant, one of them is bittern. The bittern is a by-product of the production of salts, namely '*air tua*'. The major content of the bittern is magnesium, which is $MgSO_4$ and $MgCl_2$. Other compounds are NaCl and KCl [19]. Bittern has the ability to coagulate substances in liquid waste both inorganic and organic. Implementation of the bittern as coagulant has been applied to wastewater treatment of fish flour industry [20]. It was used to reduce TSS levels in paper industry wastewater [21], reducing levels of TSS, COD, and BOD in fish processing wastewater [22].

This research will be conducted by the combination of electrocoagulation and addition of coagulant bittern which be applied to wastewater treatment of the soft drink industry where it is not done yet. The combination of this method is expected to maximize the removal percentage of contaminants in wastewater result in the soft drink industry.

2. Materials and Methods

2.1. Materials

This research, wastewater used was obtained from the soft drink industry in East Java, Indonesia, and bittern from Garam Indonesia Company. The materials could be used without further processing. For electrode, material used is iron as anode and aluminium as a cathode.

2.2. Procedure

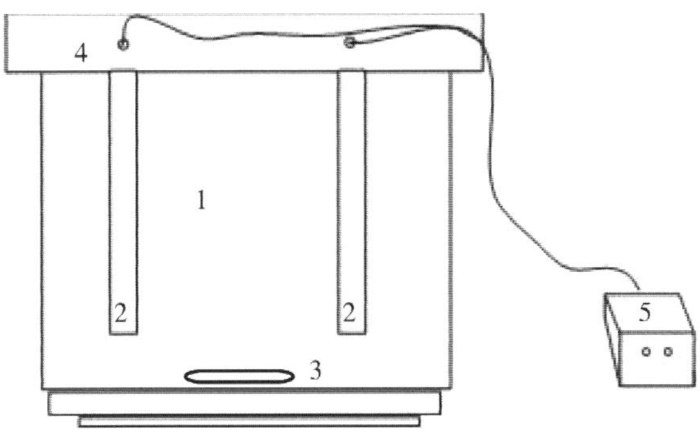

1. Tank
2. Electrode
3. *Stirrer*
4. Electrode support
5. Electric source

Figure 1. Schematic diagram of electrochemical device

Implementation of this research is begun by agitation of 5 litres of wastewater using a magnetic stirrer and be added some of the bittern in the mixture according to a certain concentration in a tank electrochemical device. It was started by 100 rpm of agitation for 3 minutes and then the agitation was decreased to 50 rpm for the certain time. The filtrate was separated and analysed COD and BOD. COD and BOD removal percentage were calculated by the following equation.

$$\text{COD Removal percentage } (\%) = \frac{(COD_{initial} - COD_{finsl})}{COD_{initial}} \times 100\% \quad (1)$$

$$\text{BOD Removal percentage } (\%) = \frac{(BOD_{initial} - BOD_{finsl})}{BOD_{initial}} \times 100\% \quad (2)$$

3. Result And Discussion

Wastewater used was obtained from the soft drink industry in East Java, Indonesia, which taken for five times. Each of sampling is analysed to obtain COD and BOD contents. Analysis result can be shown in Table 1.

The first step, wastewater is electro-coagulated with and without the addition of bittern. The result can be seen in Figure 2. Physically, the image shows that the electrocoagulation process with the addition of bittern gives better results compared to the electrocoagulation process without the addition of bitterns carried out at a certain same time. This result becomes the basis to establish where the research is conducted by adding bittern to get good results.

Table 1. The Analysis result of wastewater from the soft drink industry

Sample	The Analysis result	
	COD (mg/L)	BOD (mg/L)
Sample 1	4,941.15	2,800.00
Sample 2	6,176.44	3,000.00
Sample 3	5,173.00	3,000.00
Sample 4	4,506.53	2,500.00
Sample 5	4,403.59	2,300.00

(a) (b)

Figure 2. The image of The first step of this research (a) Without *Bittern* (b) With *Bittern*

In this research, electrodes used are iron as anodes and aluminium as cathodes. The reaction occurs at the anode follows equation (3) and then oxygen oxidize Fe^{2+} to Fe^{3+}as in equation (4) [23].

$$Fe_{(S)} \rightarrow Fe^{2+} + 2e^{-} \quad (3)$$

$$4Fe^{2+} + 4H^{+} + O_{2(g)} \rightarrow 4Fe^{3+} + 2H_2O \quad (4)$$

While aluminium is used as a cathode, the reaction follows equation (5) [23].

$$2H_2O + 2e^{-} \rightarrow H_{2(g)} + 2OH^{-} \quad (5)$$

Total reaction on anode-cathode become equation (6)

$$4Fe_{(S)} + 10H_2O_{(l)} + O_{2(g)} \rightarrow 4Fe(OH)_{3(s)} + 4H_{2(g)} \quad (6)$$

$Fe(OH)_3$ is a negative electrically charged particle that can tend to form colloids [7], [24], which causes coagulation in the sample solution, in this case, wastewater from the soft drink industry. The contaminants contained in the wastewater are coagulated by $Fe(OH)_3$ so that BOD and COD content in filtrate becomes smaller than wastewater. It can be seen in Figure 3 and Figure 4.

Figure 3 shows that COD removal percentage tends to increase with the increasing of electrocoagulation time. Maximum COD removal percentage achieves 98.31% at 0.50 % of bittern concentration and 6 hours electrocoagulation. Figure 4 shows that BOD removal percentage tends to decrease with the increasing of electrocoagulation time. The maximum BOD removal percentage achieves 95.00% at 0.50 % of bittern concentration and 2 hours electrocoagulation.

Both images show that the addition of 0.5% bittern shows the best COD and BOD removal percentages compared to other. This shows that COD and BOD removal percentage decrease with increasing of bittern addition. This is due to the bittern which mostly contains magnesium, which should be able to react to the base, the result of reaction on a node, cannot be deposited. Nevertheless, the solubility of magnesium is greater than iron (III) [24] so that much magnesium and other contaminants cannot be deposited so that decreasing of COD and BOD content cannot be maximized.

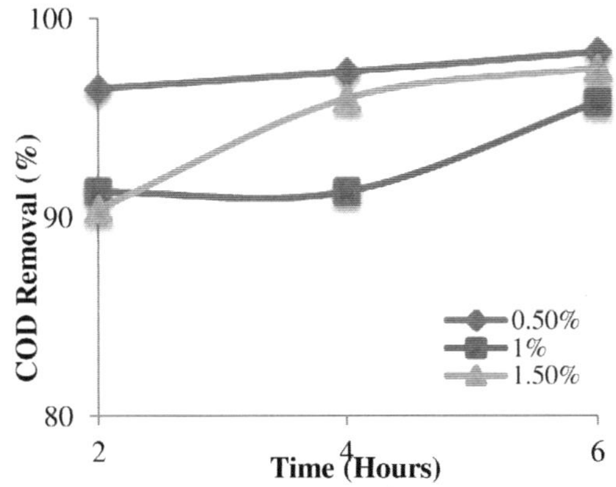

Figure 3. The Effect of Bittern Concentration on COD Removal Percentage During Electrocoagulation Process

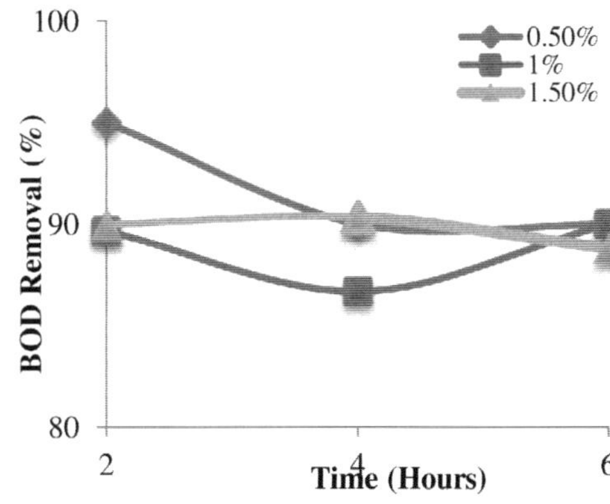

Figure 4. Effect of Bittern Concentration on BOD Removal Percentage During Electrocoagulation Process

4. Conclusions

Percent of COD removal tends to increase with the length of the electrocoagulation process and vice versa for BOD. Different things occur in percent COD and BOD removal, where the best percent removal is obtained at 0.5% bitter addition. The highest percentage of COD removal reached 98.31% at 6 hours and the highest percent of BOD removal reached 95.00% within 2 hours.

5. References

[1] Indriyati, "Pengolahan limbah cair industri minuman," *J. Tek. Lingkung.*, vol. 9, no. 1, pp. 25–30, 2008.

[2] Indriyati and J. P. Susanto, "Pengolahan limbah cair industri minuman ringan," *J. Tek. Lingkung.*, vol. 10, no. 1, pp. 85–89, 2009.

[3] A. Ryane, W. Oktiawan, and A. Syakur, "PENGGUNAAN TEKNOLOGI PLASMA DALAM MENGURANGI KANDUNGAN BOD DAN WARNA PADA LIMBAH CAIR INDUSTRI MINUMAN RINGAN," *J. Tek. Lingkung.*, vol. 4, no. 1, pp. 1–6, 2015.

[4] A. Karina, B. Zaman, and A. Syakur, "STUDI PENURUNAN COD DAN TSS LIMBAH CAIR INDUSTRI MINUMAN RINGAN DENGAN MENGGUNAKAN TEKNOLOGI PLASMA (Studi Kasus PT . Coca Cola Amatil Indonesia – Semarang)," *J. Tek. Lingkung.*, vol. 4, no. 1, p. 6, 2015.

[5] D. Y. P, S. Julaika, A. W. I. P, and R. Artikasari, "PENGARUH PENAMBAHAN KITOSAN DALAM PENURUNAN TSS PADA LIMBAH CAIR INDUSTRI MINUMAN RINGAN," in *Seminar Nasional Sains dan Teknologi Terapan V 2017 Institut Teknologi Adhi Tama Surabaya*, 2017, pp. 155–160.

[6] H. Suprihatin, "Pengolahan Limbah Cair Pembuatan Minuman Ringan Dengan Reaktor Semiaerobik, TCPS, Reactor Anaerobic Dua Tahap," *Din. Lingkung. Indones.*, vol. 5, no. 1, pp. 20–26, 2018.

[7] M. F. Ni'am, F. Othman, J. Sohaili, and Z. Fauzia, "Removal of COD and Turbidity to Improve Wastewater Quality Using Electrocoagulation Technique," *Malaysian J. Anal. Sci.*, vol. 11, no. 1, pp. 198–205, 2007.

[8] A. H. Mahvi, E. Bazrafshan, and K. A. Ownagh, "Application of electrocoagulation process using iron and aluminum electrodes for fluoride removal from aqueous environment," *E-Journal Chem.*, vol. 9, no. 4, pp. 2297–2308, 2012.

[9] H. Rubí-Juárez, C. Barrera-Díaz, I. Linares-Hernández, C. Fall, and B. Bilyeu, "A combined electrocoagulation-electrooxidation process for carwash wastewater reclamation," *Int. J. Electrochem. Sci.*, vol. 10, no. 8, pp. 6754–6767, 2015.

[10] M. Al-Shannag, Z. Al-Qodah, K. Alananbeh, N. Bouqellah, E. Assirey, and K. Bani-Melhem, "Cod reduction of baker's yeast wastewater using batch electrocoagulation," *Environ. Eng. Manag. J.*, vol. 13, no. 12, pp. 3153–3160, 2014.
[11] R. H. Al Anbari, J. Albaidani, S. M. Alfatlawi, and T. A. Al-Hamdani, "Removal of Heavy Metals From Industrial Water Using Electro-Coagulation Technique," in *Twelfth International Water Technology Conference*, 2008, pp. 1–18.
[12] G. M. Densmore, "Electrocoagulation for sulfate removal," 2015.
[13] N. Tyagi, S. Mathur, and D. Kumar, "Electrocoagulation process for textile wastewater treatment in continuous upflow reactor," *J. Sci. Ind. Res. (India).*, vol. 73, no. 3, pp. 195–198, 2014.
[14] M. Kobya, O. T. Can, and M. Bayramoglu, "Treatment of textile wastewaters by electrocoagulation using iron and aluminum electrodes," *J. Hazard. Mater.*, vol. 100, no. 1–3, pp. 163–178, 2003.
[15] A. Yulianto, L. Hakim, I. Purwaningsih, and V. A. Pravitasari, "PENGOLAHAN LIMBAH CAIR INDUSTRI BATIK PADA SKALA ELEKTROKOAGULASI," *Jurnal*, vol. 5, no. 1, pp. 6–11, 2009.
[16] F. Ozyonar and B. Karagozoglu, "Operating cost analysis and treatment of domestic wastewater by electrocoagulation using aluminum electrodes," *Polish J. Environ. Stud.*, vol. 20, no. 1, pp. 173–179, 2011.
[17] A. F. Ardhani and D. Ismawati, "Penanganan limbah cair rumah pemotongan hewan dengan metode elektrokoagulasi," 2007.
[18] E. Wardhani, M. Dirgawati, and K. P. Valyana, "Penerapan Metode Elektrooagulasi dalam Pengolahan Air Limbah Industri Penyamakan Kulit," in *Seminar Ilmiah Nasional, Penelitian masalah Lingkungan di Indonesia 8*, 2012, pp. 1–16.
[19] A. Rasmito, N. P. Asri, and J. Suwarno, "Pembuatan Kristal Epsomite Dari Air Tua," in *Seminar Rekayasa Kimia dan Proses*, 2010, pp. 1–5.
[20] K. A. Nugraha, P. Wesen, and M. Mirwan, "PEMANFAATAN BITTERN SEBAGAI KOAGULAN ALTERNATIF PENGOLAHAN LIMBAH TEPUNG IKAN," *J. Ilmu Tek. Lingkung.*, vol. 8, no. 1, pp. 1–9, 2018.
[21] Sutiyono, "PEMANFAATAN BITTERN SEBAGAI KOAGULAN PADA LIMBAH CAIR INDUSTRI KERTAS," *J. Tek. Kim.*, vol. 1, no. 1, pp. 36–42.
[22] D. Y. P, S. Julaika, A. Malik, and J. L. Goa, "PENGARUH PENAMBAHAN BITTERN PADA LIMBAH CARI DARI PROSES PENCUCIAN INDUSTRI PENGOLAHAN IKAN," *J. IPTEK*, vol. 21, no. 1, pp. 43–50, 2017.
[23] C. C. Barrera-Díaz, G. Roa-Morales, P. Balderas Hernández, C. M. Fernandez-Marchante, and M. A. Rodrigo, "Enhanced electrocoagulation: New approaches to improve the electrochemical process (Review)," *J. Electrochem. Sci. Eng.*, vol. 4, no. 4, pp. 1–14, 2014.
[24] J. R. A. Day and A. L. Underwood, *Quantitative Analysis*. 1998.

IOP Conf. Series: Materials Science and Engineering **462** (2019) 012034 doi:10.1088/1757-899X/462/1/012034

The Effect of Raja Banana Peel Extract on Acid and Peroxide Numbers in Bulk Frying Oil

D Y Purwaningsih[1], D R Zuchrillah[1] and I Nurmala[1]

[1]Chemical Engineering Department, Institut Teknologi Adhi Tama Surabaya, Indonesia

dianyp@itats.ac.id

Abstract. Bulk frying oil has been generally under the standardised quality in SNI 3555-1998 since it is not packed and is easy to be oxidised. The research was to apply raja banana peel extract that contained natural oxidant such as flavonoid to enhance the quality of bulk frying oil. The raja banana peel extract was obtained by maceration method using methanol and ethyl acetate. The weight of extract applied to the bulk frying oil was 1; 1.5; 2; 2.5 grams, with stirring time for about 40 minutes. The samples were then analysed to identify the contents of both acid and peroxide numbers. The best result was attained by applying 2.5 grams of the extract with 40 minutes of stirring time, which resulted in the acid number of 0.27938 meq/kg and peroxide number of 1.99668 meq/kg.

1. Introduction

Frying oil has been proven essential for Indonesian due to the increase in its demand, which causes higher price periodically. Consequently, people with low economic income instead choose bulk frying oil since it is more affordable for them. However, the quality of the existing bulk frying oil is distant from its standard, which is more of bad quality and not hygiene. Since the package is not fulfilled the packaging standard, bulk frying oil is more natural to be oxidised that further produces bad smell.

There are various ways to enhance the quality of bulk frying oil. The previous study found that mangosteen (*Garcinia mangostana L.*) peel extract was additionally used to improve iodide numbers because it contained a natural substance named xanthone. Besides, the addition of antioxidant that can adjourn or prevent free radicals oxidation in lipid oxidation obtained from *Raja* banana (*Musa paradisiaca sapientum*) peel extract. Raja banana peel contained a beneficial natural antioxidant, of which the antioxidant substance was called flavonoid [9]. However, this present research used *Raja* banana peel extract to regenerate the quality of bulk frying oil on acid, iodide, and peroxide numbers. Moreover, the use of *Raja* banana peel as the main ingredient of the extract, at the same time, responded to the application of the recycle process to decrease banana peel waste. This research also proved that Raja banana peel waste could be processed into different benefits, not only becoming an animal feed.

2. Literature Reviews

2.1 Frying Oil

Frying oil consists of fat originated from the purification process of plant's or animal's part, or from a synthetic method that is further purified and used to fry foods [12]. Good frying oil is that by

Indonesian National Standards (SNI). Table 1 shows the standard of frying oil regarding SNI 3555-1998.

Table 1. Standardised Frying Oil according to SNI 3555-1998

No.	Aspects to Test	Unit	Requirement
1	Color	-	Yellow
2	Smell and taste	-	Normal
3	Acid number	Meq/kg	Max. 0.3
4	Iodide number	Meq/kg	<5
5	Peroxide number	Meq/kg	Max. 2
6	Water	%	0.3

A good frying oil should consist of more unsaturated fatty acids than the saturated ones [12]. The fatty acids contained in a frying oil determine its oil quality because the acids define chemical properties and oil stability [6]. Table 2 shows different fatty acids contained in vegetable oil [3].

Table 2 Fatty Acids Composition based on Three Types of Vegetable Oil

Fatty Acids	Number of Atom C	Palm Oil (%)	Oil Core (%)	Coconut Oil (%)
Saturated Fatty Acids:				
Octanoate	8	-	3-4	5.5-9.5
Decanoate	10	-	3-7	4.5-9.5
Laurate	12	0-0,3	46-52	44-52
Myristate	14	0.9-2,5	14-17	13-19
Palmitat	16	40-46	6.5-9	7.5-10.5
Stearat	18	3.6-5,7	1-2.5	1.0-3.0
Unsaturated Fatty Acids:				
Oleat	18	39-45	13-19	5-8
Linoleate	18	7-12	0.5-2.0	1.5-2.5
Linolenate	18	1	1-5	-

2.2 Antioxidant

An antioxidant is a molecule that can postpone, decelerate, or prevent other particles oxidation process [10]. It can decelerate oxidation reaction or stop the chain reaction of the free radicals from oxidised oil (absorption of free radicals) [7]. There is four reaction mechanism including:

1. Releasing hydrogen from antioxidant
2. Releasing electron from antioxidant
3. Adding fat to the aromatic ring on antioxidant
4. Creating complex compounds between fat and aromatic ring from antioxidant.

The following states the reaction stages of fat or oil oxidation:

Lipid Radical Initiation: $RH + *O - O* \xrightarrow{Cahaya} R* + *OOH$

Peroxide formation : $R* + *O\text{-}O* \rightarrow ROO*$

$ROO* + RH \rightarrow ROOH + A*$

Propagation : $ROOH \rightarrow RO* + OH*$

Termination : $RO* + X \rightarrow$ inactive product

The addition of antioxidants can stop oxidation reactions because the stable energy compounds are accommodated by antioxidant. The following example shows the mechanism of adding an antioxidant to the oil oxidation process:

$R* + AH \rightarrow RH + A*$

$ROO* + RH \rightarrow ROOH + R$

Notes:

ROO* : Active peroxide
RH : Oil or fat
AH : Antioxidant
R* : Unsaturated fatty acids

The addition of antioxidant (AH) with low lipids concentration can prevent oil oxidation reactions that block the oxidation reaction at the initiation and propagation stages [5].

Raja banana (*Musa paradisiaca sapientum*) is included in *Musaceae* family that has many benefits, not only the flesh but also the peel. Banana peel has high nutrition especially for *Raja* banana peel [1]. Table 3 and 4 show the nutrition, vitamin, and antioxidant consisted in Raja banana peel. Meanwhile, Table 5 shows the test result of antioxidant content in Raja banana peel extract.

Table 3 Nutrition in Banana Peel

Parameter	Concentration (%)
Organic material	91.50 ± 0.05
Protein	0.90 ± 0.25
Crude lipid, Carbohydrate, Crude fiber	1.70 ± 0.10
Organic material	59.00 ± 1.36
Protein	31.70 ± 0.25

Table 4 Vitamin and Antioxidant in Raja Banana Peel

Components	Concentration (%)
Vitamin E (mg/kg)	1.07 ± 0.06
Saponin (g/100 g)	0.12 ± 0.01
Flavonoids (mg/100 g)	5.27 ± 0.47

Table 5 Antioxidant in Raja Banana Peel Extract

Seconder Metabolite	Result (mgr/gr)
Flavonoid	90.56
Saponin	-

Generally, antioxidant compounds have a similar core structure covering the benzene ring that consists of hydroxide or amino [7]. Flavonoid is a natural antioxidant included in the phenolic compound cluster that can give one or more hydrogen atoms to free radicals caused by an oxidation process so that free radicals' activities can be oppressed [8].

3. Methods

This research began with analysing the quality of bulk frying oil regarding the acid, peroxide, and iodide numbers. Afterwards, making *Raja* banana peel extract was conducted by using the maceration method equipped with methanol solvent and ethyl acetate. Figure 1 shows the procedure of employing the maceration method while Figure 2 depicts free radicals absorption utilizing the addition of a natural antioxidant.

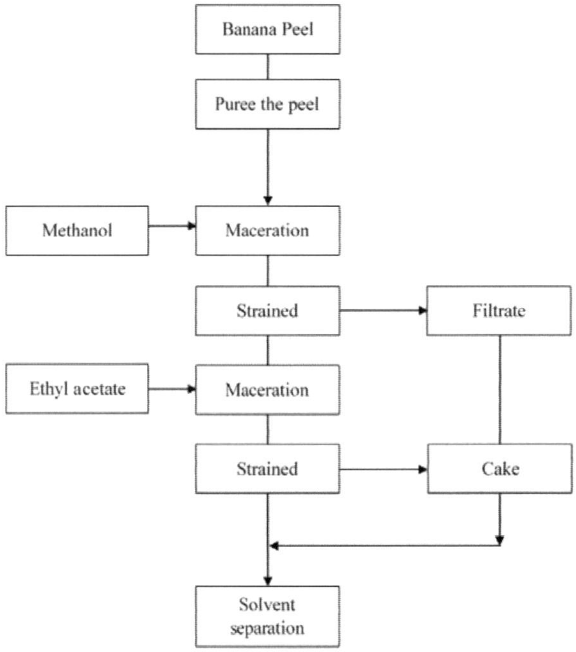

Figure 1 Block Diagram of Banana Peel Extraction

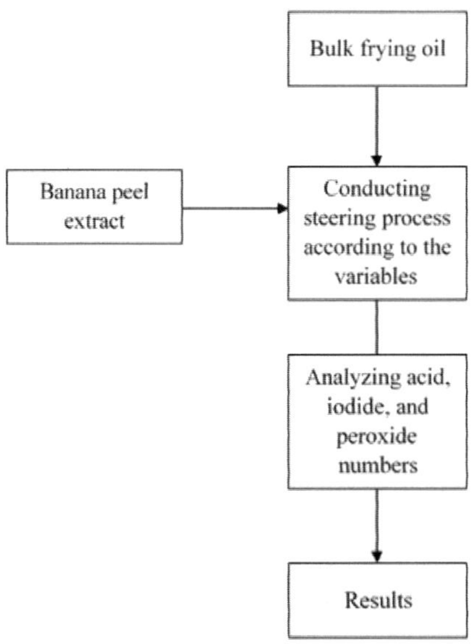

Figure 2 Block Diagram of Free Radicals Absorption using the Addition of Natural Antioxidant

4. Results And Discussion

Revealing acid, peroxide, and iodide numbers were significant to investigate the quality of frying oil. This research aimed at examining how *Raja* banana peel was able to regenerate the quality of bulk frying oil since it contained flavonoid as many as 90.56 mgr./gr. [3].

Table 6 Analysis Results of Bulk Frying Oil

Parameter	SNI	Result
Acid number	Max. 0.3 meq/kg	0.71836
Iodide number	<5 meq/kg	1.76192
Peroxide number	Max. 2 meq/kg	2.12660

In the beginning, this research made *Raja* banana peel extract and analysed bulk frying oil to reveal whether its quality met the SNI 3555-1998 or not. Table 6 shows that the analysis results of bulk frying oil did not achieve the standardised quality of SNI 3555-1998 so that the addition of antioxidant was necessary to regenerate the quality.

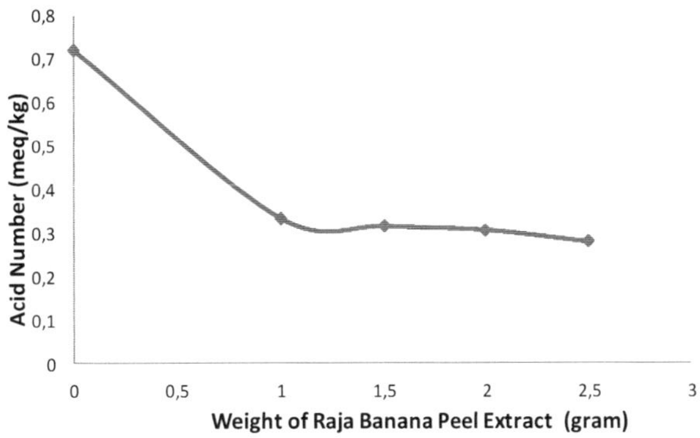

Figure 3. The relationship between the weight of Raja banana peel extract with the acid number

The acid number in the bulk frying oil showed that the oxidation process caused the present free fatty acids. By the high acid number contained, the frying oil's quality decreased or was damaged. According to Table 6, the acid number of the bulk frying oil remained unstandardized namely SNI + 0.71836 meq/kg. After being given *Raja* banana peel oil amounted to 1 gram, the acid number decreased up to 0.333 meq/kg. Figure 3 confirms the more the peel given to the bulk frying oil; the more decreased the acid number. For instance, 2.5 grams of Raja banana peel extract could decrease 61.2% of the acid number.

Besides the acid number, it was also essential to look at the peroxide number. This number was to determine the degree of the damaged oil or the fat. Unsaturated fatty acids might bind oxygen in the double bond that produced peroxide. Figure 4 shows the relationship between *Raja* banana peel extract and the various exciting time variables.

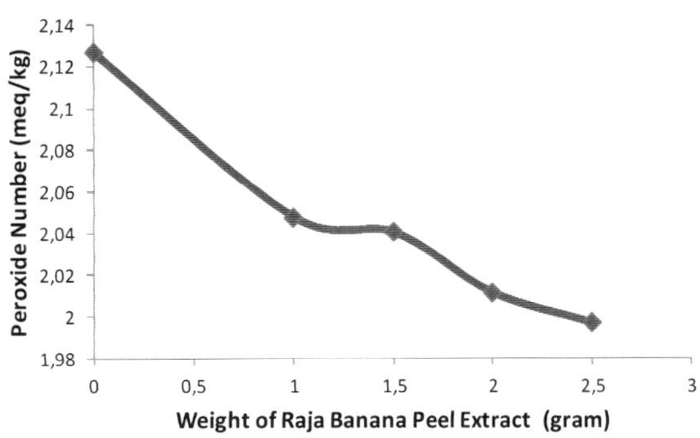

Figure 4. The relationship between the weight of Raja banana peel extract and peroxide number

Analysis result portrayed that the initial peroxide number of bulk frying oil before the addition of *Raja* banana peel extract was 2.12660 meq/kg. However, when 1 gram of Raja banana peel extract was added to the bulk frying oil, there was a decrease of the peroxide number amounting 2.0764 meq/kg. That is, the addition of *Raja* banana peel extract could decrease the peroxide number contained in bulk frying oil. The created peroxide compounds could be oppressed by using natural antioxidant from *Raja* banana peel extract. For instance, 2.5 grams of the extract could decrease the peroxide number from 2.12600 meq/kg to 1.9966 meq/kg. Thus, the decrease was about 6.24%.

The decrease of the acid and peroxide numbers were caused by the activity of antioxidant produced by *Raja* banana peel extract. Phenolic compounds contained in the extract could oppress the creation of free fatty acids caused by the oxidation process. The phenolic compounds also could prevent the oxidation process by using atom H that bonded peroxide groups, which resulted in more stable compounds.

5. Conclusion

The addition of Raja banana peel extract on bulk frying oil can enhance the oil quality. The best result is by giving 2.5 grams of *Raja* banana peel extract with the acid number 0.27938 meq/kg or decreased peroxide up to 1.997 meq/kg.

6. References

[1] Anhwange, B., Ugye, T. & T. Nyiaatagher. 2009. "Chemical Composition of Musa Sapientum (Banana) Peels". Electronic Journal of Environmental, Agricultural, and Food Chemistry.
[2] Chauhan, A., Anshita N., Kumud B., dan Yas S. 2016. "Comparative Study of Different Parts of Fruits of Musa Sp. on the Basis of their Antioxidant Activity". Scholar Research Library
[3] Eckey, S.W.1955."Vegetable Fat and Oil. Reinhold Publishing Corporation". New York.
[4] Gordon, I. 1994. "Functional Food, Food Design, Pharmafood". New York: Champman and Hall.
[5] Haryono. 2010. "Pengolahan Minyak Kelapa Sawit Bekas Menjadi Biodiesel Yogyakarta: Seminar Nasional Teknik Kimia "Kejuangan"
[6] Ketaren, S. 1986. "Minyak dan Lemak Pangan". Jakarta: Penerbit Universitas Indonesia (UI Press).
[7] Kuncahyo, I., dan Sunardi. 2007."Uji Aktivitas Antioksidan Belimbing Wuluh (Averrhoa Blimbi, L) Terhadap 1,1-Dipheyl-2-pichrylhidrazyl (DPPH)".Seminar Nasional Teknologi
[8] Pane, Elfira.2013."Uji Aktivitas Senyawa Antioksidan dari Ekstrak Metanol Kulit Pisang Raja (Musa Paradisiaca Sapientum)".Palembang: IAIN Raden Fatah
[9] Schuler P. (1990), "Natural Antioxidant Exploited Commercially", di Husdont BJF, Food Antioxidants, New York: Elsevier Applied Science
[10] Sheng, Zhan Whu et al. 2010. "Investigation of dietary fiber, protein, vitamin E and other nutritional compounds of the banana flower of two cultivars grown in China".
[11] Sitepoe, M. 2008. "Corat-Coret Anak Desa Berprofesi Ganda". Cet. 1. Jakarta: Kepustakaan Populer Gramedia.Hlm. 15-18.
[12] Standart Nasional Indonesia. 1998. "Cara Uji Minyak Dan Lemak". SNI 3555-1998

Image Compression and Encryption Using DCT and Gaussian Map

W M Rahmawati[1] and F Liantoni[1]

[1] Department of informatics Institut Teknologi Adhi Tama Surabaya

wenymistarika@gmail.com

Abstract. The use of the image in daily life keeps increasing as information technology is constantly developing. Consequently, the demand for a form of security in sending images through the internet is unavoidable. The encryption algorithm is one of the solutions. Encryption will make any image become only readable by those authorized to access. A chaos-based encryption algorithm is the most commonly used image encryption method because of its sensitivity to the parameter value. In addition to security, image transmission speed is also on demand. Images with bigger size than texts require longer transmission time. For this issue, image compression is the answer. There are many kinds of algorithm that can be used for image size compression. One of which is DCT. DCT algorithm is known to transform spatial domain to frequency. Image encryption and compression are two different processes. Nevertheless, this study combined both processes into one scheme so that they become inseparable. This study has two findings. First, in the area of compression, the scheme produced cipher image with considerable compression ratio. Second, in the area of encryption, the produced cipher image's histogram became completely different from the original histogram, indicating that the encryption algorithm used was immune to histogram analysis.

1. Introduction

Nowadays, most data transmission is done online or via the internet. The internet users begin to look for a way to improve the security and the speed of online data transmission. Many studies on data compression have been conducted [1-5]. All because smaller data size means less bandwidth needed, and the data can be transferred in a little amount of time. Many researchers also developed encryption techniques to ensure that data transferred online will not be accessible if the data ever fell to the wrong hands.

One type of data used by internet users is an image. The image has been widely used in many fields, such as medicine, transportation, military, etc. Moreover, some of those fields require high security when it comes to data transfer because most of their data are confidential. However, security alone is not enough. Data transfer is preferred to be quick as well.

Image data have different characteristics from text data. The image usually has a much bigger data size and high level of redundancy compared to text. Therefore, image compression is performed to reduce image data size, so that image's transfer takes less time. There are many algorithms available for image compression.

Image compression is classified into lossless compression and lossy compression. In lossless compression, the decrypted image has precisely the same components as the original image (before compression). That means there is no information lost from the original image. The examples of lossless compression algorithms are Huffman, LZW, etc. On the other hand, lossy compression produced a decrypted image that is different from the original image. There is information being discarded during the process[6]. However, the data being lost is not much and not recognized by naked eyes. Some examples of lossy compression algorithms are FFT, DCT, & DWT. DCT as a lossy compression algorithm, has the property of strong energy compaction. DCT can transform an image from spatial domain to frequency domain.

Image encryption is performed to ensure that unauthorized individuals can not access the image. An encryption algorithm is good if the key plays a significant role in the encryption and decryption process. This means that if a little mistake occurs when inputting the key, the produced decrypted image will be completely different from the original image. Many algorithms can be used for encryption. Some of those are DES, AES, SHA, chaos-based algorithm. In the field of image encryption, most researchers use a chaos-based method. A chaos-based method is considered as the most suitable method for image encryption because of the fact that image characteristics are different from text characteristics[7], [8]. Gaussian map as one of the chaos-based method types with a property of generating random numbers well. When it is combined with substitution and permutation, the result is an encryption method more resistant to hacking attempts[9].

Compression and encryption are two algorithms meant for different goals. Some research proposed some methods to combine compression and encryption so that both processes cannot be separable[10-13]. The proposed method is considered effective and saving more time. When the wrong hands obtain data, decompression cannot be done without decryption, and decryption cannot be done without decompression as well.

This research proposed a method of combining compression and encryption of image data. DCT algorithm was used for compression because of its strength and its convenience to compress data. A Gaussian map was used for chaos-based encryption because of its high sensitivity to an input parameter to produce random number. The combination of these methods will produce strong scheme of compression and encryption.

2. Literature Review

2.1 *Discrete Cosine Transform (DCT)*

Discrete Cosine Transform is one of lossy compression algorithm used to transform an image from spatial domain to frequency domain. DCT is mostly used in transforming image and video domain. DCT makes image or video into low and high-frequency form. With the limitations of humans' sight that is difficult to distinguish high frequency, the results of DCT having high frequencies can be suppressed. To suppress DCT results, a quantization process is required after the DCT process [14]. Quantisation is performed on the chrominance and luminance matrix.

Before performing compression, an image will be divided into 8x8 blocks. Therefore, DCT for 2 dimensions will be used with the formula that can be seen in equation (1).

$$F_k(u,v)=\frac{c(u)c(v)}{4}\sum_{i=0}^{7}\sum_{j=0}^{7} f_k(i,j)\cos\left(\frac{(2i+1)u\pi}{16}\right)\cos\left(\frac{(2j+1)v\pi}{16}\right) \quad \text{with} \quad c(e)=\begin{cases}\frac{1}{\sqrt{2}} & e=0\\ 1 & e\neq 0\end{cases} \tag{1}$$

$F_k(u,v)$ is the result of DCT in the (u,v) position, while $f_k(i,j)$ is the initial image pixel value in the (i,j) position before being transformed to DCT.

To restore the compression result using DCT, decompression is done by employing Inverse Discrete Cosine Transform function using equation (2).

$$f_k(i,j)=\frac{1}{4}\sum_{u=0}^{7}\sum_{v=0}^{7} c(u)c(v)F_k(u,v)\cos\left(\frac{(2i+1)u\pi}{16}\right)\cos\left(\frac{(2j+1)v\pi}{16}\right) \quad \text{with} \quad c(e)=\begin{cases}\frac{1}{\sqrt{2}} & e=0\\ 1 & e\neq 0\end{cases} \tag{2}$$

$f_k(i,j)$ is the IDCT result in the (i,j) position, while the DCT result in the $F_k(u,v)$ position is the result value of DCT in the (u,v) position.

2.2 *Huffman algorithm*

David A. Huffman firstly developed the Huffman Algorithm in 1952. The Huffman Algorithm is one of lossless compression because there is no lost information from its decompression results. This algorithm is widely used in compression since it is easy and straightforward by utilising the frequency of data occurrence. Frequently appear data symbols will be replaced by the shorter bits compared to appear data symbols rarely. The coding method in the Huffman algorithm is stored in a dictionary in the form of a tree called a Huffman tree. To decode the data resulted from the Huffman algorithm, the same Huffman tree for data compression is needed. That Huffman tree will give the meaning of the original symbol from the code used to decode.

2.3 *Chaotic Gaussian Map*

A chaotic map is an algorithm that is widely used in the encryption because it is difficult to guess despite its simplicity. A chaotic map will generate pseudorandom numbers which will be used for the encryption process. Pseudorandom results made by chaotic maps depend on the parameters input[9]. There are several chaotic maps can be used including Tent map, Circle map, Gaussian map, etc. To generate pseudorandom from Gaussian map, equation (3) can be used.

$$x_{n+1} = \exp(-\alpha x_n^2) + \beta \tag{3}$$

Where α and β is the input parameter that will significantly affect the Gaussian map results. A Gaussian map will generate some random values and form a sequence. This sequence will be used to randomise other sequences with the same length.

2.4 *SHA*

SHA is an encryption algorithm created by NIST. SHA has several versions, namely SHA-1, SHA-256, SHA-384, and SHA-512. The output of the SHA algorithm has the same length even though the input length is different. The maximum input and output length generated depends on the used SHA version. In SHA-1, with an input, less than 2^{64} bits long will generate a fixed period of 160 bits[15]. SHA Algorithm is widely used because it has high sensitivity. A little difference in the input will generate a completely different ciphertext result although with the same length. SHA algorithm is a reversible encryption algorithm so that there is no possibility to return the input value generated by SHA. Because of these properties, this algorithm is widely used as a part of a very secure encryption algorithm.

3. The Method Used

In the method section, it was divided into two parts, namely the compression and encryption section as well as the decompression and decryption section. In the first section, the scheme of the process to carry out a combination of compression and encryption will be explained. The input of this algorithm was an uncompressed grayscale image. The steps to do a full image compression and encryption combination are explained as follows:
Step 1: the input image was divided into 8x8 blocks. Each block through the compression process with DCT using equation (1). DCT results were then quantised. The quantisation was carried out by dividing the DCT results with chrominance and luminance matrix. The quantisation results in the form of blocks were made sequences by reading it in an indirect way so that each block generated a sequence with 64 lengths. The first three values of sequence in each block were combined to form sequence A. While the rest was combined to form sequence B.

Step 2: Sequence A was combined with all of the used keys. In this algorithm, the key consisting of $\alpha, \beta, \alpha', \beta', a, b.$ Sequence A that has been combined with the keys became input from SHA-1. In addition, sequence A was also carried out by diffusion process to increase the sensitivity. To do the diffusion process in sequence A, equation (4) was used. While sequence B was encoded using the Huffman algorithm.

$$v_1 = cyc\left[\left(t_1 + v_{i-1} + t_{i+1}\right)\bmod\left(2^{LD} - 1\right), LSB_y\left(v_{i-1} + t_{i+1}\right)\right] \quad (4)$$

Where,

$$y = \begin{cases} 1, & LD \le 3 \\ 3, & 3 < LD \le 7 \\ 7, & 7 < LD \end{cases}$$

Where t_1 is DCT value, v_1 is the result of diffusion, $cyc[x, y]$ turned the sequence x to the right as many as y the bit. $LSB_y(s)$ is the least y value significant bit off s, while LD is a representation of the number of DCT bit.

Step 3: The result of SHA in step 2 was 160 bits used to disrupt the value of Gaussian map input parameter. Initially, 160-bit SHA result was divided into five equal lengths of d_0 to d_4. Furthermore, the Gaussian input parameter was changed by SHA result with equation (5) and (6).

$$\alpha' = \left(\alpha + \left(d_0 + d_1 + U\left(d_2\right)\right) / 2^{32}\right) \bmod 1 \quad (5)$$

$$\beta' = \left(\beta + \left(L\left(d_2\right) \oplus d_3 \oplus d_4\right) / 2^{32}\right) \bmod 1 \quad (6)$$

Where $U(A)$ are the upper 16 bits and $L(A)$ are the lower 16 bits.

Step 4: Doing Gaussian map iteration using α' and β' resulted from step 3 as the input parameter. The result of the Gaussian map was a random sequence used to mask the results of Huffman encoding resulted in step 1. The masking process can be done using the following equation:

$$c_i = \left(r_i + m1_{c_{i-1} + r_{i+1} \bmod L} + c_{i-1} + r_{i+1}\right) \bmod 2^{32} \quad (7)$$

Where ml is the result of the Huffman algorithm and m1 is the sequence of Gaussian map results.

Step 5: The diffusion results of sequence A from step 2 was mutated using cat map. The mutated result encoded using the Huffman algorithm. Next, as in step 4 generated the new Gaussian value to be masked with Huffman Algorithm result using equation (7).

Step 6: Combining the masking results in step 4 and step 5 as a cipher image.

Meanwhile, to carry out decompression and description, the steps were the opposite of step 1 to 6.

4. Result and Analysis

The results of the program done were cipher images in the form of random images in which the original form cannot be recognized, while the input was an uncompressed grayscale image. As can be seen, Figure 1.a is the example of the input image, while Figure 1.b is the result of the chiper image.

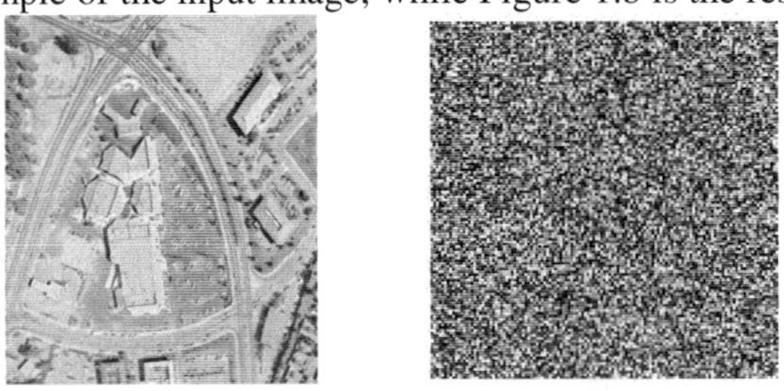

(a) (b)

Figure 1. (a) Input image (b) Cipher image

The analysis section was divided into the compression process and part of encryption. In the compression section, the result of compression can be seen from the compression ratio formulated in equation (8).

$$ComprRatio = \frac{SizeofPlain\,\mathrm{Im}age}{SizeofChiper\,\mathrm{Im}age} \qquad (8)$$

The higher the compression ratio, the better the compression performance. The result of compression ratio calculation can be seen in table 1. From table 1, it can be seen that the cipher image generated, on average, had four up to five times smaller than the original image.

Table 1. The result of image compression ratio

Image name	Compression ratio	Image name	Compression ratio
Aerial	3.90793083	Bridge	4.1282519
Alarm	5.31948052	Goldhill	4.5893
Baboon	4.16101587	Pepper	4.78784
Barb	4.82876510	Pirate	4.513498
Bird	5.67116649	Zelda	5.178255

While from the encryption section, it was analysed using histogram analysis test. It can be seen in Figure 2 the histogram of the original result was completely different from the histogram of the cipher image and did not have a pattern similar to the histogram of the original image. The histogram of the cipher image that did not have pattern in which the original image was unpredictable.

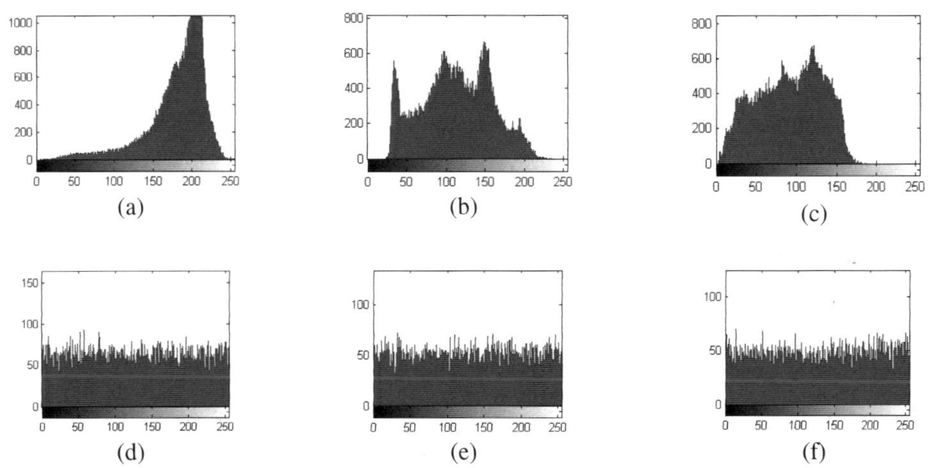

Figure 2 (a) histogram of original Aerial image (b) histogram of original Barb image (c) histogram of original Zelda image (d) histogram of cipher Aerial image (e) histogram of cipher Barb image (f) histogram of cipher Zelda image

5. Conclusion

From the conducted research on image compression and encryption put in on the scheme, the compression carried out using DCT resulted in excellent compression ratio that was 4 to 5 times smaller than the original size. While the chaos-based encryption, Gaussian map was carried out resulted in cipher image that withstood the histogram analysis attack since it resulted in cipher image in which the histogram had a completely different pattern with the original image histogram.

References

[1] A. Chu, "LZAC lossless data compression," in *Proceedings DCC 2002. Data Compression Conference*, 2002, p. 449.

[2] F. Rashid, A. Miri, and I. Woungang, "Secure image deduplication through image compression," *J. Inf. Secur. Appl.*, vol. 27–28, pp. 54–64, Apr. 2016.

[3] H. Wu, X. Sun, J. Yang, W. Zeng, and F. Wu, "Lossless Compression of JPEG Coded Photo Collections," *IEEE Trans. Image Process.*, vol. 25, no. 6, pp. 2684–2696, Jun. 2016.

[4] M. Iwahashi, H. Kobayashi, and H. Kiya, "Lossy compression of sparse histogram image," in *2012 IEEE International Conference on Acoustics, Speech and Signal Processing (ICASSP)*, 2012, pp. 1361–1364.

[5] S. K. Roy, S. Kumar, B. Chanda, B. B. Chaudhuri, and S. Banerjee, "Fractal image compression using upper bound on scaling parameter," *Chaos, Solitons & Fractals*, vol. 106, pp. 16–22, Jan. 2018.

[6] I. M. Pu, *Fundamental Data Compression*, 1st ed. Oxford: Butterworth-Heinemann, Elsevier, 2006.

[7] E. Yavuz, R. Yazıcı, M. C. Kasapbaşı, and E. Yamaç, "A chaos-based image encryption algorithm with simple logical functions," *Comput. Electr. Eng.*, vol. 54, pp. 471–483, Aug. 2016.

[8] Y. Zhou, L. Bao, and C. L. P. Chen, "A new 1D chaotic system for image encryption," *Signal Processing*, vol. 97, pp. 172–182, Apr. 2014.

[9] A. Sahay and C. Pradhan, "Gauss iterated map based RGB image encryption approach," in *2017 International Conference on Communication and Signal Processing (ICCSP)*, 2017, pp. 0015–0018.

[10] C.-H. Yuen and K.-W. Wong, "A chaos-based joint image compression and encryption scheme using DCT and SHA-1," *Appl. Soft Comput.*, vol. 11, no. 8, pp. 5092–5098, Dec. 2011.

[11] Y. Zhang, B. Xu, and N. Zhou, "A novel image compression–encryption hybrid algorithm based on the analysis of sparse representation," *Opt. Commun.*, vol. 392, pp. 223–233, Jun. 2017.

[12] M. Brindha and N. Ammasai Gounden, "A chaos-based image encryption and lossless compression algorithm using hash table and Chinese Remainder Theorem," *Appl. Soft Comput.*, vol. 40, pp. 379–390, 2016.

[13] H. Zhu, C. Zhao, and X. Zhang, "A novel image encryption–compression scheme using hyper-chaos and Chinese remainder theorem," *Signal Process. Image Commun.*, vol. 28, no. 6, pp. 670–680, Jul. 2013.

[14] D. Salomon and G. Motta, *Handbook of Data Compression*, Fifth. New York: Springer, 2010.

[15] W. Stallings, *Cryptography and network security : principles and practice*, Fifth. Prentice Hall, 2011.

Acknowledgement

The authors would like to thank Institut Teknologi Adhi Tama that has given the facility to conduct this research. The authors would also like to thank the Directorate General of Higher Education that has provided a chance and financial support so that this research can be well-conducted.

IOP Conf. Series: Materials Science and Engineering **462** (2019) 012036 doi:10.1088/1757-899X/462/1/012036

Inventory and Transportation Models for Multi-item Single-Supplier through Purchasing Consortium for the Fishpond Manager

Nur Rahmawati[1], Ika Widya Ardhyani[2] and Sinta Dewi[3]

[1]Industrial Engineering Department, Institut Teknologi Adhi Tama Surabaya, Jl. Arief Rahman Hakim 100, Surabaya, Indonesia
[2]Industrial Engineering Department, Universitas Maarif Hasyim Latif Sidoarjo, Indonesia
[3]Jurusan Teknik Industri, Institut Teknologi Telkom Surabaya, Indonesia

rahmawatinur1987@gmail.com

Abstract. Purchasing Consortium (PC) is the merger of two or more organisations with separate legal entities or independently or through the third party using division and union of the purchase volume, information, and resources owned. This merger aims to increase the bargaining power so that the product or service purchase price can be lower than a supplier. This research tried to conduct modelling to minimise the annual total cost of several buyers doing multi-item through PC at single supplier offering a fixed discount rate by shipping union. In this study, an analytical method was employed to solve the problems using the optimisation approach. From numerical experiment results, by making purchases through PC with shipping unification, the total cost generated was better compared to the purchase without going through PC or PC without shipping union.

1. Introduction

Purchasing Consortium (PC) is a strategy in purchasing in which the merger of two or more organisation with separate legal entities either independently or through the third party [4]. The merger was done using division and union of the purchase volume, information, and resources owned [7]. By doing the union, it is expected that the bargaining power can be increased so that the benefits from a supplier can be obtained [1].

There are several benefits obtained by purchasing through PC. These benefits include reducing the cost of purchasing products, reducing the risk of shipping, reducing the transaction cost and increasing productivity, flexibility of inventory, better access to the resources or markets, providing high quality products, reducing logistics cost, workload and transaction cost, establishing better relationship with the supplier through commitment to make contracts, sharing experience and information in order to improve product and service quality [9]. In addition to these benefits, there are other benefits from PC establishment including the maximisation of existing expert staffs [2] and a decrease in shipping lead time [3].

Gresik is one of the areas in East Java where most of the population earns a living as farmers, especially in Duduk Sampean District. The fishpond managers always do the activity of purchasing fertiliser, especially urea and TSP as well as other medicines. However, usually, they still purchase individually in small quantities. Therefore, this research compiled inventory and transportation model to minimise the annual total cost from each member on multi-item purchases with fixed discount scheme [6]. The form of PC adopted from research [7] and [5] was first buying in concentration phase. In this type of PC membership, all members got the same work portion. Analytic method was employed to generate a minimum annual total cost. The components of value analysed including the cost of purchasing, ordering, storage, PC operating, and transportation.

2. Method

In Figure 1, the conceptual model used in this research is presented. Some buyers who need similar items formed PC to be able to meet the discount quota set by the supplier (K). Every j-th buyer had i-th demand item of R_{ij}. Consortium collected the demand for each piece for all buyers. After that, the total purchasing quota of i-th item (Q_{ij}) was calculated to be ordered from the supplier. Since the purchase was made in large quantities, the discount quota from a supplier can be fulfilled. The supplier gave a discount of r from the total purchase. The price for i-th item from supplier is P_i. The i-th item proportion given by the j-th buyer is Q_{ij} with the cost of i-th item is P_i. Also, each buyer received saving obtained from the collaboration based on each buyer contribution. Item shipping was done separately.

Notation

The notations used in this research model include:

P_i	= Item cost per unit of purchase for items i-th
R_{ij}	= Annual demand of i-th item of buyer j
C_{od}^{PCS}	= Ordering cost from PC to supplier
C_{od}^{RPC}	= Ordering cost from buyer to PC
C_{od}^{RS}	= Ordering cost from buyer to supplier
C_{op}^{PC}	= PC operating costs

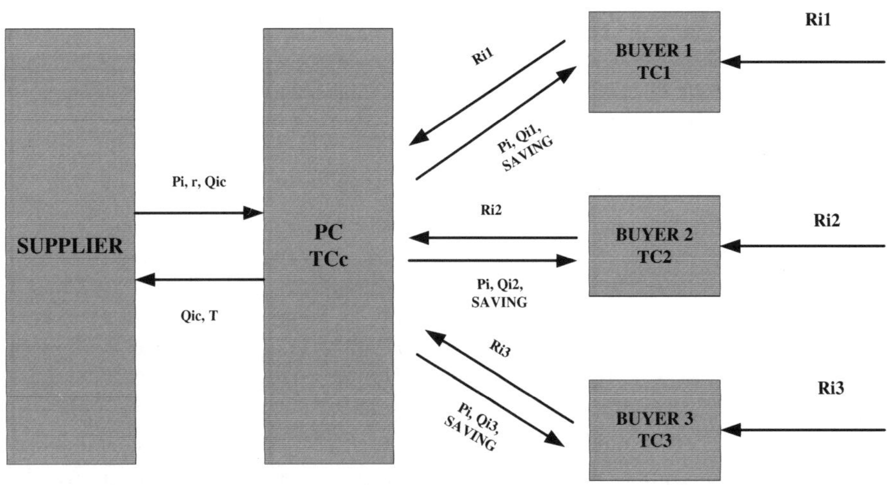

Figure 1. Conceptual model of single supplier multi-item purchase on PC

Q_{ic}	= Order quantity in units per year for i-th item by PC
Q_{ij}	= Order quantity in units per year for i-th item by PC
F	= Storage cost fraction per year
m	= R_i/Q_i = Order frequency in one year
T	= 1/m = Time between orders
d_{oj}	= Distance from supplier to j-th buyer without shipping union
n	= Number of items
l	= Number of buyers
K	= Minimum number of purchases with discount from supplier in IDR

nrd_j	= Vehicle rental duration (days) on the j-th buyer
r	= Discount offered by the supplier for purchases nominal ≥ K
ϕ_{j_beli}	= Saving obtained by j-th buyer due to joining PC in making a purchase
C_{tr}	= Transportation cost per distance unit
C_{rent}	= Vehicle rental fee

2.1 Assumption

The assumptions used in this research model are:

1. The relationship of purchasing quota increase with the amount of discount obtained was assumed proportional (Fixed discount rate)
2. All PC members were expected to make an order on all types of product at the same time
3. Transportation costs are considered beared entirely by the buyer
4. One supplier was supplying all items to all buyers
5. Backorder is not allowed
6. Items bought from the supplier were directly sent to the buyers
7. The capacity of transport vehicle was ignored

2.2 Mathematical Model

2.2.1 Purchasing without going through PC. The following joint replenishment model is derived from the EOQ single item concept [8].

The total costs for Buyer j as follows:

TC_j = Purchase cost + ordering cost + storage cost + transportation cost

$$TC_j(m) = \sum_{i=1}^{n} P_i R_{ij} + C_{od} m + \frac{F \sum_{i=1}^{n} P_i R_{ij}}{2m} + m\big((ndr_j * C_{rent}) + 2C_{tr} d_{0j}\big) \quad (1)$$

By lowering TC toward m, the optimum order frequency for each buyer will be obtained are as follows:

$$\frac{\partial TC}{\partial m} = 0$$

$$C_{od} - \frac{F \sum_{i=1}^{n} P_i R_{ij}}{2(m^*)^2} + \big(C_{rent} + 2C_{tr} d_{0j}\big) = 0$$

$$m^* = \sqrt{\frac{F \sum_{i=1}^{n} P_i R_{ij}}{2\big(C_{od} + (ndr_j * C_{rent}) + 2C_{tr} d_{0j}\big)}} \quad (2)$$

By inputting m* to the $TC_j(m)$, the maximum total cost of each buyer will be obtained. The limitation used in this scenario is:

$$\sum_{i=1}^{n} P_i \frac{R_{ij}}{m} < K \quad untuk\ semua\ j$$

$$2\big(C_{od} + C_{rent} + 2C_{tr} d_{0j}\big) > 0$$

The first limitation is the discount quota limitation. Equation 1 and 2 are only applicable if the total item purchase is less than K. While the second limitation is used to guarantee that equation 2 has a result.

Total Cost of the system is:

$$TC_S(m) = \sum_{j=1}^{l} TC_j(m) \quad (3)$$

2.2.2 Purchasing through PC. The following joint replenishment model is derived from EOQ single item concept [8].

IOP Conf. Series: Materials Science and Engineering **462** (2019) 012036 doi:10.1088/1757-899X/462/1/012036

The total costs experienced by the entire system include:
TCs = Purchase Cost + Ordering Cost + PC Operational Cost+ Storage Cost + Transportation Cost

$$TC_S = (1-r)\sum_{i=1}^{n}\sum_{j=1}^{l} P_i R_{ij} + (C_{od}^{PCS} + C_{od}^{RPC})m + C_{op}^{PC} + \left(\frac{F(1-r)\sum_{i=1}^{n}\sum_{j=1}^{l} P_i R_{ij}}{2m}\right) + m\left((\sum_{j=1}^{l} ndr_j * C_{rent}) + C_{tr}\sum_{j=1}^{l} 2d_{0j}\right) \quad (4)$$

By derivated TC toward m, the optimum order frequency on the PC will be obtained as follows:

$$\partial TC/\partial m = 0$$

$$(C_{od}^{PCS} + C_{od}^{RPC}) - \frac{F(1-r)\sum_{i=1}^{n}\sum_{j=1}^{l} P_i R_{ij}}{2(m^*)^2} + \left((\sum_{j=1}^{l} ndr_j * C_{rent}) + 2C_{tr}d_{0j}\right) = 0 \quad (5)$$

$$m^* = \sqrt{\frac{F(1-r)\sum_{i=1}^{n}\sum_{j=1}^{l} P_i R_{ij}}{2(C_{od}^{PCS} + C_{od}^{RPC} + (\sum_{j=1}^{l} ndr_j * C_{rent}) + C_{tr}\sum_{j=1}^{l} 2d_{0j})}} \quad (6)$$

Condition:

$$2(C_{od}^{PCS} + C_{od}^{RPC} + (\sum_{j=1}^{l} ndr_j * C_{rent}) + C_{tr}\sum_{j=1}^{l} 2d_{0j}) > 0 \quad (7)$$

$$\sum_{i=1}^{n} P_i \frac{R_{ic}}{m} \geq K \quad (8)$$

Equation 8 is a discount quota limit. Since the purchase was made through PC, the total purchase by the PC can meet the discount quota from the supplier (K). While equation seven was used to guarantee that equation 6 has a result.

2.3 Data and Parameter

The following is the parameter and data used in the model

Table 1. Buyer Data

i-th Buyer	1			2			3		
i-th Item	1	2	3	1	2	3	1	2	3
Item price (Pi)	130.000	110.000	130.000	130.000	110.000	130.000	130.000	110.000	130.000
Demand (Rij)	100	100	15	12	9	3	6	6	-

In this research, there were three buyers making purchases on the same item with different needs as can be seen in Table 1. The parameter data used in this model can be seen in Table 2.

Table 2. Parameter Model Data

Storage cost fraction per item (F)	0.3	Vehicle rental fee (Crent)	100,000
Ordering cost from buyer to supplier (Cod_BPC)	25,000	Transportation cost per Km distance (Ctr)	10,000
Ordering fee from PC to supplier (Cod_PCS)	25,000		
The percentage of discount from the supplier	1.5%		

3. Result and Discussion

From the numerical experiments that have been conducted, the following results are obtained:

Table 3. Numerical experiment results

Scenario		m	BD	BPS	BIS	BO	Bop	Total Cost
0	Buyer 1	3.91	898,612	25,950,000	996,287	97,675	-	33,441,939
	Buyer 2	1.34	295,161	2,940,000	328,702	33,541	-	
	Buyer 3	0.94	207,602	1,440,000	230,981	23,379	-	
1	PC	2.93	1,968,947	29,875,050	1,529,450	146,499	500,000	34,019,945
2	PC	4.00	920,129	29,875,050	1,120,157	200,028	500,000	32,615,364

From the results of numerical experiments in table 4, the results show that in cases in which the number of buyers is three people, the purchase made without going through PC (Scenario 0) have a smaller total cost compared to the purchase made through PC without shipping union (Scenario 1). However, the purchase through PC with shipping union (Scenario 2) has the smallest total cost compared to the other two scenarios.

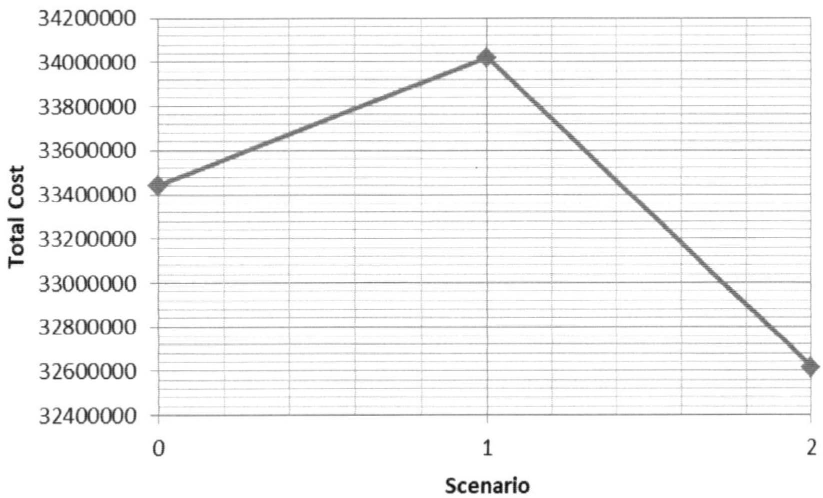

Figure 2. Cost Comparison in Three Scenarios

4. Conclusion

The conclusions that can be drawn from this study include: In this research, it was found the importance of coordination regarding inventory, ordering, and transportation policy between several buyers who make a multi-item purchase at the same supplier by sharing information to minimise the total annual cost of each buyer. In the case of a multi-item at the single supplier in three buyers offered fixed discount scheme, scenario two by joint in both purchasing consortium and transportation showed a smallest total cost.

5. References

[1] Bloch, R. E., Perlman, S. P., & Brown, J. S. (2006). An analysis of group purchasing organizations' contracting practices under the Antitrust Laws: Myth and Reality. White Paper. Mayer, Brown, Rowe & Maw.

[2] Essig, M. (2000). Purchasing consortia as symbiotic relationships: developing the concept of "consortium sourcing". European Journal of Purchasing & Supply Management, 6(1), 13-22.

[3] Ghaderi, H., & Leman, Z. (2013). Horizontal collaboration in *purchasing*: A successful case from small and medium enterprises (SMEs). African Journal of Business Management, 7(10), 750-753

[4] Huber, B., Sweeney, E., & Smyth, A. (2004). Electronic *Purchasing* Consortia: a Procurement Direction for the Future?.

[5] Nollet, J., & Beaulieu, M. (2003). The development of group purchasing: an empirical study in the healthcare sector. Journal of Purchasing and Supply Management, 9(1), 3-10.

[6] Rahmawati, N., Baihaqi, I., and Widodo, E., 2017. Pengembangan model persediaan dan transportasi untuk pembelian multi item pada supplier tunggal pada purchasing consortium. *Proceeding SENDI_U*.

[7] Schotanus, F. (2007). *Horizontal cooperativepurchasing*. University of Twente.

[8] Tersine, R. J. (1994). Principles of inventory and materials management.

[9] Yu, W. (2012). *Cooperative purchasing* in SME's: evidence from China's retail sector.

Acknowledgement

The authors would like to thank DRPM of Higher Education that has sponsored this research. In addition, the author would also like to thank the parties who helped in conducting this research.

IOP Conf. Series: Materials Science and Engineering **462** (2019) 012037 doi:10.1088/1757-899X/462/1/012037

The Effect of variations in PVD installation distance and thickness of soft soil layer for the degree of consolidation and time of consolidation at Gunung Anyar, Surabaya

M K Wardani[1] and Naufan[2]
[1]Lecturer, Department of Civil Engineering, Institut Teknologi AdhiTama, Surabaya, Indonesia
[2]Undergaduate Student, Department of Civil Engineering, Institut Teknologi AdhiTama, Surabaya, Indonesia

milakusuma@itats.ac.id

Abstract. Gunung Anyar is one of the areas in East Surabaya which has quite a high indication of the spread of soil clay with The combined Cv is 0.0008 - 0.000125 cm^2/s , which indicates long a consolidation time. The appropriate soil improvement method to speed up the consolidation time is to use a combination of preloading and prefabricated vertical drain. Simple statistical analysis was performed to obtain prefabricated vertical drain's thickness and length in general is 3.33 mm and 100 mm. The analysis consists of the effect PVD installation distance which is 1 m, 1.5 m and 2 m of triangular and rectangular patterns and variations thickness of soft soil clay are 10 m, 15 m, and 20 m. At the same installation distance with different thickness of soft soil layer, the changes of degreethe of consolidation was only <1%. Triangular pattern with an installation distance of 1 m PVD soft soil thickness 20 m, with 90% degree of consolidation the time requirement is 11 – 25 weeks. if soft soil thickness same and the difference in installation distance of 1 m, degreethe of consolidation decreased by 18.32% - 51.05% triangular pattern and 23.75% -45.69% rectangular pattern.

1. Introduction

Zonification of the soft soil in East Surabaya obtained the thickness of the soft soil clay layer ranging 6.5-21 m with combined Cv values ranging between 0.000168 - 0.00228 cm^s/s. Gunung Anyar, a part of East Surabaya has combined CV value of 0.0008 - 0.000125 cm^s/s.[1] That has indicates long consolidation time , therefore soil improvement method that is widely used to speed up the consolidation time is using a combination of vertical drain and precompression. Vertical drain can be divided into two, namely prefabricated vertical drain (PVD) and sand column [2]. Compression time for varying PVD installation distances with a triangular pattern indicating that the thickness of the soft soil layer (Hdr) has no significant effect on the distance of PVD installation that is close together [1].

Apart from that, many designers use PVD from various suppliers where each product has a different thickness and length. The depth and length of the PVD used can affect the target time for achieving the degree consolidation. Therefore, it can be calculated to obtain the depth and length of PVD in general from some products from some PVD's Suppliers using simple statistical analysis. Then the calculation of consolidation time is based on 90% the degree consolidation with variations in distance and thickness of soft soil.

An analysis of the effect of variations installation distance on numerous variations of soft soil was carried out at Gunung Anyar. The vertical drain installation pattern used is a rectangular and triangular pattern. Changes in soft soil thickness based on zonification are 5 m, 10 m, and 20 m while variations in PVD installation distance are 1 m, 1.5 m and 2 m.

2. Prefabricated Vertical Drain (PVD)

The PVD is a thin, synthetic drainage element comprised of a drainage core wrapped in a geotextile filter [3]. Prefabricated vertical (PV) drains are commonly used to decrease the drainage path within soft soils to accelerate the time of primary consolidation [4]. Prefabricated vertical drains are displacement drains of a small volume that exhibits considerably with fewer disturbances to the soil mass than the displacement sand drains [5]. Excess pore-water pressures, created by preloading, During the compression and rearrangement of the soil structure, the excess pore pressures were maintained at higher levels similar [6].

The degree of consolidation obtained from pore pressures (Up) is consistently less than that from settlements (Us)[7]. Soil improvement methods using PVD are assumed to be slightly disturbed depending on the sensitivity of the soil [8]. This influence is referred to as smear zone PVD which is affected by the installation pattern of PVD, horizontal permeability coefficient, vertical permeability coefficient[9].

$$F_n = \ln\left(\frac{D}{d_w}\right) - \frac{3}{4} \tag{1}$$

Where d_w is the PVD equivalent diameter or in the specification is referred to thick (m) which is affected pattern and D is the installation distance. The duration of consolidation in the design of the vertical drain depends on the value of Ch whereas specified in Table 1.

Table 1. Ch / Cv value comparison parameter [10]

Characteristics of soil clay layers	C_h/C_v
Relatively homogeneous (almost no permeable layers)	1 – 1.5
Sediment clays (there are non-continuous lenses and sand layers)	2 – 4
Clay-coated *(varved day)* or clay with a layer of sand that is more or less continuous	3 – 15

The formula to calculate the average degree of consolidation for vertical water flow and radial flow use PVD are as follows [11]

$$U = 1 - [(1 - Uv)(1 - Uh)] \tag{2}$$

with a Uv value based on

$$U_v = \frac{\left(\frac{4T_v}{\pi}\right)^{0,5}}{\left[1+\left(\frac{4T_v}{\pi}\right)^{2,8}\right]^{1,79}} \tag{3}$$

$$U_h = 1 - \exp\left[\frac{-8\frac{c_h t}{D^2}}{F_n}\right] \tag{4}$$

Calculation prediction method for Ugab which is widely used in PVD planning is the elasto method - flexible finite element, that the results obtained are approximate results that are almost as real as in the implementation. [10]. Calculations for Uv and Uh values are explained in the following equation:

$$U_v = 2\sqrt{\frac{Tv}{\pi}} \times 100\% \tag{5}$$

$$U_h = \left[1 - \left(\frac{1}{e^{\left(\frac{t\times 8\times C_h}{D^2\times 2\times F_n}\right)}}\right)\right] \tag{6}$$

3. Results and Calculations

3.1 Calculation of vertical drains width and length

The collection PVD products used in this paper by collecting brochures from suppliers. Supplier selection is carried out directly from local and imported suppliers. Table 2. shows some examples of vertical drain specifications from several distributors that are distinguished by width and length, where the most differential specifications are the width range of each product.

Table 2.PVD products which are mostly used in a construction project

	Distributor	Product name	Width (mm)	Length (mm)
1	PT.Teknindo Geosystem	Vertical Wick Drain	3,0-5,0	100
2	HB Wick Drains	Membra Drain MD-xx	2,4-3,4	97-98
3	USDS Drainage	USDS Drainage 74	3,3	102
4	Sunzo	High-Performance PVD	4	100
5	Sunzo	PVD SPB (China Standard)	3,5-5,0	100
6	DAEHAN i. m.	PVD (wick drain)	3,0-4,5	105
7	Tencate	Alidrain PVD	3,0-5,0	100

From Table 2, A Simple statistical analysis is performed by grouping data into six classes and 0.4 interval classes. The results of analytical processing for PVD thickness are given in Table 3.

Table 3. Processing of PVD thickness data group statistics

data	frequency	fkk
2.4 – 2.8	1	1
2.9 – 3.2	3	4
3.3 – 3.7	3	7
3.8 – 4 1	1	8
4.2 – 4.5	1	9
4.6 – 5 0	3	12
Total	12	

Table 3 shows that the minimum thickness that often occurs in some PVD products is 3.3 - 3.7 mm and the overall vertical drain length based on Table 2 is 100 mm. The median value of the PVD thickness data in Table 3 is 3.36 mm, and the mean value is 3.74 mm.

3.2 Calculation of time consolidation with PVD

The calculation of time 90% degree of consolidation used $Tv_{90\%}$ value = 0848 [13]. The calculation with a triangle and rectangle installation pattern with variations in distance (D) and thickness of the soft soil layer (Hdr) based on theory [12] and [9]. The Ch / Cv ratio assumption was 1.5 assuming a homogeneous clay soil layer. The target time used in the calculation analysis was limited to 30 weeks.

Table 4. Time consolidation with 90% degree of consolidation

	$t\ (U_{gab} = 90\%)$ (week)											
	Hansbo						Terzaghi					
Hdr	10 m		15 m		20 m		10 m		15 m		20 m	
Installation pattern	Rect.	Tri.	Rect.	Tri.	Rect.	Tri.	Rect.	Tri.	Rect.	Tri.	Rect.	Tri.
S = 1.0 meter	25	30	25	>30	25	>30	13	11	13	11	13	11
S = 1.5 meter	>30	>30	>30	>30	>30	>30	>30	28	>30	29	>30	29
S = 2.0 meter	>30	>30	>30	>30	>30	>30	>30	>30	>30	>30	>30	>30

In general, if the PVD installation distance is the same as the thickness of soft soil, time for degree consolidation is the same. Table 4 shows that a distance of 1 m with variations in soft soil thickness was faster in triangular patterns with a time between 11-25 weeks while 13-30 weeks in rectangular patterns. Whereas with the distance installation PVD 2 m with all variation soft soil layers to gain 90% degree of consolidation need more than 30 weeks

3.3 Effect of distance on the degree of consolidation

An analysis of the degree of consolidation based on variations distance and pattern of PVD installation on the thickness of soft soil clay layer (Hdr) has been done from the time target planning which was 30 weeks.

Table 5. The degree of consolidation with variations distance PVD and Soft soil layer

The degree of Consolidation (%)												
	Hansbo						Terzaghi					
Hdr	10 m		15 m		20 m		10 m		15 m		20 m	
Installation pattern	Rect.	Tri.	Rect.	Tri.	Rect.	Tri.	Rect.	Tri.	Rect.	Tri.	Rect.	Tri.
S = 1.0 meter	90.66	89.83	90.24	89.33	90.03	89.07	91.14	90.64	90.86	90.38	90.72	90.24
S = 1.5 meter	68.64	61.29	67.07	59.35	66.29	58.38	88.69	90.36	88.12	90.68	87.84	90.46
S = 2.0 meter	48.27	42.35	45.68	39.46	44.39	38.02	69.22	73.88	67.68	72.58	66.91	71.92

In general. At the same installation distance with different thickness of soft soil layer, the changes of the degree of consolidation were only <1%. Degree of consolidation 90% achieved at 1 m installation distance, for example from the calculation results with a distance of 1 m with a soft soil thickness of 20 m in the triangular pattern obtained 89.07% [9] and 90.24% [12].

The difference of distance between PVD installations is getting farther, then the degree of consolidation achieved is smaller with the same time target. The smallest degree of consolidation when the distance of 2 m PVD installation with a thickness of 20 m soft soil layer from the calculation of the degree of consolidation that can be achieved is 38.02% [9] and 71.92% [12]. If soft soil thickness same and the difference in installation distance of 1 m, the degree of consolidation decreased by 18.32% - 51.05% triangular pattern and 23.75% -45.69% rectangular pattern.

4. Conclusion

The general PVD thickness products are between 3.3 - 3.7 mm and the overall vertical drain length is 100 mm. The median value for PVD thickness is 3.36 mm, and the average thickness value is 3.74 mm. Triangular patterns PVD with a time between 11-25 weeks while 13-30 weeks in rectangular patterns. Whereas with the distance installation PVD 2 m with all variation soft soil layers to gain 90% degree of consolidation need more than 30 weeks. In general. At the same installation distance with different thickness of soft soil layer, the changes of the degree of consolidation were only <1%. If soft soil thickness same and the difference in installation distance of 1 m, the degree of consolidation decreased by 18.32% - 51.05% triangular pattern and 23.75% -45.69% rectangular pattern.

5. References

[1] Baihaqi D H, Lastiasih Y and Kumalasari T, Zonafikasi Parameter Pemampatan Tanah Lunak Dan Analisa Terhadap Waktu Dan Variasi Jarak Pemasangan Di Wilayah Surabaya Timur, JURNAL TEKNIK POMITS

[2] Mochtar, Indrasurya B . 2000. Teknologi Perbaikan Tanah dan Alternatif Perencanaan pada Tanah Bermasalah (Problrmatic Soils). Surabaya: Jurusan Teknik Sipil FTSP-ITS.

[3] Holtz, Bergado and Balasubramaniam. 2011. Prefabricated Vertical Drains in Soft Soil Clay : A Case Study of the New Bangkok International Airport Project. Canadian Journal Geotechnic

[4] Sukla A S and Kambekar A R 2013 Working Of Prefabricated Vertical Drain – A Case Study, IJIRSET Vol 2, ISSUE 8, August 2013

[5] Hansbo, S 1979 Consolidation of clay by band-shaped prefabricated drains Ground Eng., 12(15), 16-25.

[6] Bergado, DT, Long PV, and Balasubramaniam, A S 1996a. Compressibility and flow parameters from PVD improved soft Bangkok clay. Geotechnical Engineering Journal, 27(1): 1–20.

[7] Holtz, R.D., and Broms, B B. 1972. Long term loading tests at Ska-Edeby, Sweden. In Proceedings of the Specialty Conference on the Performance of Earth and Earth-Supported Structures, Lafayette, Ind., Vol. 1, pp. 435–464.

[8] Rowe, K.R., Taechakumthorn C. (2008). Combined effect of PVD's and reinforcement on embankments over rate sensitive soils. Geotextiles and geomembrane journal. 26 (2008) 239-247.

[9] Barron R A 1948 Consolidation of fine-grained soils by drain wells, Trans ASCE, 133 (paper 2346), 718-724.

[10] Jamiolkowski, M Lancellotta, R dan Wolski, W 1983 Precompression andSpeeding Up Consolidation Proceedings of the 8th. EuropeanConference on Soil Mechanics andFoundation Engineering, Vol. 3., hal.1201-1226, Helsinki

[11] Terzaghi K. 1943. Theoritical Soil Mechanics. New York . John Willey & Sons, inc

[12] Aspar N A W dan Fitriyani N E 2016 pengaruh jarak dan pola prefabricated vertical drain (PVD) pada perbaikan tanah lempung lunak. MPI Vol.10 No.1, April 2016.

[13] Das, Braja M 1998 Mekanika Tanah: Prinsip–Prinsip Rekayasa Geoteknik. Diterjemahkan oleh Noor Endah

Effect of Flow Rate and Temperature on Erosion Corrosion Rate of Crude Palm Oil Against Elbow A53 Grade B Carbon Steel Material

Budi Prasojo[1], Hendri Budi Kurniyanto[1], Tarikh Azis R[1], Subagio So'im[1], Asfiem Rahmat Haqin[1]

[1] Shipbuilding Institute of Polytechnic Surabaya, Surabaya, Indonesia

budiprasojo1968@gmail.com; budi.hb97@gmail.com; tarikh@ppns.ac.id; bagiosoim@gmail.com; asfiemrahmat@gmail.com

Abstract. In Center Port Nilam there is a construction of storage tanks and pipelines that will be used for the distribution and storage of Crude Palm Oil (CPO). CPO will be transported to the storage tank through a pipeline using carbon steel A53 grade B. In the pipeline there is a pipe bride as high as 8.6 meters, and under the pipe bride, there is an elbow that has the potential to erosion corrosion. The experiment is to test the effect of flow rate and temperature to the corrosion-erosion rate of CPO in elbow material of carbon steel A53 grade B. For calculations use the DNV RP O501 standard and for testing using the ASTM G31-72 standard. This research use variation of the fluid speed of 1 ms, 2 m/s, 3 m/s and temperature of 40°C, 50°C, 60°C. From the results using manual calculations for the flow rate of 1 m/s, 2 m/s, 3 m/s erosion corrosion rate successively is 0.0030 mmpy, 0.0440 mmpy, and 0.1946 mm / year. The erosion-corrosion rate successively is 0.0315 mmpy, 0.0450 mmpy and 0.0531 mmpy for flow rate of 1 m/s, erosion corrosion rate of 0,0596 mmpy, 0,0781 mmpy, 0,0950 mmpy for flow rate of 2 m/s and 0.1044 mmpy, 0.1274 mmpy and 0.1497 mmpy for flow rate of 2 m/s for the test results using testing method at a temperature of 40 °C, 50 °C, 60 °C. It can be concluded that the greater flow rate and temperature can lead to greater erosion corrosion rate.

1. Introduction

Erosion corrosion is one thing that can cause failure in a piping system. All production companies or contractors significantly avoid erosion corrosion. The occurrence of erosion-corrosion due to fluid flowing in the order has a corrosive nature, and there is a substantial element in the fluid that flows [1]. There is the construction of storage tanks and pipelines in Central Nilam Port, Surabaya, Indonesia. Storage tanks and pipes will be used to store and transport Crude Palm Oil containing sand [2]. Crude Palm Oil will be transported from ship to storage tank through 8-inch diameter pipeline. At the time of discharge, Crude Palm Oil temperature is maintained between 40°C – 60°C. On the pipeline, there is an 8.6-meter pipe bridge. Under the pipe bridge, there is an elbow that is potentially exposed to erosion corrosion.

Noviadam [4] investigated erosion-corrosion in the pipeline system. It was found the way to analyse the rate of erosion-corrosion on ST 60 steel material using seawater fluid. Investigated about erosion-corrosion in the elbow in the pipeline [3-5]. It was found the way to calculate the rate of erosion on elbow manually concerning standard DNV RP O501 2007/2015. Based on the problem, this research will analyse the effect of flow rate and temperature on crude palm oil erosion rate to A53 grade B carbon steel elbow. This research will be done by erosion corrosion testing of weight loss method which refers to standard ASTM G31 - 72 and manual calculations referring to DNV standard - RP O501 [6-8].

2. Methods

2.1. Flow diagram

This research is using weight loss method which refers to standard ASTM G31-72, and manual calculation of erosion rate refers to DNV RP O501 2007/2015 standard. This research investigated the rate of Crude Palm Oil erosion (CPO) on elbow using carbon steel material A53 Grade B. The variables used in this erosion corrosion rate testing are velocity of flow rate and temperature, where the rate of flow rate is 1 m/s, 2 m/s, 3 m/s and the test temperature is 40°C, 50°C and 60°C. Also, this research also calculates the erosion-corrosion rate manually referred to DNV RP O501 2007/2015 standard. The speed of the flow rate used for manual calculations is 1 m/s, 2 m/s and 3 m/s. This research method is shown in the flow diagram in Figure 1 below.

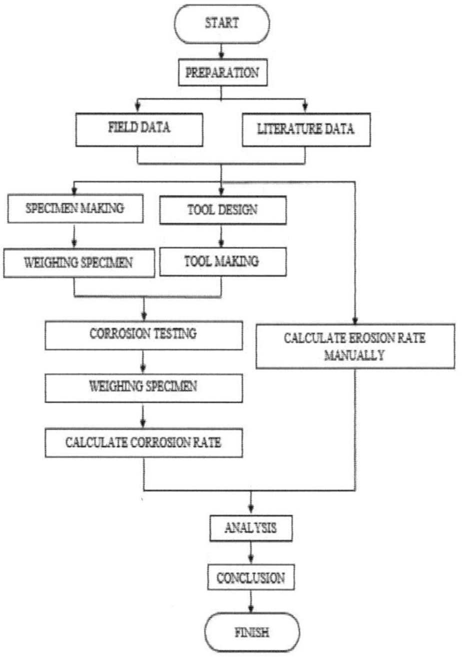

Figure 1. Flow diagram of this research

2.2. Calculation formula

2.2.1. Erosion rate

Base on DNV RP O501, to calculate the erosion rate on elbow required the following steps.

- *Calculation for the characteristic impact angle (α)*

$$\alpha = \arctan\left(\frac{1}{\sqrt{2}.\ R\ \text{curvature}}\right) \quad (1)$$

Where:
α = Impact Angle (°)
R curvature = radius of curvature (-)

- *Calculation for the dimensionless parameter group (A)*

$$A = \frac{\rho_m^2 .\tan(\alpha).\ Up\ .\ ID}{\rho_p\ .\ \mu_m} \quad (2)$$

Where:
A = Dimensionless Parameter Group (-)
ρ_m = Fluid Mixture Density (kg/m^3)
α = Impact Angle (°)
ID = Inside Diameter (m)
ρ_P = Particle Density (kg/m^3)
μ_m = Fluid Mixture Viscosity (kg/m.s)
Up = Impact Particle Velocity (m/s)

- *Determination the value of critical particle diameter*

$$\gamma_c = \frac{\rho_m}{\rho_p . [1.88 . \ln(A) . -6{,}04]}, \gamma_c < 0{,}1 \tag{3}$$

Where:
γ_c = Critical Particle Diameter (-)
ρ_m = Fluid Mixture Density (kg/m^3)
ρ_p = Particle Density (kg/m^3)
A = The Dimensionless Parameter Group

- *Calculation for the particle size correction function (G)*

$$G = \begin{cases} \frac{\gamma}{\gamma_c}, \gamma < \gamma_c \\ 1, \gamma \geq \gamma_c \end{cases} \tag{4}$$

Where,
G = The Particle Size Correction (-)
γ_c = Critical Particle Diameter (-)
γ = dp/ID (-)

- *Calculation for the characteristic pipe bend area exposed to erosion*

$$A_t = \frac{\pi}{4 . \sin(\alpha)} \tag{5}$$

Where,
A_t = Area Exposed to Erosion (m^2)
ID = Pipe Inside Diameter (m)
α = Impact angle (°)

- *Determination the value of the function F(α) by using the angle (α)*

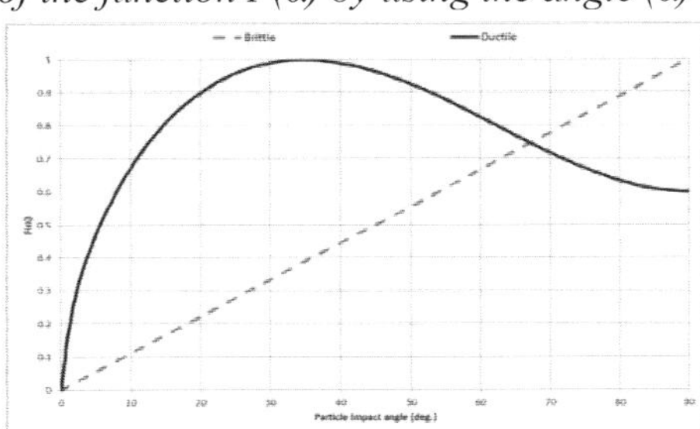

Figure 2. Function *F(α) by using the angle (α)*

- *Calculate erosion rate*

$$EL = \frac{K.F(\alpha).\sin(\alpha).Up^n}{\rho_t.A_t}.G.C_1.GF.\dot{m}_p.C_{unit} \tag{6}$$

Where:
EL = Erosion Rate (mmpy)
K = Material Constant [$(m/s)^{-n}$]
α = Impact Angle (^{0})
U_p = Impact Particle Velocity (m/s)
n = Velocity Exponent
ρ_t = Density Material (kg/m^3)
A_t = Area Exposed to Erosion (m^2)
G =The Particle Size Correction (-)
C1 = Model/Geometry Factor (-)
GF = Geometry Factor (-)
$\dot{m}$ = Mass Flow Particle (kg/s)
C_{unit} = Unit Conversion Factor (-)

2.2.2. Weight loss testing method

Base on ASTM G31 – 72, to obtain the erosion-corrosion rate using the following formula below

$$Cr = \frac{K.W}{D.A.T} \tag{7}$$

Where,
Cr = Corrosion Rate (mmpy)
K = Constant ($8{,}76 \times 10^4$)
W = Weight Loss (gram)
D = Material Density ($gram/cm^3$)
A = Area (cm^2)
T = Time Exposure (hours)

3. Results

3.1. Erosion corrosion rate by manual calculation

The manual calculation of erosion-corrosion rate base on DNV RP O501, From the calculation of erosion-corrosion rate, that each flow rate velocity has different erosion rate values, it can be seen the effect of flow rate velocity to the rate of erosion of carbon steel material A53 Gr B in Figure 2. From the results obtained, it can be concluded that the higher of flow rate, can cause greater the rate of erosion.

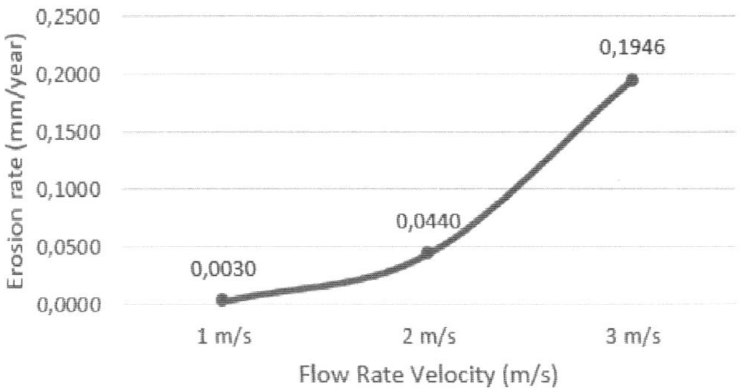

Figure 3. Results of the erosion-corrosion rate by manual calculation

3.2. Erosion corrosion by the testing method

The method used in corrosion test is weight loss which refers to ASTM G31-72 standard. This test uses a variation of flow rate velocity and temperature. The velocity of flow rate used is 1 m/s, 2m/s, 3m/s and the temperature used is 40°C, 50°C, and 60°C. The following corrosion test results are shown in Table 1.

Table 1. Result of erosion *corrosion* rate based on ASTM G 31-72

Flow Rate (M/S)	Temperature (Celcius)	Specimen	W1 (g)	W2 (g)	ΔW (g)	A (cm^2)	CR (mmpy)	Average CR (mmpy)
	40	A1	49.0666	49.0624	0.0042	30.974	0.0315	
	40	A2	52.1152	52.1106	0.0046	33.809	0.0316	0.0315
	40	A3	51.6551	51.6506	0.0045	33.209	0.0315	
	50	B1	50.8292	50.8230	0.0062	33.764	0.0426	
1 m/s	50	B2	51.7867	51.7803	0.0064	32.679	0.0455	0.0450
	50	B3	50.6658	50.6591	0.0067	33.209	0.0468	
	60	C1	50.2644	50.2569	0.0075	32.099	0.0542	
	60	C2	49.2396	49.2310	0.0086	33.209	0.0598	0.0562
	60	C3	48.3933	48.3857	0.0076	32.099	0.0547	
	40	D1	51.2448	51.2357	0.0091	34.319	0.0614	
	40	D2	51.1802	51.1719	0.0083	34.939	0.0551	0.0576
	40	D3	50.4658	50.4575	0.0083	34.319	0.0563	
	50	E1	47.4521	47.4416	0.0105	32.099	0.0756	
2 m/s	50	E2	49.6780	49.6679	0.0101	32.654	0.0716	0.0759
	50	E3	49.2330	49.2215	0.0115	33.209	0.0805	
	60	F1	49.3783	49.3652	0.0131	32.654	0.0934	
	60	F2	50.2092	50.1955	0.0137	33.209	0.0958	0.0926
	60	F3	51.2686	51.2557	0.0129	33.834	0.0885	
	40	G1	51.4288	51.4122	0.0166	33.764	0.1142	
	40	G2	53.2954	53.2804	0.0150	35.559	0.0979	0.1044
	40	G3	48.6913	48.6771	0.0142	32.609	0.1011	
	50	H1	48.3124	48.2938	0.0186	33.209	0.1300	
3 m/s	50	H2	51.6048	51.5853	0.0195	34.319	0.1319	0.1276
	50	H3	47.5486	47.5322	0.0164	31.519	0.1208	
	60	I1	50.5969	50.5729	0.0240	34.409	0.1619	
	60	I2	47.1310	47.1125	0.0185	31.519	0.1363	0.1497
	60	I3	47.4284	47.4079	0.0205	31.519	0.1510	

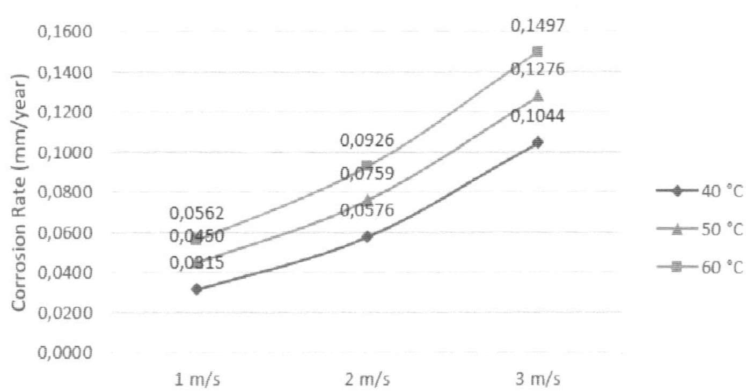

Figure 4. Effect of flow rate velocity and temperature to the erosion-corrosion rate

From the calculation of erosion rate, that the effect of flow rate velocity and temperature has different erosion rate values. From the table above, it can be seen the impact of flow rate velocity and temperature to the rate of erosion of carbon steel material A53 Gr B shown in graphical form in Figure 3. From the results obtained, it can be concluded that the higher of flow rate, can cause greater the rate of erosion. Moreover, also the higher of temperature can cause greater the rate of erosion.

IOP Conf. Series: Materials Science and Engineering **462** (2019) 012038 doi:10.1088/1757-899X/462/1/012038

4. Results

Based on result and analysis, it can be concluded that the higher velocity of flow rate can cause the higher value of the erosion-corrosion rate. From the results of manual calculations, the lowest erosion corrosion rate is found at a velocity 1 m/s in the amount of 0.003 mmpy, while the highest is at a velocity of 3 m/s of 0.1946 mmpy. From the test results obtained the lowest erosion corrosion rate at the fluid flow rate of 1 m/s at temperature 40°C with a value of 0.0315 mmpy while the highest value is at the fluid flow rate of 3 m/s at temperature 60°C with a value of 0.1497 mmpy. Increasing the flow rate of fluid in the pipe will increase the erosion-corrosion rate and if the fluid temperature increase, the corrosion rate will be increased.

5. References

[1] American Society of Testing and Material G31 – 72 2004 *Laboratory Immersion Corrosion Testing of Metals*

[2] Det Norske Veritas and Germanischer Lloyd RP – O501 (2015) *Managing Sand Production and Erosion*

[3] Herman Pratikno, H. S. 2017 *Effect of Heat Treatment and Types of Bacteria on Corrosion Resistance of Steel Materials* (Surabaya: 10 November Institute of Technology)

[4] M. Riki Noviadam A. M. 2016 *Analysis of Corrosion Rate of Erosion on Steel ST60 in Various Medium Sea Water* (Surabaya: Surabaya State University)

[5] Muhammad, F. A. 2016 *Effect of Plastic Deformation Level 304 L Stainless Material on Micro Structure and Sulfur Corrosion Rate Temperature 120°C* (Surabaya: Shipbuilding Institute of Polytechnic Surabaya)

[6] Rizki Ornelasari, M. 2015 *Analysis of Corrosion Rate in 304 Stainless Steel Using ASTM G31-72 Method in Nira Media* (Surabaya: Surabaya State University)

[7] Rizkyawan. 2017 *The Influence of Holding Time Variations in Normalizing the A53 Carbon Steel Corrosion Rate with Mud Media* (Jember: Jember State University)

[8] Yunan, A. M. 2017 *Analysis of Erosion Rate and Lifetime Calculation of Stainless Steel Material 304, 316 and 317 on Reject Flow 1ST Cleaner to 2ND Cleaner OCC Line Voith Unit SP 3-5 AT PT. X* (Surabaya: Shipbuilding Institute of Polytechnic Surabaya)

IOP Conf. Series: Materials Science and Engineering **462** (2019) 012039 doi:10.1088/1757-899X/462/1/012039

The Influence of the Piston Head Shape on the Performance of a Single Cylinder Diesel Engine: An Experimental Study

D Khusna[1], A Susilo[2], Sudarto[2], A Suharyanto[2]

[1] Department of Mechanical Engineering, Faculty of Industrial Engineering, Institut Teknologi Adhi Tama, 60111 Surabaya, Indonesia
[2] Brawijaya University, 65145 Surabaya, Indonesia

d.khusna2017@gmail.com

Abstract. Air pollution due to motor vehicles encouraged this research to develop proper combustion in the combustion chamber. The combustion in motor vehicles produces not only environmentally friendly combustion but also better engine performance. In a diesel engine, diesel fuel is put into the combustion chamber using injection through a nozzle. This fuel spray is subjected to break up and atomization before getting mixed with air and experiencing combustion in the combustion chamber. Diesel engine combustion chamber with different piston head will affect the combustion process and eventually affect the diesel engine performance. This research used the experimental method with the piston head shapes being tested were a low standard, semi-spherical, and multispherical. While the engine performance sought as a result of changes in the piston head shape are power (N), the use of specific fuel (Sfc), and thermal efficiency (ηth).

1. Introduction

A diesel engine is one type of engine or internal combustion engine with diesel fuel. In the beginning, this engine was discovered by Rudolf Diesel who designed a diesel engine using coal as fuel. However, in the future, the use of liquid oil results in more efficient engine performance. In the history of its application, the Packard diesel engine was used as an aircraft propulsion engine in early 1927, and Charles Lindbergh flew the Stinson SM1B with Packard diesel engine in 1928. The journey of cars with a diesel engine was completed on 6 January 1930. That journey was started from Indianapolis to New York City - as far as 1,300 km. Today, the diesel engine has been widely used in the field of transportation.

The diesel engine is called as Compression Ignition Engine or CI Engine because the combustion process occurs due to compression at high pressure and temperature. The fuel is put into the combustion chamber using injection through a nozzle. The fuel spray from this nozzle is subjected to fogging and atomization [1]. This fogging process is the first step before getting mixed with air and undergoing combustion due to high pressure and temperature.

The excellent mixing process of air and fuel will result in proper combustion. The mixed air and fuel to the combustion chamber happen when the piston goes to the top dead point. As a result, there is a combined flow in the swirl, the flow in the radial direction from the top of the piston crown (squish), and tumble. The flow of this air and fuel mixture is closely related to the shape of the piston head. Therefore, the shape of the piston head affects the combustion occurred [1,2]. On this basis, this research was prepared to determine the extent of the influence of the piston head shape on diesel engine performance.

2. Theory

2.1 Diesel Engine Performance
The most discussed performances of the diesel engine are power, torque, average effective pressure (BMEP), specific fuel consumption (SFC), and thermal efficiency. Power on a diesel engine is a measure of diesel engine ability to do a work per unit of time. The tool used to measure power is dynamometer. This research used electric dynamometer in which the engine power was measured by the electrical power produced by the generator to ignite the lamp load on the electric circuit. Specific fuel consumption is a measure of the fuel amount needed per unit of power per hour. While the thermal efficiency is defined as the efficiency of using heat from fuel converted into mechanical fuel (shaft). This research did not discuss the average effective pressure and torque because the value can be obtained from the known power.

2.2 Mathematical Model
A mathematical model for calculating generator power, engine power, specific fuel consumption, and thermal efficiency used equations (1), (2), (3) and (4) as follows:

$$N_g = V.I.Cos(\varphi) \tag{1}$$

$$N_e = \frac{N_g}{\eta_g} \tag{2}$$

$$SFC = \frac{3{,}6\dot{m}_f}{N_e} \tag{3}$$

$$\eta_{th} = \frac{10^6 N_e}{\dot{m}_f.LHV} \tag{4}$$

where: V = electrical voltage (Volt); I = electric current (Ampere); $\cos(\varphi)$ = electrical power coefficient; N_g = generator power (watt); N_e = engine power (watt); η_g = generator efficiency; SFC = specific fuel consumption (kg/(watt.jam)); $\dot{m}_f$= fuel (consumption) rate (g/s); η_{th} = engine thermal efficiency; LHV = lower heat fuel value (kJ/kg)

3. Method

3.1 Test Material
Materials tested here were three (3) types of the piston with different piston head shapes, namely flow standard, semi-spherical, and multi-spherical as shown in figure 1.

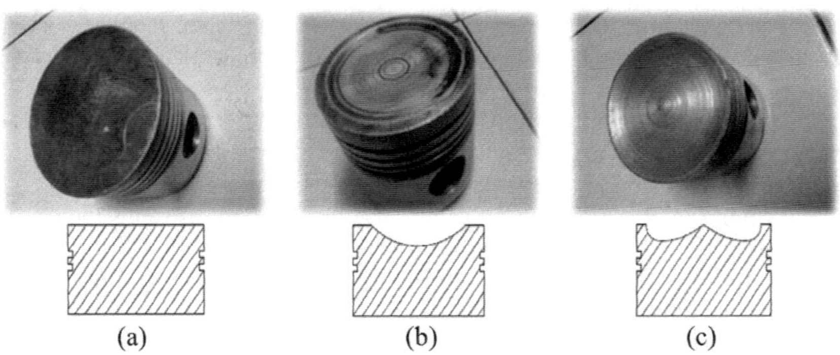

Figure 1. Piston types (a) flat standard, (b) semi-spherical, (c) multi-spherical

3.2 Test Material
The equipment used for this research are listed as follows:

Table 1. Data of Equipment used

No	Equipment	Description/Specification
1	Diesel Engine	7 HP number type of MD175A, one horizontal cylinder block, four steps, 75 mm bore, 80 mm stroke, the compression ratio of 21±1, the standard power of 4.41 kW at 2200 rpm, an injection pressure of 14±0.5 Mpa, water fluid cooling, fuel advanced angle of 21 ± 1^0CA
2	Generator	1 phase, 3 kW power, Cos θ = 1, max rpm of 1800 rpm, 50/60 Hz, 0.8 efficiency, 230/115 V voltage, 13/26 A current at 1500/1800 rpm, 70 kg weigh
3	Electric Load Circuit	(for generator loading when measuring engine power)
4	Measuring tube	(for measuring fuel consumption)
5	Amperemeter	(for measuring the electric current when loading)
6	Voltmeter	(for measuring the voltage when loading)
7	Tachometer	Fuji Kogya brand, min/max 10/10.000 rpm, 1 rpm accuracy
8	Stopwatch	Chrono brand (made in Jepang), 0,1 s accuracy
9	Toolkit	(for unloading the engine when changing the piston)

The equipment circuit when testing is shown in figure 2.

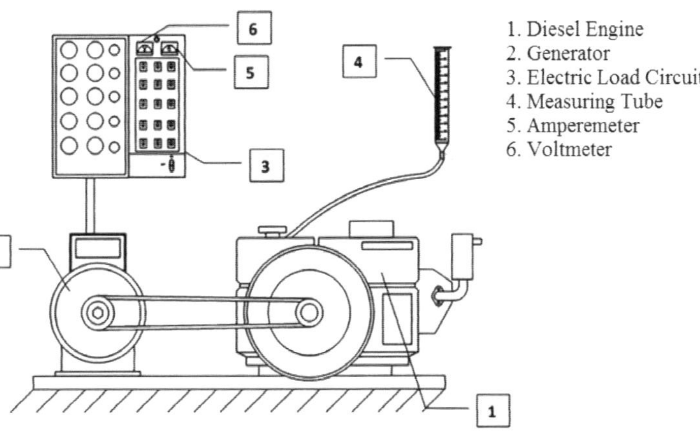

Figure 2. Test equipment circuit

4. Result and Discussion

4.1 Engine Power

Engine power is the power resulted from combustion process in the combustion chamber. The greater the engine speed, the more combustion cycles occurred so that the engine power will increase as well. The increase in engine power with the increase in engine speed can also be predicted theoretically from the mathematical equation of the relationship between power and engine speed.

In the electrical load circuit, the addition of electric load will burden the generator driven by the engine so that the engine rotation will tend to decrease. Therefore, it takes additional fuel to maintain constant engine rotation. Thus, the increase in electricity load will be accompanied by an increase in fuel consumption and vice versa.

In this research, it was found that the use of piston type 2 (semi-spherical) and type 3 (multi-spherical) provided greater power than type 1 (flat standard). The combustion process occurred in diesel engine began with a fogged fuel spray. In this fogging process, there was fuel breakup and atomization, as well as getting mixed with air (oxygen) causing combustion due to high pressure and temperature. In semispherical and multi-spherical piston type, the after chamber in the combustion chamber allowed the air-fuel mixture to swirl, squish, and tumble causing the mixture to be more homogeneous and resulted in better combustion. Better combustion generated better power.

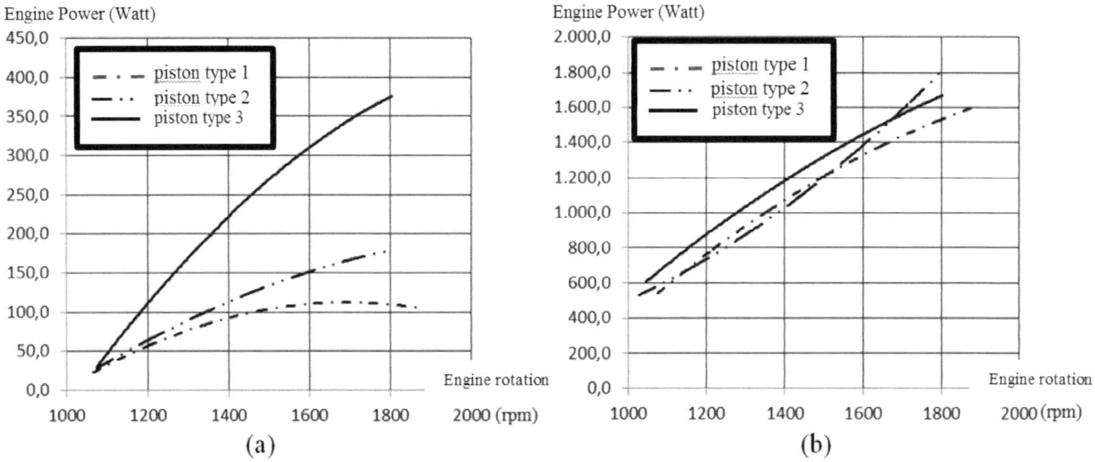

Figure 3. Engine power at (a) 200 W electric load, and (b) 1000 W electrical load

In this research, it was found that at small electrical load (200 W), the percentage of increased power due to the use of piston type 2 and 3 was greater than in greater electric load (1000 W). As previously described, small electric load consumed small amount of fuel. This small fuel consumption caused fuel spray easy to be subjected to fogging which led to good combustion and fast increase of engine power.

4.2 Specific fuel consumption (SFC)

Specific fuel consumption is fuel consumption per unit of power per hour. By using piston type 2 and 3, the increase in power might reduce the specific fuel consumption. The use of piston type 2 and 3 in this research gave a greater percentage of SFC reduction at small electric load (200 W) compared to in greater electric load (1000 W) in the same rotation, as shown in Figure 4. This can be explained since small electric load (200 W) had a higher percentage of power increase. It could also mean that at the same rotation and load, the use of semi-spherical and multi-spherical piston resulted in greater fuel consumption savings compared to the flat standard piston.

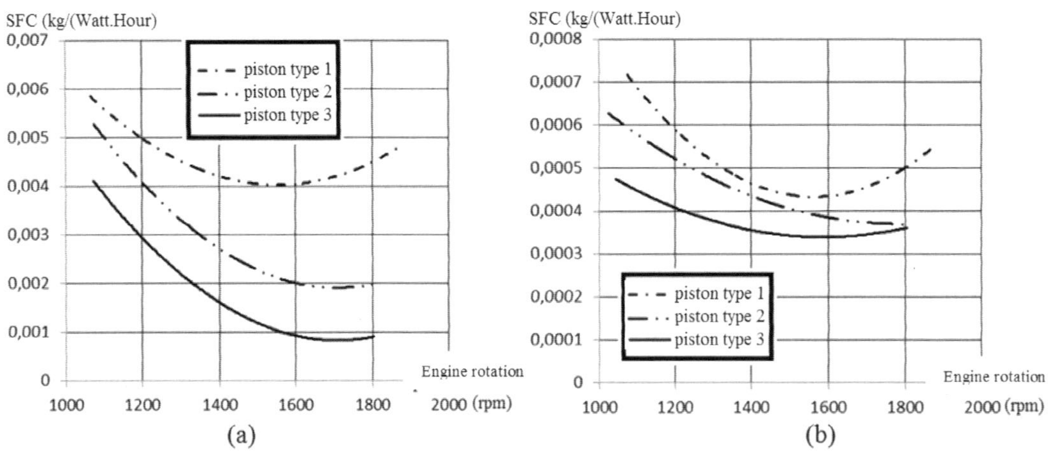

Figure 4. SFC graph at (a) 200 W electric load, and (b) 1000 W electric load

4.3 Thermal Efficiency

As in other internal combustion engines, in this research, thermal efficiency showed a low value on small engine rotation then getting higher with the increase in engine rotation. The low value of thermal efficiency at small electric load (200 W) as in Figure 5 (a) was due to the low output power to rotate the generator. Therefore, with fixed fuel consumption, the effect on efficiency value became low, as if the heat energy from fuel was only used to rotate the engine and generator without any energy (electricity) being used.

The use of semispherical and multi-spherical piston type gave greater efficiency effect since the engine output power was greater. This explanation was similar to the explanation of engine power that in these piston types, fuel spray had a greater chance for breakup in the after chamber of the combustion chamber so that the atomization became more complete and resulted in better combustion.

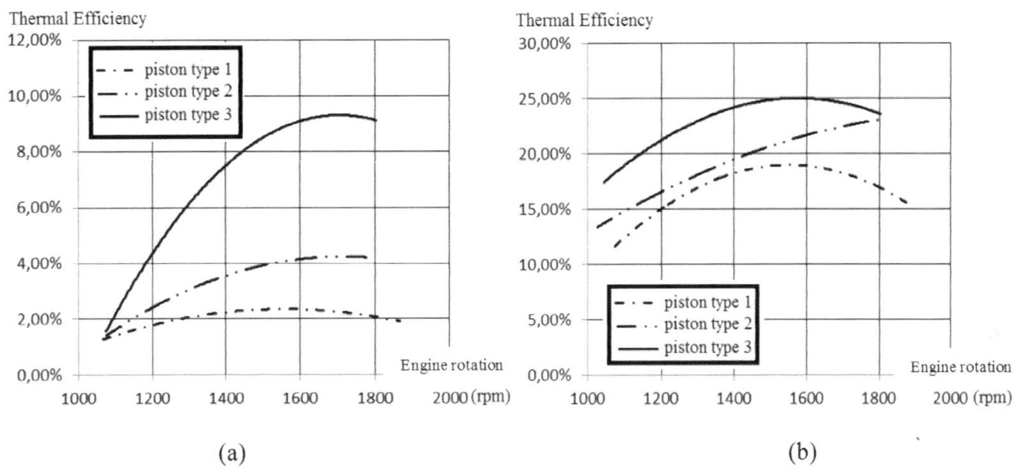

Figure 5. Efficiency graph at (a) 200 W electric load, and (b) 1000 W electric load

5. Conclusion

From this research, it can be concluded that, the use of semi-spherical and multi-spherical piston provided increased power and efficiency as well as reducing specific fuel consumption. The effect of increased power and efficiency, as well as reduced specific fuel consumption, was greater at low electricity load.

6. References

[1] Morris C J and Dent J C 2016 *Proc. Instn. Mech. Engrs.* (Pennsylvania: Pennsylvania State University) Vol 190 41/76
[2] Prasad S L V et al 2011 *Int. J. of Engr. Sci. and Technol. (IJEST)* **3:2** 1571-75
[3] Congbo Y, Zhendong Z, Yuedong S, Tao S and Renjie Z 2016 *J. Engr. Appli. of Comp. Fluid Mech.* **10:1** 311-29
[4] Olumide AT and Ademola AD 2018 *Europ. J. of Engr and Technol.* **6:1** 13-25
[5] Peeraiah M et al 2015 *Inter. J. of Emerg. Trends in Engr. Research (IJETER)* **Vol. 3 No.6** 499-510
[6] Chandrashekharapua R R et al 2012 *J. Transact. of the Canad. Soc. for Mech. Engr.* **Vol. 36 No. 4** 429-438
[7] A. Necati Ö 2017 *Int. J. of App. Sci. and Technol.* **vol.** 1, NO:1, 22-8
[8] Heywood J B 1988 *Internal Combustion Engine Fundamentals* McGraw-Hill, New York p 491.
[9] Hiroyasu H, Kadota T and Arai M 1989 *Development and Use of a Spray Combustion Model to Predict Diesel Engine Efficiency and Pollutant Emission* Bulletin of the JSME vol. 26, no. 24, Paper 214-12, pp. 569-575.

[10] Khusna D 2009 *Karakteristik Semprotan Menumbuk Dinding Piston Type Mexican-Hat Dengan Bahan Bakar Biodiesel Dalam Ruang Bakar Bertekanan* Digital Library ITS Master Theses Rekayasa Konversi Energi Surabaya p 73-9
[11] Wjayanti F and Dadan I 2014 *Analisis Pengaruh Bentuk Permukaan Piston Terhadap Kinerja Motor Bensin* Jurnal Ilmiah Mesin Universitas Islam 45 Bekasi, Vol.02 No.01, 2014
[12] Khusna D and Dhofir M 2016 *Pengaruh Penggunaan Bahan Bakar Biodiesel Hasil Pengolahan Minyak Jelantah dan Solar B0, B20 dan B40 Terhadap Performa dan Emisi Gas Buang Pada Mesin Diesel 2 tak untuk Kapal* Jurusan Teknik Mesin ITATS Surabaya

Acknowledgement

The authors would like to thank the Department of Mechanical Engineering ITATS that has facilitated this research.

IOP Conf. Series: Materials Science and Engineering **462** (2019) 012040 doi:10.1088/1757-899X/462/1/012040

Numeric Simulation of the Effect of Varying Velocities on Catalytic Converter and Exhaust Gas Emission

Suheni[1], R Sunoko[2], S Wahyudi[2] and A S Leksono[1]

[1]Departement of Environmental Engineering, Brawijaya University, Malang, Indonesia
[2]Departement of Mechanical Engineering, Brawijaya University, Malang, Indonesia

irsuheni@gmail.com

Abstract. Fossil fuel-powered vehicles are the most significant contributor to air pollution nowadays. They emit substances that are harmful to both human and environment. Carbon monoxide (CO) and Hydrocarbon (HC) is the most common harmful exhaust gases found in the environment. One of the technologies used for reducing CO emissions is catalytic converters. The catalytic converters usually installed in the exhaust system. The numeric study was conducted on the interaction of flow, reaction, and thermal effects in the form of a 2D model of the catalytic converter. Simulation using Fluent software was then used to find out the extent of the catalyst effects on CO gas absorption when velocities were 6.3 m/s, 15 m/s, and 20 m/s. The simulation revealed that CO mass fraction and temperature around the exhaust channel increased as the velocity increased.

1. Introduction

Carbon monoxide (CO) and Hydrocarbon (HC) are pollutants mostly produced by vehicles. The accumulation of CO in the air may cause health problems and even deaths. On the other hand, HC polluted air may trigger cancer cells form in the human body. When pollutant concentration in the air is getting higher, deaths caused by respiratory tract problems or cardiovascular problems increas[1]. Catalytic converters can be a solution to this. When installed in the vehicles exhaust systems, catalytic converters are known to reduce CO and HC emitted by vehicles[2-5].

Uneven flow during cold start of the catalytic converters has a significant effect causing differences in ignition characteristic among channels, particularly in the exhaust channel with lower temperature [6]. G. Bella *et al.* [7] and M. Badami *et al.* [8] conducted studies on the effect of uneven flow distributed from oxidation inlet to catalytic converters. By varying catalysts used in catalytic converters such as quartz glass, alumina ceramic, and copper, Junhu Zhou et al. improved micro-combustor that had low stability [9]. The result was that combustion chamber stability increased when those three materials were used as a catalyst. Moreover, the chamber was able to function in extreme equivalence ratio. The other finding was that the use of copper as a catalyst caused lower temperature distribution in the combustion chamber when compared with the other materials [9].

In previous studies, analysis on the effect of the speed of fluid passing through catalytic converters is still rare, so the objective of this study was to find out the result of fluid velocity in catalytic converters on gas emission when copper was used as a catalyst. The researcher expected that this study might help in the overcoming of pollution problems.

2. Research Method

In this research, catalytic converters in the vehicle were placed after the exhaust manifold system as an exhaust gas controller (to convert CO and HC into CO^2 and water vapour). The flow inside the catalytic converter was simulated using the Fluent software. The flow inside the catalytic converter was varied in 3 different velocities: 6.3 m/s, 15 m/s, and 20 m/s. The fluid used in this research was carbon monoxide (CO). The other conditions were 300°C inlet temperature condition, 1 atm (atmospheric pressure), 1.1233 kg/m^3 density, and 1.75x10^{-5} Kg/m.s viscosity.

The catalytic converter geometry in this research is shown in figure 1. Meanwhile, figure 2shows that the number of cells used was 33,102, the number of faces was 67,095, and the number of nodes was 33,982.

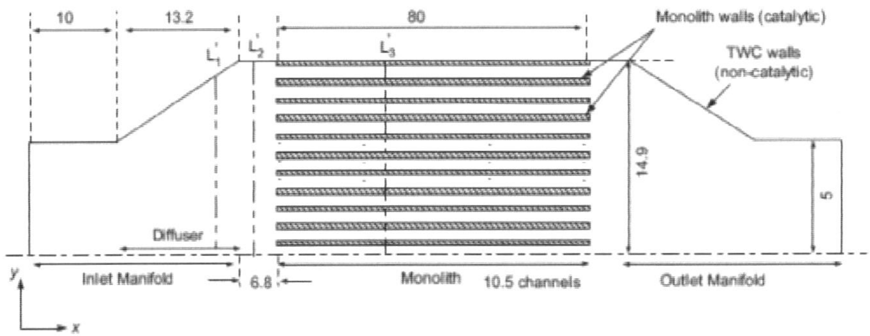

Figure 1. Catalytic Converter Geometry.

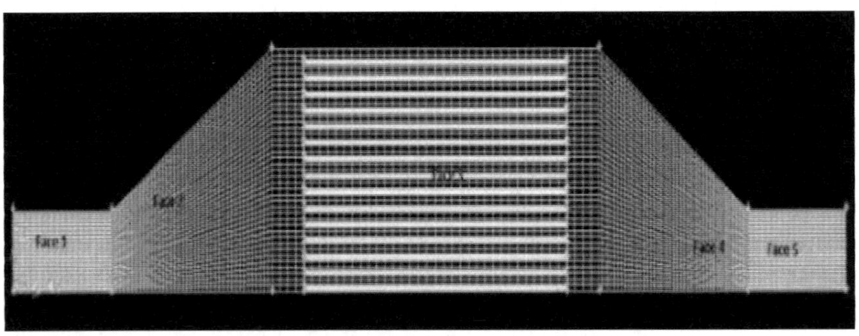

Figure 2. Meshing for Catalytic Converter.

3. Result And Discussion

3.1 Validation

This research result was validated by comparing to the effect of research by Gaurav et al. [10]. The validation was performed for the effect of the 6.3m/s velocity simulation. Table 1 shows that the result of this research was almost similar to that found by Gaurev et al [10]. This research found that the magnitude of velocity in the exhaust area was 1.26x10^1, while in the study conducted by Gaurev et al. was 1.30x10^1. The difference that was only 3.17% (less the 10%) means that the simulation validity was good.

Table 1. Validation.

Velocity	Present study	Gaurav et. al.	Distance
6.3 m/s	$1.26x10^1$	$1.30x10^1$	3.17%

3.2 Velocity Contour

The results of the simulation at three different velocities showed that there was a pattern in the magnitude of the velocity. The pattern showed that high recirculation occurred in inlet diffuser, characterised by a flow rotating at the upper part and lower part induced by the existence of sudden area expansion in all 3 inlet velocities as shown in Figure 3. The flow became non-uniform because of this condition.

Transversal velocity component was lower than the axial velocity component at the centre part. This means that the fluid shifted to the upper part and the lower part. Also, the pressure dropped along the channel because of friction. Friction increased drastically because of the inlet velocity within every channel. When the flow speed is high, the friction resistance becomes high as well.

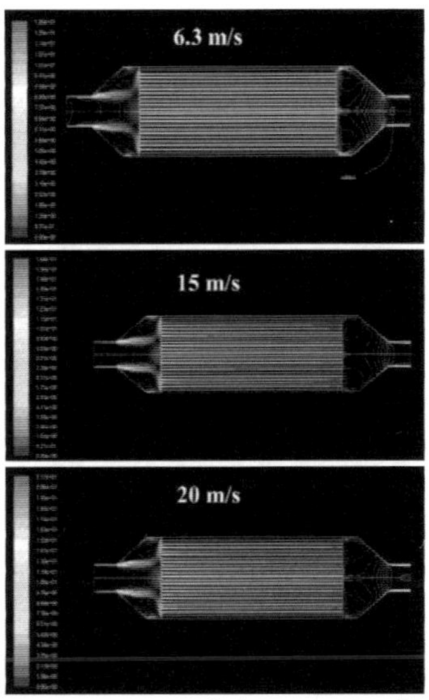

Figure 3. Different contours of the magnitude of the velocity.

3.3 CO Mass Fraction Contour

The analysis result of carbon monoxide fluid showed that mass fraction distributed within the catalytic converter had a high concentration in the centre of the channel. The mass fraction distribution is represented by red colour in Figure 4. Mass fraction decreased while flowing to the outlet channel.

As the velocity was increasing, the area of mass fraction decrease was getting wider and more directed toward the direction of the outlet. This was caused by temperature in the peripheral channel that became lower than that in the centre channel because of the heat loss to the environment.

The flaw of 2D geometric planar was that it could not capture transversal heat shift through the monolithic solid structure. Mass fraction decreased as CO moved to peripheral channel and was multiplied in the last peripheral channel. The initial decrease of CO mass fraction increased as CO flowed further from temperature loss-zone to environment responsible for lower conversion in the previous two channels in the monolithic catalyst. Qualitatively, the same result was observed for other operational conditions and full-scale geometry. This result was obtained for rather simple kinetics for CO oxidation (under low activation energy). For more complex kinetics and higher energy activation, conversion decrease within the peripheral channel is expected to be higher.

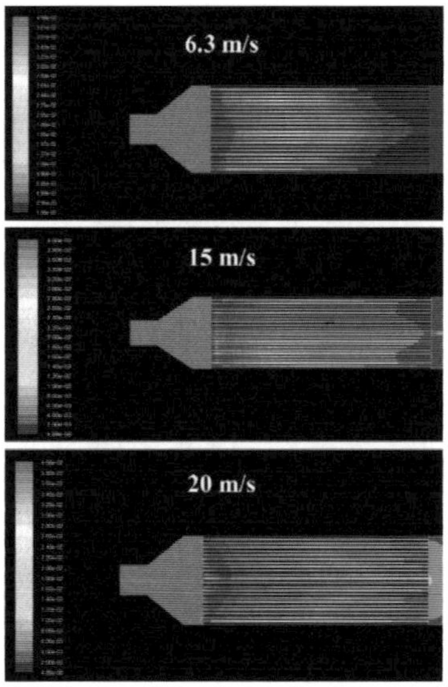

Figure 4. CO mass fraction contours comparison.

3.4 Temperature Contour

Simulation result of temperature contour showed that the temperature around the channel was lower than that in the centre of the channel. As shown in figure 5, this result occurred in all three velocities. As velocity increased, temperature around the channel more significantly decreased. This could be observed from the heat distribution area in various inlet area (red colour) that got narrower when velocity was 20 m/s.

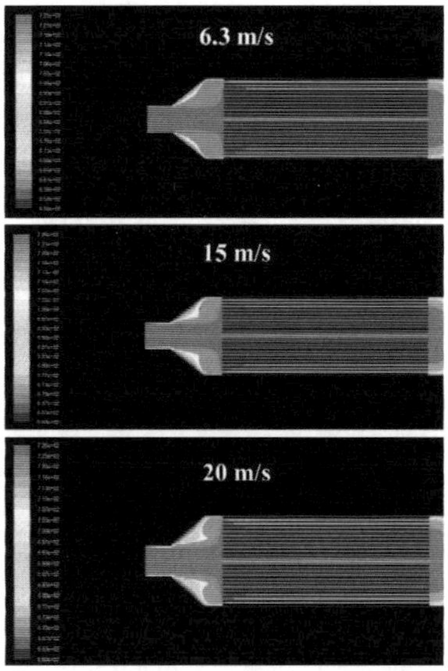

Figure 5. Temperature contours (K) comparison

4. Conclusions

The data analysis of the simulation result showed the phenomenon and the flow characteristic of the fluid that flowed through the catalytic converter. The data also the effects of varying velocities. Overall, the conclusions drawn from the study are as the following: Highest CO mass fraction was in the centre of the channel (inlet) for 6.3 m/s velocity. As the velocity increases, the flow in the area where mass fraction decreased was getting wider and going toward the bay. As speed reached 20 m/s, the temperature decrease around the channel became greater. Finally, it is expected that the result of this research will contribute to any efforts directed toward overcoming pollution, by the use and the approach of engineering technology.

5. References

[1] J. Huang, X. Pan, X. Guo, and G. Li, "Impacts of air pollution wave on years of life lost: A crucial way to communicate the health risks of air pollution to the public," *Environment International,* vol. 113, pp. 42-49, 2018/04/01/ 2018.

[2] F. Zeng, J. Finke, D. Olsen, A. White, and K. L. Hohn, "Modeling of three-way catalytic converter performance with exhaust mixtures from dithering natural gas-fueled engines," *Chemical Engineering Journal,* vol. 352, pp. 389-404, 2018/11/15/ 2018.

[3] R. M. Bagus Irawan, P. Purwanto, and H. Hadiyanto, "Optimum Design of Manganese-coated Copper Catalytic Converter to Reduce Carbon Monoxide Emissions on Gasoline Motor," *Procedia Environmental Sciences,* vol. 23, pp. 86-92, 2015/01/01/ 2015.

[4] S. Ramalingam, S. Rajendran, and P. Ganesan, "Performance improvement and exhaust emissions reduction in biodiesel operated diesel engine through the use of operating parameters and catalytic converter: A review," *Renewable and Sustainable Energy Reviews,* vol. 81, pp. 3215-3222, 2018/01/01/ 2018.

[5] A. Ghofur, Soemarno, A. Hadi, and M. D. Putra, "Potential fly ash waste as catalytic converter for reduction of HC and CO emissions," *Sustainable Environment Research,* 2018/09/07/ 2018.

[6] V. K. Chakravarthy, J. C. Conklin, C. S. Daw, and E. F. D'Azevedo, "Multi-dimensional simulations of cold-start transients in a catalytic converter under steady inflow conditions," *Applied Catalysis A: General,* vol. 241, pp. 289-306, 2003/02/20/ 2003.

[7] G. Bella, V. Rocco, and M. Maggiore, "A Study of Inlet Flow Distortion Effects on Automotive Catalytic Converters," *Journal of Engineering for Gas Turbines and Power,* vol. 113, pp. 419-426, 1991.

[8] M. Badami, F. Millo, A. Zuarini, and M. Gambarotto, "CFD Analysis and Experimental Validation of the Inlet Flow Distribution in Close Coupled Catalytic Converters," 2003.

[9] J. Zhou, Y. Wang, W. Yang, J. Liu, Z. Wang, and K. Cen, "Combustion of hydrogen–air in catalytic micro-combustors made of different material," *International Journal of Hydrogen Energy,* vol. 34, pp. 3535-3545, 2009/05/01/ 2009.

[10] G. Agrawal, N. S. Kaisare, S. Pushpavanam, and K. Ramanathan, "Modeling the effect of flow mal-distribution on the performance of a catalytic converter," *Chemical Engineering Science,* vol. 71, pp. 310-320, 2012/03/26/ 2012.

Risk Analysis of Musculoskeletal Complaints with Rula Method in Chemical Company

B A Aziz[1], **L Handoko**[2] **and A I Juniani**[2]

[1]Study Program of Safety Engineering, Marine Engineering Department, Shipbuilding Institute of Polytechnic Surabaya, Jl. Teknik Kimia, Kampus ITS, Sukolilo, Surabaya 60111, Indonesia
[2]Study Program of Design Manufacture, Marine Engineering Department, Ship building Institute of Polytechnic Surabaya, Jl. Teknik Kimia, Kampus ITS, Sukolilo, Surabaya 60111, Indonesia

lukmanhandoko@gmail.com

Abstract. Office workers seem trivial, not many people have aware of occupational diseases risk, office workers who most are computer users in their daily work. Often the symptoms are ignored, most people assume that it is due to ordinary fatigue whereas things it is the beginning of the onset of occupational diseases. Muscle receive static load repeatedly in a long time, will be able to cause complaints of damage to joints, ligaments, and tendons. In this study, the method is RULA that shows a very high and moderate risk level. From the results, such assessments are used as a basis for applying improvements facilities at work. Design through to a table and a chair by analysing anthropometry, it is anticipated that analysis ergonomics and analysis of the redesigning. They used this application for reasonable enough CATIA. That is in pop up based on observation, so it can be an alternative redesigning and possible to reduce the potential of any disease, from behind their backs and purchased operator local control room, publicly listed heavy duty equipment operators affected by the policy and office workers.

1. Introduction

This research was conducted on office workers, where the work looked normal, but it turned out that the impact was huge. In this study using a combination of RULA and CATIA and using the Indonesian population as a sample. Where this theme is rarely discussed because office work looks ordinary and does not have a high level of work risk so that it is considered a job that is free from occupational diseases.

Musculoskeletal Complaints are complaints on the part of the skeletal muscle that is felt by a person from very mild complaints to very painful. If the muscle receives constant static load for a long time, it can cause complaints in the form of damage to the joints, ligaments, and tendons.

Office workers are more comfortable to be exposed to discomfort and pain at work. The occupational risk factors are the most significant factors to this health problem, and it can be found in many industries.MSDs can degrade the health of the workers thus reduces the workers' performance in completing their tasks. Several kinds of literature had suggested that the effect of low performance by the workers can have significant economic and social consequences.

Rapid Upper Limb Assessment (RULA) is a method used to measure musculoskeletal disorders risk factors in the neck and upper body. Mc Atamney and Corlett developed the RULA from the University of New York Institute of Occupational Ergonomics, the United Kingdom in 1993 [1].

RULA itself is used to calculate ergonomic risk factors in jobs where the work is doing much work in a sitting position or standing without any movement. RULA is also used to calculate risk factors in the form of posture, power or load, static work and repetitions carried out at work.

The Nordic Body Map Method (NBM) is different from the RULA method. This method is a method used to assess the severity of the occurrence of a skeletal muscle disorder or injury. In its application, the Nordic Body Map method is done by using a worksheet in the form of a body map that is very simple, easy to understand, inexpensive, and requires a short time for each. The Observer can directly interview or ask the respondent about which skeletal muscles are experiencing dysentery or pain.

Anthropometry comes from "anthro" which has human meanings and "metri" which has a meaning of size. Anthropometry is a study of the measurement of the body of the human dimension of bone, muscle, and adipose or fat tissue. Anthropometric data is used for various purposes, such as designing workstations, work facilities, and product design.

2. Research

The first phase of the study was the initial identification phase with literature study and field study. Then the problem identification stage is carried out and determines the purpose of the research. After that, continue with the data collection stage. Data collection is done by taking data and observing directly. There are two types of data obtained in this study, primary data and secondary data. The primary data collection in this study was obtained from the results of interviews, structured questionnaires, checklists, and sampling and measurement. Whereas, secondary data is data collected from the literature and references relating to the issues discussed, data collected from PT. Chemistry, in the form of company profiles, and data on the number of workers. After the data processing stage is complete, then analyse and draw conclusions and suggestions.

3. Result and Discussion

Observation data was to data were by using RULA worksheet data to measure work posture by dividing observations of administrative body personnel into two groups, A and B. The activity carried out was to determine the RULA score from the results of a combination of A and B scores so that a grand score was found. After that, determine the level of action from the RULA calculation on the work posture. Then continued by assessing the results of the NBM questionnaire which was filled in by the administrative staff. Based on the results of data processing with a calculation that aims to get the RULA worksheet and NBM questionnaire assessment, it will obtain various levels of action against work posture. From the level of action, RULA was identified and analyzed to be associated with the NBM questionnaire. From the results of the analysis, after doing their work activities can be minimised by change work posture.

Figure 1. Office Employee

3.1 Group a posture

- A posture of the upper arm forms an angle > 20 given a score of 2
- Forearm posture forms an angle of 60-100, score 1

- Wrist section <15 is given a score of 2
- Wrist rotation is in the midline with a score of 1 Activity score
- Activities carried out repeatedly, more than four times/minute with a score = 1
- Load score
- Load <2 kg with a score = 0 The total score for group A is 3 + 1 = 4

3.2 Group B posture

- Neck posture (neck) The neck forms an angle of <20 given a score = 2
- Body posture (trunk) The trunk forms an angle of 0-20 given a score = 2
- Group B posture score is = 2Activity score
- Activity score, Activities are carried out repeatedly, more than four times/minute with a score = 1
- Load Score, Load <2 kg with a score = 0 The total score for group A is 3 + 1 = 4

Table 1. The Calculation Result

Group A			Rula Score Sheet			Group B	
	Right	Left					
Upper Arm	2	2				2	Neck
Lower Arms	1	1	Use Table A		Use Table B	2	Trunk
Wrist	2	2	3	3	3	1	Legs
Wrist Twist	1	1					
			+		+		
			Muscle Use		Muscle Use		
			1		1		
			+		+		
			Force/ Load		Force/ Load		
			0		0		
			=		=		
Score C			4	4	4	Score D	
4			Grand Score (Score C & D)			4	
			R: 4		L: 4		

The final score for office staff activities with sitting posture based on Table is = 4. Based on the score, the level of risk of office employee activity with sitting posture is in the medium risk level category, and further investigation is required, there may be a change in the improvement of work attitude. After knowing the level of risk in the form of a score value on the RULA method for office workers at PT. Chemistry. Risk control is needed in the form of improvement of work postures carried out by workers. These improvements are carried out by adding or redesigning work facilities in the workspace. Risk control in the work area is a process to eliminate or reduce risk factors that have been identified and assessed using the RULA method. The improvements that will be made to the work area aim to reduce the risk of injury to the lowest acceptable level without a risk that can cause injury or accident.

The risk control that will be provided refers to the risk control hierarchy, elimination, substitution, design (engineering control), administrative control, and Personal Protective Equipment (PPE). The controls that will be applied by the conditions of the work area are the control of Engineering (engineering Control) and Administrative (Administrative Control).

The application in analysing RULA is used as a tool to carry out an initial analysis that can determine how far the risk of workers to be affected by the factors that cause injury (posture, static muscle contraction, repetitive movements, and style). The application used is CATIA, in which there is a picture of a visual appearance that can be used to find the best work posture.

Figure 2. Determination of manikin in CATIA

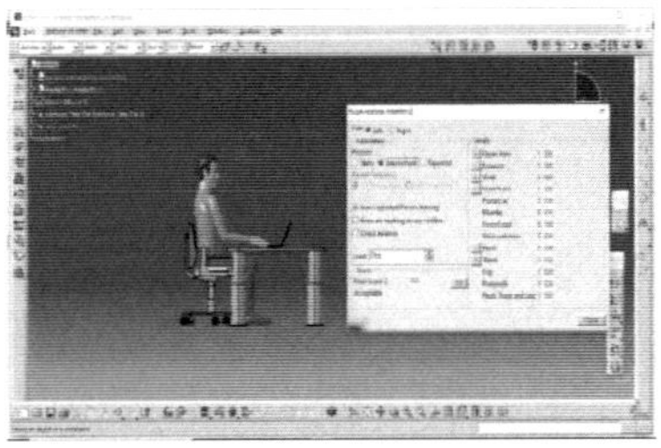

Figure 3. CATIA results Look right.

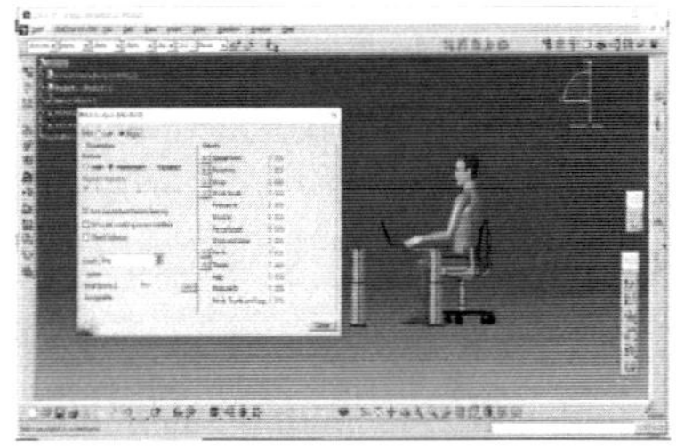

Figure 4. CATIA results Look left

From the results of the RULA meeting using simulation Catia on the redesign, calculations can be known that the RULA obtained in number 2 means that it is safe to be used by workers to do their work. For the redesign, results can be seen in the picture below. Previously it can be seen in table 1 that having a RULA 4 can be reduced to 2, by using the redesign that has been applied.

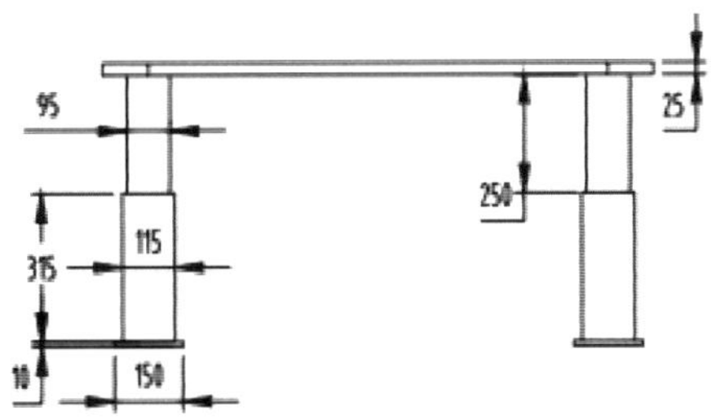

Figure 5. Redesign of workbench

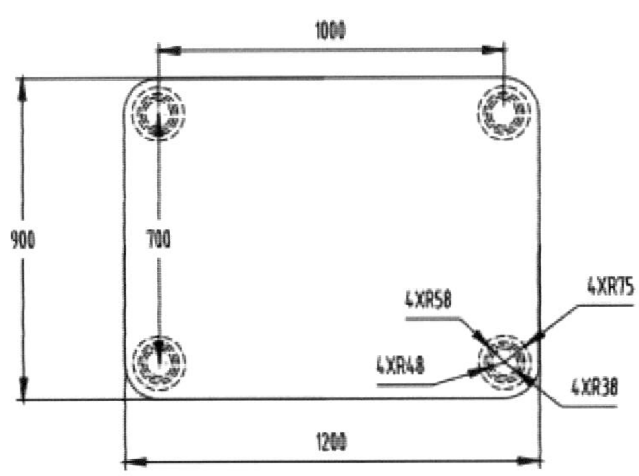

Figure 6. Detail drawing

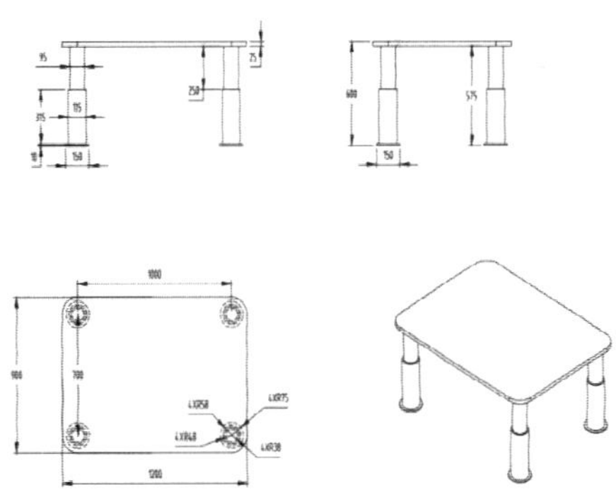

Figure 7. Detail drawing

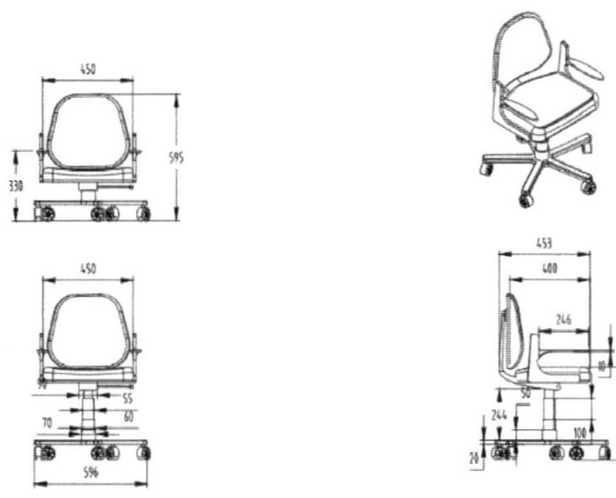

Figure 8. Redesign of work chairs

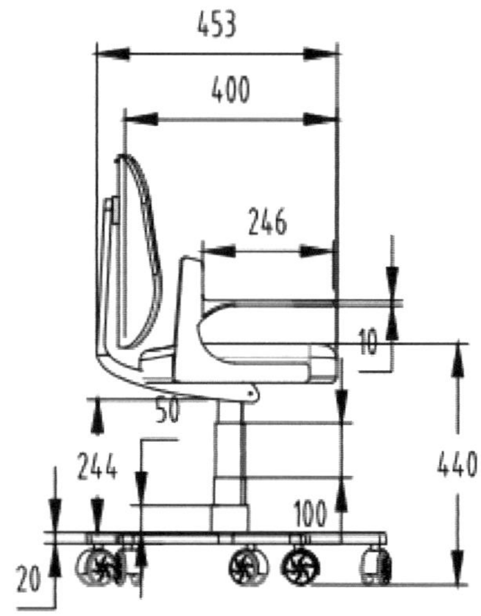

Figure 9. Detail of the redesign of work chairs

From the results of the redesign, it can be said that the results of RULA can go down from the initial value of 4 to 2, of which value 2 is a safe value for work.

The following is an explanation of the advantages of the new design work facilities in the form of tables and chairs after the application is applied to Local Control Room Operators, Office Workers, and Heavy Equipment Operators:

The advantages of the new design of tables and chairs are re-designed by the standards of the Permenaker, and the mannequin in Catia uses the 95th percentile which is average for Asians. For a workbench, there is an addition to adjusting the height of the table with a height of 58 cm to 68 cm. Moreover, the high change from the work table is beneficial to improve the worker's body position. So with the design, the RULA score is initially valued at 4, now it has been reduced to 2 which means the work posture of the workers has been excellent.

4. Conclusions

The recommended changes to the ergonomic desk and chair design at PT Kimia is to adjust to the anthropometry of the human body, as well as provide a soft upholstery on the back and seat backs and provide additional support for heavy equipment operator seats. The recommendation of the chair and table design in this study is feasible to use because the ergonomics can be tested using posture analysis and RULA test on CATIA applications.

5. References

[1] Ahmadi, Abu (2012) *Psikologi Belajar*. Jakarta. Rineka Cipta

[2] Bhaskara, W. R. (2011). *Studi Postur dan Gerak Dinamis Menggunakan Permodelan dan Simulasi Gerak 3D Berbasis Skeletal System.* Surabaya: Tugas Akhir Jurusan Teknik Industri Institut Teknologi Sepuluh Nopember (ITS).

[3] NHIS. (2008). *Musculoskeletal Disorders. Centers For Diseases Control and Prevention* (CDC) http://www.cdc.gov/nchs/nhis.htm (Sitasi Juni 2014)

[4] Notoatmodjo. (2005). *Metode Penelitian Kesehatan. Jakarta* : Rineka Cipta

[5] Nurmianto, Eko. (2004). *Ergonomi, Konsep Dasar dan Aplikasinya*. Surabaya : Prima Printing

[6] OSHA. 2011. *Prevention of Musculoskeletal Disorders in the Workplace*. U.S. Departement of Labor. https://www.osha.gov/SLTC/ergonomics/ (Sitasi Februari 2014)

[7] Suma'mur PK. (2009). *Higiene Perusahaan dan Kesehatan Kerja*. Jakarta: CV. Haji Massagung.

IOP Conf. Series: Materials Science and Engineering **462** (2019) 012041 doi:10.1088/1757-899X/462/1/012041

[8] Suyatno. (1985). *Meningkatkan Produktivitas Dengan Ergonomi*. PT. Pustaka Binaman Pessindo. Jakarta.
[9] Tarwaka, P. (2010). *Ergonomi Industri Dasar-Dasar Pengetahuan Ergonomi dan Aplikasi di Tempat Kerja*. Surakarta: Harapan Press.
[10] Tarwaka, P. (2004). *Ergonomi untuk keselaamatan, kesehatan kerja dan produktivitas kerja*. Surakarta: Uniba Press.

An Analysis of Lead (Pb) Levels in the Urine of Gas Station Operators Based on Individual Characteristics (A Case Study at Kali Rungkut and Panjang Jiwo Gas Station Surabaya)

J Caroline[1], S Choiriyah[1] and G A Cristata[2]

[1]Department of Civil Engineering, Institut Teknologi Adi Tama Surabaya, Indonesia
[2]Department of Environmental Engineering, Institut Teknologi Adi Tama Surabaya, Indonesia

jennycar@itats.ac.id

Abstract. Improved means of transportation have several positive and negative impacts. The most-occurred adverse effects is a decrease in environmental quality. Emissions from vehicles will continue to increase every single day and will be the dominant source of air pollutants. Pb or lead is one of the contaminants of vehicle mobility. Gas station operators are at risk of having high Pb levels in their bodies. This happened because the gas station operators are exposed to Pb for 8 hours/day from motor vehicle emissions as well as gasoline fumes when consumers refuel. This study aims to analyse and determine the lead (Pb) levels in the urine of gas station operators based on their characteristics at Panjang Jiwo and Kalirungkut gas station Surabaya. This research is quantitative research with a descriptive method that compares the results of measurements in the field with the Decree of the Minister of Health of the Republic of Indonesia Number 1406 / Menkes / SK / XI concerning the Examination Standard of Black Lead Levels in Human Biomarker Specimen. Measurements were made on the urine of twelve gas station operators at two gas stations. The results of the study showed that the analysis of Pb levels in the urine of twelve respondents was below the threshold value of 0.1 mg/l. There is also a tendency that the longer the service life of the operator and the increasing age, the higher the Pb level in the urine of gas station operators.

1. Introduction

The increased means of transportation has negative impacts since it can reduce the environmental quality. Emission from vehicles will continue to increase every single day and will be the dominant source of air pollutants. In big cities like DKI Jakarta, the contribution of pollutants from motor vehicles is around 70%. The primary contaminants found in the emission of motor vehicles are carbon monoxide (CO), various hydrocarbon compounds, different nitrogen oxide (NOx) and sulfur (SOx), as well as particulate matters and dust including lead (Pb) [1]. According to Mifbakhuddin, lead is usually used as a mixture of gasoline fuel which serves to improve lubrication power and combustion efficiency so that motor vehicle performance increases [2]. These chemical substances are burned along with gasoline in the engine, and the remaining ±70% comes out with the emission from combustion. Librawati added that every litre of gas in 88 octane number contains lead is usually used as a mixture of gasoline fuel which serves to improve lubrication power and combustion efficiency so that motor vehicle performance increases 0.70 g of Tetraethyl Pb and 0.84 g Tetramethyl Pb compounds [3]. Every one litre of gasoline that was burned will emit 0.56 g Pb to the air. Gas station operators are one of the objects affected by Pb from vehicles. Every day they breathe pollutants from vehicle exhaust gases in their workplaces. Increased accumulation of Pb in the human body will cause gastrointestinal syndrome, decreased consciousness (cognitive effect), anaemia, kidney damage, hypertension, neuromuscular and psychological consequences, as well as central nerve damage, behavioural changes, and inhibited erythrocytes formation [2].

The gas station is one of the places with high-level Pb exposure and pollution because it is not only due to gasoline containing Pb but also due to many vehicles come in and out to refuel. The gas station operators are at risk of having high Pb level in their bodies. This due to the operators is exposed to Pb for 8 hours/day from the motor vehicle emission and gasoline fumes during consumer refuel process. Also, most gas station operators do not use any PPE (Personal Protective Equipment) while working [3]. This study aims to analyse and determine the lead (Pb) levels in the urine of gas station operators based on their characteristics study case at Panjang Jiwo and Kalirungkut gas station Surabaya. Therefore, research on the relationship between the period of employment and Pb levels in the gas station operator bodies through urine analysis was conducted. This research aims to find out and analyse lead (Pb) levels in the urine of gas station operators based on their characteristics at Panjang Jiwo and Kalirungkut gas station.

2. Literature Review

Pb in the atmosphere comes from incomplete combustion of regular and premium gasoline. Premium gasoline consists of 2-4 grams Pb per gallon with an average of 2.8 grams. Regular gasoline consists of 2.3 grams Pb in average per gallon. On average 70-80% Pb in gasoline is dispensed from the exhaust pipe as particulate. As an air pollutant, Pb is the result of gasoline combustion in the vehicle and lead emission in Co2 smelting [4]. Pb is already in gasoline fuel. Therefore, Pb becomes an element that is released freely in the air due to the incomplete excretion process in vehicle engine [5]. Pb enters the human body through various means, including through inhalation, gastrointestinal tract, and even dermal contact duct. However, breath it the main pathway of Pb exposure. For more stable pollutants such as Pb, some hydrocarbons-halogen, and polyaromatic hydrocarbons can fall to the ground along with rainwater or settle with dust, and contaminate soil and water. The compound can enter the food chain which eventually enters the human body through raw food [1]. The entry of Pb into the body of gas station operators can be through inhalation. Vehicle emissions are containing Pb and gasoline fumes while refuelling enters the gas station operator body and then distributed by blood to the soft tissues (nervous system, kidneys, liver) and hard tissues (bones, nails, hair, teeth) as well. That 90% of Pb entering the body will be stored in bone, and only a small portion is stored in other tissues [3]. Urine is a residual metabolic fluid produced by the kidney and excreted from the body through urinating. Urine consists of water and materials dissolved in it. These dissolved materials are in the form of body metabolic waste such as urea, evaporated salt, and other organic matters [6].

3. Research Method

This research is quantitative research with the descriptive method by comparing the results obtained from measurements in the field with the Decree of the Minister of Health of the Republic of Indonesia Number 1406/Menkes/SK/XI/2002 concerning Examination Standard of Black Lead Levels in Human Biomarker Specimen. This research was conducted at two gas stations (Panjang Jiwo and Kalirungkut) in Surabaya with urine as the test materials from 12 respondents; 6 respondents from Kalirungkut gas station and 6 respondents form Panjang Jiwo gas station.

4. Result and Discussion

The research was started by doing interviews about respondent characteristics, including age and the period of employment.

Table 1. Respondent Distribution Based on the Age of Gas Station Operators

No	Age (years old)	Frequency (people)	Percentage (%)
1	20-30	6	50
2	31-40	4	33.33
3	> 40	2	16.67
	Total	12	100

Based on the research results in Table 1, it is known that 50% of respondents aged between 20-30 years old (6 people) and only 16.67% (2 people) aged more than 40 years old.

Table 2. Respondent Distribution Based on the Period of Employment at Gas Station

No	Period of Employment (year)	Frequency (people)	Percentage (%)
1	1-10	5	41.7
2	11-20	7	58.3
	Total	12	100

Based on the research results in Table 2, it is known that 58.3% of respondents have been working for more than ten years (7 people) and only 41.7% have been working for less than ten years (5 people).

Pb Levels in the Urine of Gas Station Operators

Table 3. Pb Levels in the Urine of Gas Station Operators Based on Their Age and Period of Employment

No.	Gas Station	Sample Code	Age (years old)	Period of Employment (year)	Pb Levels in the Urine (mg/l) (SNI<0.150)
1	Kalirungkut Gas Station	A1	24	6	0.096
2		A2	21	4	0.078
3		A3	41	20	0.114
4		A4	48	20	0.128
5		A5	25	7	0.082
6		A6	30	11	0.091
7	Panjang Jiwo Gas Station	B1	31	13	0.102
8		B2	34	16	0.118
9		B3	36	18	0.132
10		B4	22	5	0.108
11		B5	35	15	0.135
12		B6	20	2	0.071
	The lowest		20	2	0.071
	The highest		48	20	0.135

According to Table 3, all 12 respondents have Pb levels below the threshold value of Permenkes No. 1406/Menkes/SK/XI/2002 which is 0.150 mg/l, with the lowest Pb level of 0.071 mg/l and the highest Pb Level of 0.135 mg/l.

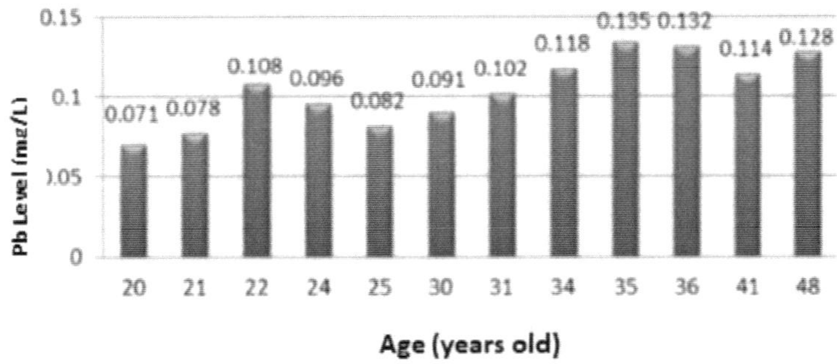

Figure 1. The Relationship between Gas Station Operators' Age and Pb Level in the Urine

According to Figure 1, the oldest respondent is 48 years old with Pb level of 0.128 mg/L and the youngest with 0.071 mg/L. However, the highest Pb level is in the 35-year-old respondent. This difference shows that the longest the exposure time or period of employment may affect the accumulated Pb in the respondent's body. High Pb level may be affected by how much exposure they have got and operators' age is potential to increase Pb levels in the body. Pb content in adults differs due to environmental and geographical factors in which they are living [1-2].

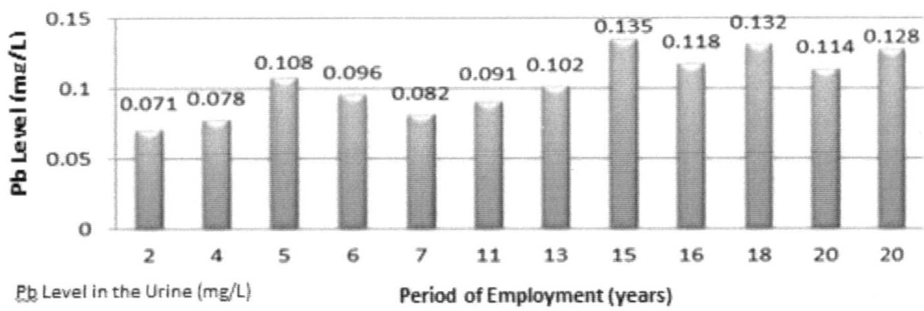

Figure 2. The Relationship between Gas Station Operators' Period of Employment and Pb Level in the Urine

Figure 2 shows the highest Pb level (0.132 mg/L) is in the operator who has been working for 15 years, while those who have been working for 20 years have Pb levels of 0.128 mg/L and 0.114mg/L respectively. According to Wahyudi [1] in Mifbakhuddin [2] the function of the immune system in the body decreases according to the age. The biological half-life of Pb in human bones is estimated to be 2-3 years. Pb in the blood can be detected in the half-life of 20 days, whereas the total Pb excretion in the body occurred within a half-life of about 28 days. From the blood and the deposit place, Pb then excreted through urine, faeces and sweat [3], while Pb excretion through kidney is through glomerular filtration. This explained that the longer the exposure time or period of employment, the more accumulated Pb in the gas station operators' body. Pb content in the body is used for occupational exposure. In general, Pb excretion runs very slowly [4].

The analysis results showed that all 12 respondents had Pb level of less than 0.150 mg/L, but over time, the source of exposure could be more significant. The length and amount of exposure and the increasing age might increase Pb accumulation in the body. Thus, the gas station executives need to take precautions to minimise the entry of Pb into the operators' organisation such as providing knowledge about the source of exposure, hazardous effects, control of Pb exposure prevention, and the use of mask while working. The gas station executives should review the policy of not allowing the use of cover while working since the effects of accumulated Pb may cause poisoning and fatal consequences [5].

5. Conclusion
Pb levels in the urine of 12 respondents were under the threshold value of Kepmenkes No. 1406/Menkes/SK/XI/2002 concerning Examination Standard of Black Lead Levels in Human Biomarker Specimen which is 0.150 mg/L. There was a tendency in which the longer the period of employment, the higher Pb levels in the urine of gas station operators. There was a tendency in which the older the operator, the higher Pb levels contained in the urine.

6. References
[1] Wahyudi, Mohamad Tri. 2012. Sistem Gas Buang Sepeda Motor. Yogyakarta : PT Pustaka Insan Madani
[2] Mifbakhuddin, et al. 2010. Hubungan Antara Paparan Gas Buang Kendaraan (Pb) Dengan Kadar Hemoglobin Dan Eritrosit Berdasarkanlama Kerja Pada Petugas Operator Wanita SPBU Di Wilayah Semarang Selatan. Jurnal. Fakultas Kesehatan Masyarakat Universitas Muhammadiyah Semarang : Vol. 6 No. 2 Tahun 2010
[3] Khairunisa, et al. 2016. Analisis Kadar Timbal (Pb) Dan Hemoglobin (Hb) Pada Operator Spbu Di Kota Pekanbaru Dan Potensinya Sebagai Rancangan Handout Dalam Pembelajaran Biologi Di SMA. Skripsi. Program Studi Pendidikan Biologi Fakultas Keguruan dan Ilmu Pendidikan Universitas Riau
[4] Riyadina, Woro. 1997. Pengaruh Pencemaran Pb Terhadap Kesehata. Pusat Penelitian Penyakita Tidak Menular. Badan Penelitian dan Pengembangan Kesehatan Depkes RI. Media Litbang : Vol. VII No. 03 & 04 / 1997
[5] Rosyidah, Hesti. 2010. Hubungan Antara Kadar Pb Dalam Darah Dengan Kejadian Hipertensi Pada Operator SPBU Di Kota Yogyakarta. Jurnal. Fakultas Kesehatan Masyarakat, Universitas Ahmad dahlan, Yogyakarta : Vol. 4.No. 2, JUNI 2010 : 76 – 143. ISSN : 1978-0575
[6] Retnaningtyas, Iklimahika. 2012. Sistem Ekskresi. http: //skp.unair.ac.id/repository/GuruIndonesia/SistemEkresi_iklimahikaretnaningtyas_14335.pdf

IOP Conf. Series: Materials Science and Engineering **462** (2019) 012043 doi:10.1088/1757-899X/462/1/012043

An Analysis of Concrete Test Weight with Different Water Cement Factors Using Histogram in Quality Management

S Choiriyah[1] and J Caroline[1]
[1]Department of Civil Engineering, Institut Teknologi Adhi Taman Surabaya, Surabaya 60117, Indonesia

siti.choiriyah@itats.ac.id

Abstract. Concrete is a product with the composition of binding medium in which there is an aggregate pigments or particles. The binder is formed by cement and water mixture. Cater cement factor affects the compressive strength result. In producing concrete, a quality control is needed to obtain performance or process variation. In order to obtain the quality, statistical analysis using histogram method in the form tables and graphs is needed. From the measurement analysis results of 30 concrete test objects with different cement water factors getting uneven and unpredictable distribution results. The greater the cement water factor, the more dispersed the distribution of the cement will be but the weight of the concrete will decrease.

1. Introduction

Concrete is naturally a material that is a basic form of modern social life and is a composite material consisting of coarse and fine aggregates mixed with water and cement as binder and filler between the aggregates [1]. The weight of concrete test object is also strongly influenced by water/cement ratio [2]. To get the performance or variation process, it is necessary to have a quality control. Quality management is aspects of the overall management function that implements and carries out the quality of a company / organization, in order to meet customer needs and time requirements with a economical and economical budget, a project manager must enter & conduct quality management training [3]. The measurement process for concrete quality can be obtained from concrete testing, namely compressive strength, weight, volume and specific gravity [4]. For measuring the difference in the concrete mixture process, histogram method was employed. Histogram is a tool or method for making data summary to be easily analyzed [5]. The presentation of the bar diagram is depicted with cells that have the same area of data between cell 1 and the other with no distance, because the data includes continuous data that continues so that the one with the next one must coincide [6]. In the manufacture of concrete there are several things that need to be considered in concrete quality control with variations in concrete production, namely variations in cement water factors [7].

2. Literature Review

Quality control techniques so that data is easily analyzed is using a histogram method that has the benefits of: obtaining process stability, performance or process variation, testing, evaluating, developing and monitoring process improvement [8]. Quality control techniques to ease data analysis process was by using histogram method with the benefits including: obtaining the process stability, performance or process variation, testing, evaluating, developing and monitoring process improvement [9]. Concrete is a complicated composite material, can be made easily even by those who have no understanding of concrete technology, but this misconception of simplicity often results in problems with products, including the reputation of concrete as building material [10].The weight of concrete is also a very decisive factor since the measurement of concrete test results are obtained from compressive strength, weight, and volume. The water cement ratio is the ratio of the weight of water to the weight of cement used in a concrete mix. A lower ratio leads to higher strength and durability, but may make the mix difficult to work with the use of plasticizers or super plasticizers [11].

3. Method

The data source used in the histogram analysis was test/measurement results of concrete test object from the laboratory as many as 30 specimens with Water Cement Factor (WCF) of 0.4, 0.5 and 0.6 at 28-day-old concrete.

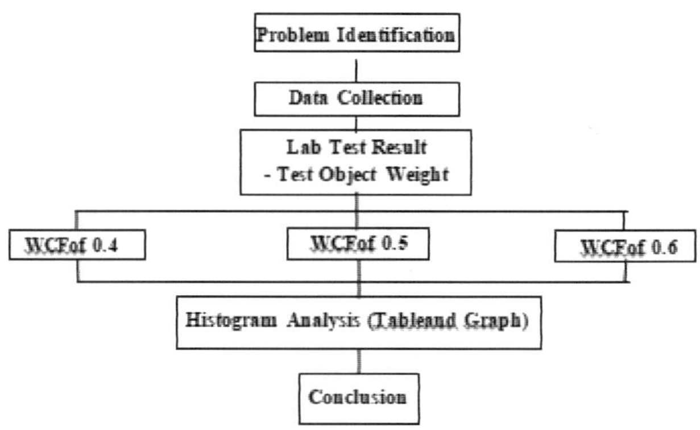

Figure 1. Flow Diagram

The several steps requird to contructs histogram beginning with computing the range (R) of data [12] :
Step 1 : compute the range (R) of the data. R = Xmax – Xmin
Step 2 : Compute the appropriate number of classes (k): K = √n, round up a k to of 6 classes.
Step 3 : Compute the appropreate class with (h): h = R/k
Step 4 : Determine the unit of measurement (m): 1,00
Step 5 : Finalize the class width (h) by rounding up to the next highest to measure, So highest h = 20
Step 6 : Compute the lower boundary of the first class (L1). L1 = (Xmin-m/2)
Step 7 : Determine th lower boundary of the reaming of the classes:

L2 = L1 + h
L3 = L2 + h
L4 = L3 + h
L5 = L4 + h
L6 = L5 + h
L7 = L6 + h

Step 8 : Using the class intervals, a frequency table is constructed shown. Note the upper boundary for L7 is also shown.
Step 9 : Making diagram number of traffic fines paid and fine paid.

4. Result and Discussion

This research results on the measurement of test object weight and histogram analysis are presented in the form of table and graph.

4.1 The weight of concrete test object with WCF of 0.4

Table 1. The weight results of 28-day concrete test object.

2.9	3.25	2.9	2.975	2.98	3.22
3.1	2.95	2.95	3.15	3.21	2.95
2.95	3.2	2.915	3.22	3.22	3.2
3.22	3.225	3.22	3.22	3.21	3.2
3.15	3.1	3.2	3.21	2.96	2.95

Computation:

A. Computing the total :
   X = 93.105
   Computing the mean :
   X’ = x/n = 93.105/ 30 = 3.10
B. Computing the range (R) of the data :
   R = X max – X min
   R’ = 3.225 – 2.90 = 0.325
C. Computing the approximate class number of classes (k)
   K = √n = √30 = 5.48 = 6
D. Computing the approximate class width :
   H = R/K = 0.325 / 6 = 0.054 = 0.1
E. M = 0.01
F. Finalizing the class width (h) by rounding up to the next highest of measure :
   h = 0.1
G. Computing the lower boundary of the first class :
   L1 = Xmin – (m/2) = 2.90 – (0.01/2) = 2.895
H. Determining the lower boundaries of the remaining :
   L2 = L1 + h = 2.895 + 0.1 = 2.995
   L3 = L2 + h = 3.895 + 0.1 = 3.095
   L4 = L3 + h = 4.895 + 0.1 = 3.195
   L5 = L4 + h = 5.895 + 0.1 = 3.295
   L6 = L5 + h = 6.895 + 0.1 = 3.395
   L7 = L6 + h = 7.895 + 0.1 = 3.495
I. Using the class intervals, a frequency table is constructed as shown.
   The upper boundary note for L7 is also shown.

IOP Conf. Series: Materials Science and Engineering **462** (2019) 012043 doi:10.1088/1757-899X/462/1/012043

Table 2. 0.4 WCF Histogram

Class	Tally	Frequency
2.895 - 2.995	IIIIIIIIIII	11
2.995 - 3.095		0
3.095 - 3.195	IIII	4
3.195 - 3.295	IIIIIIIIIII II	13
3.295 - 3.395		0
3.395 - 3.495		0

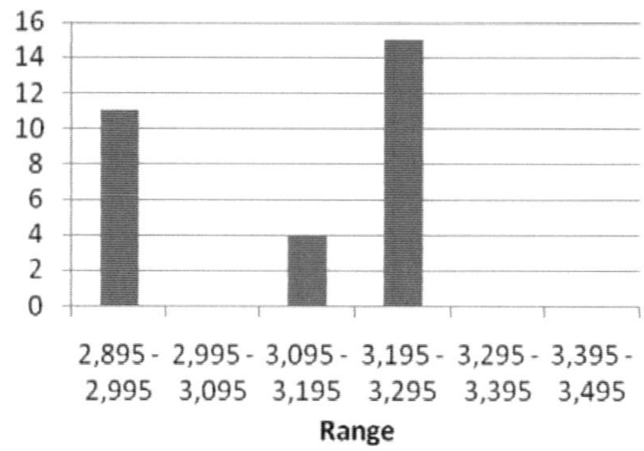

Figure 2. The Graph of 0.4 WCF Histogram

Based on histogram analysis, the weight spread of concrete test subject was uneven and 3 ranges were not fulfilled.

4.2 The weight of concrete test subject with 0.5 WCF

Table 3. The weight results of 28-day concrete test object.

2.9	3.25	2.9	2.975	2.975	3.15
3.1	2.95	2.95	3.15	3.21	2.95
2.95	3.2	2.92	3.22	3.22	3.22
3.22	3.225	3.22	3.22	3.1	3.22
3.15	3.1	3.2	3.21	2.92	2.92

Computation:

A. Computing the total:
X = 92.885
Computing the mean:
X' = x/n = 92.885 / 30 = 3.09

B. Computing the range (R) of the data:
R = X max – X min
R' = 3.22 – 2.90 = 0.32

C. Computing the approximate class number of classes (k)
K = √n = √30 = 5.48 = 6

IOP Conf. Series: Materials Science and Engineering **462** (2019) 012043 doi:10.1088/1757-899X/462/1/012043

D. Computing the approximate class width:
   H = R/K = 0.32 / 6 = 0.053 = 0.1
E. M = 0.01
F. Finalizing the class width (h) by rounding up to the next highest of measure:
   h = 0.1
G. Computing the lower boundary of the first class:
   L1 = Xmin – (m/2) = 2.90 – (0.01/2) = 2.895
H. Determining the lower boundaries of the remaining:
   L2 = L1 + h = 2.895 + 0.1 = 2.995
   L3 = L2 + h = 3.895 + 0.1 = 3.095
   L4 = L3 + h = 4.895 + 0.1 = 3.195
   L5 = L4 + h = 5.895 + 0.1 = 3.295
   L6 = L5 + h = 6.895 + 0.1 = 3.395
   L7 = L6 + h = 7.895 + 0.1 = 3.495
I. Using the class intervals, a frequency table is constructed as shown.
   The upper boundary note for L7 is also shows.

Table 4. 0.5 WCF Histogram

Class	Tally	Frequency
2.895 - 2.995	IIII IIII III	11
2.995 - 3.095		0
3.095 - 3.195	IIII II	6
3.195 - 3.295	IIII IIII IIIII	13
3.295 - 3.395		0
3.395 - 3.495		0

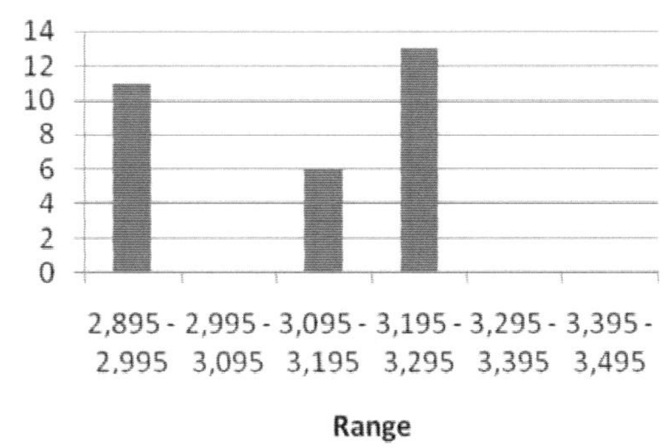

Figure 3. The Graph of 0.5 WCF

Based on histogram analysis, the weight spread of concrete test subject was uneven and 3 ranges were not fulfilled.

4.3 The weight of concrete test subject with 0.6 WCF

Table 5. The weight results of 28-day concrete test object

2.56	3.2	2.21	3.2	2.9	2.9
2.92	2.2	2.92	3.3	3.5	3.2
3.27	2.56	3.27	3.38	3.2	3.38
3.23	3.3	3.23	3.5	3.2	2.9
3.2	2.79	3.38	3.9	2.95	3.2

Computation:

A. Computing the total:
 X = 92.85
 Computing the mean:
 X' = x/n = 92.85 / 30 = 3.09

B. Computing the range (R) of the data:
 R = X max – X min
 R' = 3.90 – 2.79 = 1.11

C. Computing the approximate class number of classes (k)
 $K = \sqrt{n} = \sqrt{30} = 5.48 = 6$

D. Computing the approximate class width:
 H = R/K = 1.11 / 6 = 0.185 = 0.1

E. M = 0.01

F. Finalizing the class width (h) by rounding up to the next highest of measure:
 h = 0.1

G. Computing the lower boundary of the first class:
 L1 = Xmin – (m/2) = 2.79 – (0.01/2) = 2.785

H. Determining the lower boundaries of the remaining:
 L2 = L1 + h = 2.785 + 0.1 = 2.885
 L3 = L2 + h = 2.885 + 0.1 = 2.985
 L4 = L3 + h = 2.985 + 0.1 = 3.085
 L5 = L4 + h = 3.085 + 0.1 = 3.185
 L6 = L5 + h = 3.185 + 0.1 = 3.285
 L7 = L6 + h = 3.285 + 0.1 = 3.385

I. Using the class intervals, a frequency table is constructed as shown.
 The upper boundary note for L7 is also shown.

Table 6. 0.6 WCF Histogram

Class	Tally	Frequency
2.895 - 2.995	|||| |||| |||	11
2.995 - 3.095		0
3.095 - 3.195		0
3.195 - 3.295	|||| |||| ||||	12
3.295 - 3.395	||||	4
3.395 - 3.495	|||	3

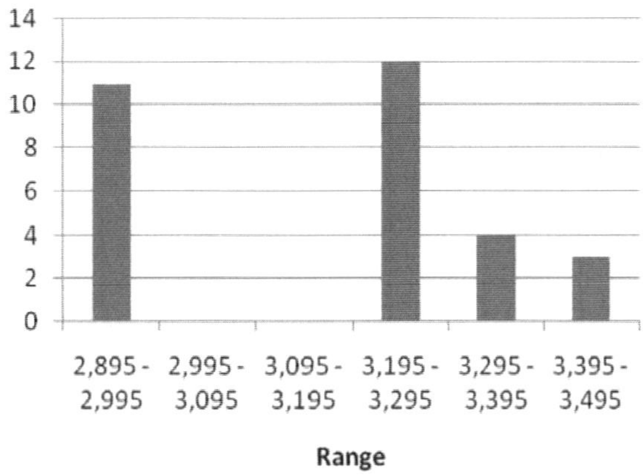

Figure 4. The Graph of 0.6 WCF

Based on histogram analysis, the weight spread of concrete test subject was uneven and 2 ranges were not fulfilled

5. Conclusion

From table 1,2,3 and images of histograms 2,3,4 on specimens with FAS (0,4) FAS (0,5) and FAS (0,6), the results are uneven distribution. Indicates that the process is unstable and unpredictable with almost the same graphical form. Then the greater the FAS, the more spread the test specimen (FAS = 0.6).

6. References

[1] Subakti, Aman, 2012. Teknologi Beton Dalam Praktek I. Surabaya: ITS press
[2] A.M. Naville; J.J.Brocks "Concrete Technology" c Logman Group Ltd. 1990
[3] Wijaya, Buletin BAPEKIN, KIMPRASWIL Edisi 4 http://huangcrop-wordpress-com.cdn.ampproject.org/v/s/huangcorp.wordpress.com/2008/04/29/manajemen-kualitas-quality-management/amp/?amp_js_v=a2&amp_gsa=1&usqp
[4] Sadji. 1982. Teknologi Beton. Surabaya: ITS
[5] Sajiyo. 2012. Quality Management. Surabaya: Untag
[6] Purwono, Rachmat. 2010. Pengendalian Mutu Beton Sesuai SNI, ACI dan ASTM. Surabaya : ITS press
[7] Sajiyo. 2012. Quality Management. Surabaya: Untag
[8] Hendra Poerwanto G https://sites.google.com/site/kelolakualitas/Histogram
[9] Paul Nugraha Antoni. 2007. Teknologi Beton Dari Material, Pembuatan, ke Beton Kinerja Tinggi. Surabaya: Penerbit Andi
[10] http://en.m.wikipedia.org/wiki/Water-cement_ratio
[11] Budi Kho, "Pengetahuan Tentang Manajemen Produksi & Operasional, Manajemen SDM & Manajemen Kualitas', 1 Juli 2016

Implementation IMO Regulation of Ballast Water Management at Inaport 2nd Jakarta Based Environmental Risk Assessment

Minto Basuki[1], Lukmandono[2], Maria Margareta Zau Beu[1].

[1] Naval Architecture and Vessel Building Engineering Department, Institut Teknologi Adhi Tama Surabaya, Jln Arif Rachman Hakim 100, Surabaya, Indonesia.
[2] Industrial Engineering Departement, Institut Teknologi Adhi Tama Surabaya, Jln Arif Rachman Hakim 100, Surabaya, Indonesia

mintobasuki@yahoo.co.id

Abstract. This study aims to analyse the application of the rules of IMO (International Maritime Organization) relating to ballast water management to vessel back at the port in the operational area of Inaport 2nd Jakarta. IMO rules governing vessel ballast water management, are applied internationally starting September 8, 2017. Data samples are taken from the number of vessel visits, especially international vessel routes that visited in 2012-2016 and disposed of ballast water at the operational port of Inaport 2nd. The analytical method uses qualitative and deterministic based on the formulation of IMO MEPC 56/23 ANNEX 2. The volume of discarded ballast water is varied for 10% of vessel visiting up to 60% of the vessels visiting, with the increment of 5 to 25 tons of ballast water which is disposed of by each vessel. In 2016, for 60% of vessel that visited and disposed of 25 tons, the waters of Inaport 2nd would receive 48,741.06 kL of ballast water. The increase in the number of vessel and the amount of ballast water that is disposed of will increase the potential of invasive follow-up species to local streams and increase marine pollution in the port area.

1. Introduction

The vessel is a floating building that moves dynamically with humans and goods transported and moved from one place to another. As part of a huge quantity of products and human transport systems, the vessel must be safe and secure until the goal is to move the cargo. The vessel as a means of transportation must be protected starting from the shipbuilding process to the operational stage of the vessel so that it is seaworthy. Likewise, environmental protection due to vessel operations. Protection of people, goods and the environment due to vessel operations requires rules, both nationally and internationally. Non-compliance with national and international regulations resulted in legal risks, including the prohibition of sailing, the detention of vessels and crew, fines.

International and national vessel companies must comply with international and domestic rules; these rules are in the form of law and regulation. This rule is internationally grouped, promulgated and organised by an international maritime institution, namely the International Maritime Organization (IMO). This organisation is a combination of several countries in managing maritime and then generally regulated in IMO. IMO, in carrying out its duties and functions to form several committees, as well as a combination of several committees with different objectives, the different purposes are based on differences in their interests, including: (i) the MSC (Marine Safety Committee) this committee deals with and engages in safety, both vessel safety, people and the environment. This committee has also produced several rules, either in the form of conventions or codes, one of which is a convention on life safety at sea or known as the SOLAS (Safety of Life At Sea) convention. (ii) the MEPC (Marine Environment Protection Committee) working area of this committee is specifically tasked and focused on environmental protection due to the operation of the vessel and other floating equipment; this committee has produced a regulation on environmental protection called MARPOL (Marine Pollution) in 1973, then an amendment in 1978.

Some research that has been done by people related to Environmental Risk Assessment in shipyard companies, among others, uses several methods. Green shipbuilding and recycling related analysis uses literature review, and focuses on the by-products of shipyard companies, in the form of water, air and soil pollutants, to increase socio-economic factors [10]. Evaluating the chemical content of 12 metal concentrations in sediments on the coast of Karachi, Pakistan, this study uses the Numerical Quality Sediment Guidelines approach, due to the operation of shipyard companies [6]. Risk assessment on the transport of gas hydrate using the semi-quantitative risk method to determine the risk index of 80 risks identified, three risks cannot be accepted, 28 risks can be accepted but in the ALARP risk level [9]. Assessment of environmental pollution due to the operation of shipyard companies in South Korea, the characteristics of pollutants include dust, odour and noise [3]. Conducted research on the effects of contaminants, especially water pollution near shipyard operations and other scrap industries around West Bengal India and their impact on the DNA of workers involved in shipyard companies [7]. Study of absence rates due to illness in the private sector in Greece, by taking data samples on shipyard company workers and data from insurance [4].

Research into environmental pollutant factors, especially the effect on the DNA of workers exposed to paint particles in the industry [2], this study compared workers exposed to paint particles and workers not exposed to paint particles within a week of working time. An Ecological Risk Assessment study to assess risks to the environment using the Relative Risk Model combined with the Monte Carlo method [1]. Analysis of environmental influences in the form of pollutant material due to the ship breaking process of animal and plant life in the waters around the ship scraping process. The analysis was carried out based on recorded data statistics, published data and direct observation data using remote sensing [5].

Referring to the regulations of IMO (International Maritime Organization) which has been classified since 2004, that all types and types of vessels with sizes above 400 GT (Gross Tonnage) and have non-domestic vessel routes, must undergo inspection and have certificates relating to BWM (Ballast Water Management). Vessels with a size below 400 GT (Gross Tonnage) can follow the inspection requirements and have a document in their own country, where the vessel is registered. These provisions can be adopted by the government of the Republic of Indonesia, domestic (domestic) vessels can follow the rules that will be made by the Government of the Republic of Indonesia itself [12].

The adverse impact of the commercial ship of deballasting on the aquatic and economic ecosystems of the Tanjung Emas Port in Semarang has been analyzed by Tjahjana [13]. Based on the study, ballasting and deballasting to vessels that were disposed of containing invasive species, so that further processing was necessary. The presence of invasive species, proposed a model of ship water management in Inaport 3th can be carried out by accommodating vessel's water in a shelter, then processing it with a capacity of 51,090 m3 per month or 81,744 kL per month. Based on data from the IMO (International Maritime Organization), in one year, for vessels sailing in the waters around the world, the vessels contain 10 billion tons of water ballast. Thousands of marine species will also contain the vessel's ballast water with micro size in it and included in the voyage until the destination port. Based on data from the IMO

(International Maritime Organization) also, that per day, in one-hour duration of 7,000 species moved from one water location to another due to ballasting and deballasting processes [11].

2. National and International Rules of Ballast Water Management

As a State with a maritime basis, the Government of the Republic of Indonesia has regulated the territorial waters owned by Law Number 6 of 1996 concerning Indonesian Waters. Law Number 6 of 1996 concerning Indonesian Waters as an integral part of Law Number 17 of 1985 concerning the Ratification of the United Nations Convention on the Law of the Sea (United Nations Convention on the Law of the Sea). The Government of the Republic of Indonesia has also regulated the process and procedures for vessel operations in the form of Law Number 17 of 2008 concerning shipping. Support for Law Number 17 of 2008 concerning ship is Law Number 43 of 2008 concerning State Areas. In line with these laws and regulations, the government has also issued Government Regulations. Government Regulations that regulate include: (i) Government Regulation Number 19 of 1999 concerning Control of Sea Pollution and/or Destruction, (ii) Government Regulation Number 21 of 2010 concerning Maritime Environmental Protection [14].

Internationally the legislation relating to ballast water management is regulated by the IMO (International Maritime Organization). These laws and regulations were born from an international convention, which was agreed upon at a Diplomatic conference held at the headquarters of the IMO (International Maritime Organization) in London, England on February 13, 2004. IMO enacted regulations relating to ballast water management (International Maritime Organization) aims to minimise the risk of new invasive species entering other waters. The IMO (International Maritime Organization) regulation which regulates ballast water management, is divided into two parts. Standard D-1 (Ballast Water Exchange), which is still valid to date is carried out by rinsing the water three times in the sea which is more than 200 nautical miles from the beach with a depth of more than 200 meters. The next standard is Standard D-2 (Ballast Water Treatment). This standard requires the treatment of ballast water which is found to contain more than ten microorganisms per cubic meter which is more than or equal to 50 microns. With this water treatment, there will be no more microorganisms that will escape to the new environment, so that environmental damage can be prevented.

3. Ballast Water as Marine Pollution in Inaport 2nd

Every day, an estimated 10,000 marine species are transported around the world in ballast water. This high level of invasive species being transported and discharged into inhabitable waters has called for legal action to take place. Conducting IMO MEPC 56/23 ANNEX 2 document, the overall probability of a successful invasion also depends in part on the number of organisms and the frequency with which they are introduced over the entire period of the exemption. Therefore, it is recommended that a risk assessment should consider estimates of at least the following four factors: (i) the total volume of water discharged, (ii) the amount of water discharged in any event (voyage), (iii) the total number of discharge events and (iv) the temporal distribution of discharge events [8].

The amount of water returned by vessels that were dumped at the Port under the auspices of Inaport 2nd from 2012 to 2016 was used for analysis. The ballast water discharged by the vessel in 2016 as in Figure 1 and Figure 2 is analysed based on International vessel visit data. This was done, because vessels with international routes were suspected to have had invasive follow-on species in the dumped water that was thrown away. Formulation of ballast water discharged by the vessel also includes a factor of 95% confidence level. Based on Figure 1 and Figure 2, the ballast water discharged by the vessel is increment for each vessel of 5 tons to 25 tons. The number of vessels that dispose of ballast is also propagated between 10% and 60% of the total vessel visits.

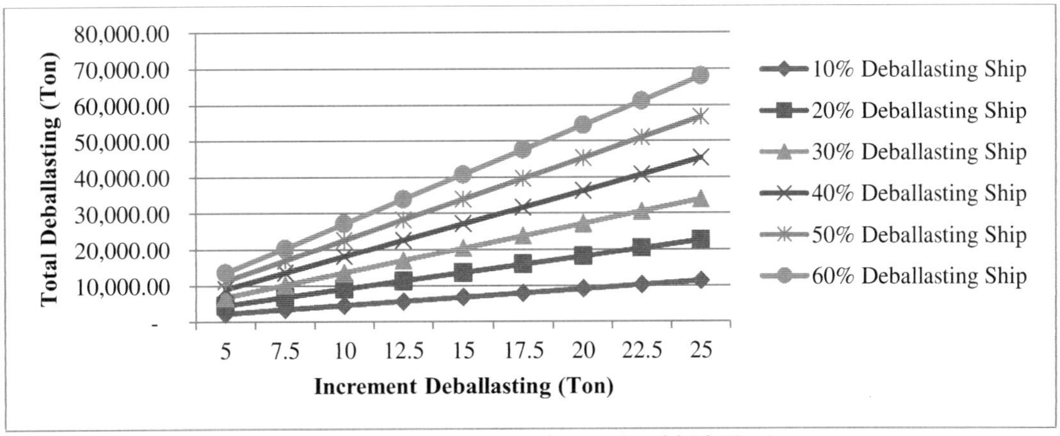

Figure 1. Deballasting of Vessel at 2016 (Ton)

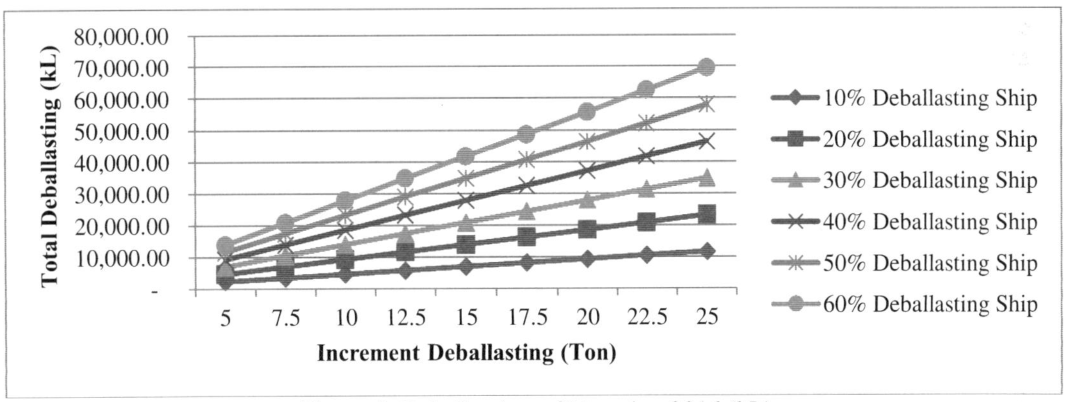

Figure 2. Deballasting of Vessel at 2016 (kL)

4. Result and Discussion.

Based on figures 1 through 2 which take data samples of international route vessel visits, the potential increase in pollutants due to the disposal of ballast water and sediment discharged by the vessel will continue to increase. This indication can be seen with the continued increase in world trade and will have a logical impact on the environment at the port of destination. Increased maritime technology and the development of world trade will also be followed by the development of shipbuilding technology and will lead to a diversity of vessel types and an increase in the main size of vessels. The increase in the size of the main vessel built will be followed by an increase in the volume of water returned by the vessel. This will lead to an increase in the amount of water replaced by vessels exchanged and disposed of at the destination port. So that the potential for increased pollutants from ballast and sediment-water, will be even more enormous, this is what the Government and port managers must anticipate.

Regarding rules, that the Government of the Republic of Indonesia has just ratified the IMO rules relating to the Ballast Water Management in November 2015. Stretching of ballast water management in ports managed by Inaport is still not much done by the port operator. This indication can be seen from the human resources of port managers in the effort to implement Ballast Water Management; not all understand the international rules. Regarding regulations in government legislation, it has not yet been

compiled, so it needs hard work again in its realisation. Socialization and implementation need to be carried out immediately, both by the government as regulator and by the port operator.
Continuing to increase vessel visits, especially international route vessels, there is an excellent potential for increased water to be returned to vessels that are disposed of at Inaport 2nd in Jakarta. This increase also has the potential to increase the risk of pollutants, especially invasive follow-up species and the potential for sediment to be discharged from the vessel. We have to capture these opportunities, as an advantage, in managing the vessel's water ballast. By the results of the MEPC 71 trial on 03-07 July 2017 where the implementation schedule of the Ballast Water Management Treatment System (BWMTS) was a vessel's water treatment system before being disposed of by sea, due to the vessel being added some loads to the vessel. This ballast water must be treated before being discharged into the sea again, because it is indicated that invasive species are included in the ballast water. This invasive species will become a predator at the port or sea where the water is discharged from the vessel. Ballast Water Management Treatment System (BWMTS) in the form of a set of equipment to process water in vessels. This can be seen from the potential for water to be returned by vessels that are disposed of in port waters totalling 47,552.25 tons, equivalent to 48,741.06 kL based on 60% of vessels that discharge water and assume that each vessel throws 25 tons of deballasting

5. Conclusion
Based on the results of the discussion above, some conclusions can be drawn from the results of this study, including:
1. Although it has ratified the IMO rules relating to Ballast Water Management, the implementation of the regulation has not been entirely carried out, including in terms of the readiness of human resources.
2. Increased vessel visits, especially international routes vessels will also increase pollutant potential due to ballasting and deballasting activities. This solution is in the form of invasive and sedimentary species due to water discharge from vessels.
3. In 2016, there was a potential for water to be returned by vessels that were disposed of 47,552.25 tons, equivalent to 48,741.06 kL into the waters of Inaport 2nd. The ballast water are calculated based on 60% of the vessels that dispose of ballast water and the assumption that each vessel discharged 25 tons of deballasting.

Reference
[1] Hayes, E.H., and Landis, W.G., 2004, Regional Ecological Risk Assessment of a Near Shore Marine Environment: Cherry Point WA, *Human and Ecological Risk Assessment*, **10** (2), pp. 299-325.
[2] Cassini, C, Calloni, C, Bortolini, G, Garcia, S. C., Dornelles, M.A., Henriques, J.A.P., Erdtmann, B., and Salvador, M., 2011, Occupational Risk Assessment of Oxidative Stress and Genotoxicity in Workers Exposed to Paints During a Working Week, *International Journal of Occupational Medicine and Environmental Health*, **24** (3), pp. 308 – 319
[3] Chung, J.W., Lee, M.E., and Lee, H.D., 2011, Characteristics of environmental pollution related with public complaints in an industrial shipbuilding complex, Korea, *Environmental Monitoring Assessment Journal*, **177**, pp 73–84.
[4] Alexopoulos, E.C., Merekoulias, G., Tanagra, D., Konstantinou, E.C., Efi Mikelatou, E., and Eleni Jelastopulu, E., 2012, Sickness Absence in the Private Sector of Greece: Comparing Shipyard Industry and National Insurance Data, *International Journal of Environmental Research and Public Health,* **9**, pp. 1171-1181.

[5] Abdullah, H.M., Mahboob, M.G., Banu, M.R., and Seker, D.Z., 2013, Monitoring the drastic growth of ship breaking yards in Sitakunda: a threat to the coastal environment of Bangladesh, *Environmental Monitoring Assessment*, **18**, pp. 3839–3851
[6] Mashiatullah, A., Chaudanry, M.Z., Ahmad, N., Javed, T., and Ghaffar, A., (2013), Metal pollution and ecological risk assessment in marine sediments of Karachi Coast, Pakistan, *Journal Environmental Monitoring Assessment*, **185**, pp. 1555-1565.
[7] Singh, M., Das, A., Singh, D., Maiti, P., Md. Shabbir. Md., and Das, A., 2014, the High genotoxicity of vessel yard contaminants on Allium cepa and calf thymus DNA, *Environmental and Chemistry Lett*, **12**, pp. 321–327.
[8] Werschkun, B., Banerji, S., Basurko, O.C., David, M., Fuhr, F., Gollasch, S., Grummt, T., Haarich, M., Jha, A.,N., Kacan, S., Kehrer, A., Linders, J., Mesbahi, E., Pughiuc, D., Richardson, S.D., Schwarz-Schulz, B., Shah, A., Theobald, N., Urs von Gunten, Wieck, S., and Höfer, T., 2014, Emerging risks from ballast water treatment: The run-up to the International Ballast Water Management Convention, *Chemosphere*, **112**, pp. 256–266.
[9] Kim, K., Kang, H., and Kim, Y., 2015, Risk Assessment for Natural Gas Hydrate Carriers: A Hazard Identification (HAZID) Study, *Journal Energies*, **8**, pp. 3142-3164.
[10] Rahman, A., and Karim, Md. M., 2015, Green Shipbuilding and Recycling: Issues and Challenges, *International Journal of Environmental Science and Development*, **6** (11), pp 838-842.
[11] Thahir, S., 2017a, *IMO BWM Convention dan Peluang Jasa Pengolahan Air Ballast Eksternal*, Majalah Lautindo.
[12] Thahir, S., 2017b, *PENTING! Aturan "Ballast Water" Kapal Untuk Operasional Dalam Negeri*, Majalah Lautindo.
[13] Tjahjana, A., A., 2017, Analisis Pengelolaan Air Ballast Kapal Niaga Berbasis Lingkungan Di Pelabuhan Tanjung Emas Semarang, Disertasi, Pasca Sarjana Ilmu Lingkungan, Universitas Diponegoro, Semarang.
[14] Basuki, M., Lukmandono, Margareta, M.Z.B., 2018, Ballast Water Management Berbasis Environmental Risk Assessment di Perairan Indonesia, *Prosiding Simposium Nasional Kelautan Ke V*, Fakultas Kelautan dan Perikanan, Universitas Hasanudin, Makasar.

Acknowledgement
Authors gratefully thanks to Indonesia Ministry Research Technology and Higher Education through research grant for funding support in this research (PDUPT scheme).

IOP Conf. Series: Materials Science and Engineering **462** (2019) 012045 doi:10.1088/1757-899X/462/1/012045

Pile Foundation Analysis on High – Rise Building using Finite Element-Spring Method on Sandy Clay Soil

D K Fitriyah[1], J Propika[1], L L Lestari[1], H Istiono[1], D Pertiwi[1], and R Sekartadji[1]
[1]Institut Teknologi Adhi Tama Surabaya

Abstract. It is a common thing used by a planner to get the pile foundation needs with the finite element program. The output generated from the finite element program is then processed to produce the required number of piles on the planned structure. The finite element program that is often used is SAP2000. From several cases in the field, there were differences in the results of the number of piles needed between the analysis using the manual calculation method with the Nakazawa method. So, in this paper a comparison of the need for piles based on the manual calculation method with Nakazawa method will be compared, which is devoted to soft soil material according to soil data. The results showed that the number of pile requirements based on the Nakazawa method (n = 244 pieces) was much smaller than the manual calculation method (n = 320 pieces). This shows that the analysis of pile needs is more effective with the use of the Nakazawa method.
Keyword : *high rise building, pile foundation, spring*

1. Introduction

1.1. Background

High rise building can stand firm because it is supported by strong under construction of buildings. Foundation is the most important part of building construction. The foundation serves to hold all the loads in the building and forward them to the ground. [1]

So, considering the needs of the piles is most important. Commonly, to calculate the need for piles, only use manual calculations that only consider axial resistance. Whereas in some cases, it is also necessary to calculate the piles which requires consideration of the resistance of elastic forces due to soil conditions and the loading that occurs. Seeing the conditions that have happened, it is necessary to have a comparative analysis of the need for piles based on the Nakazawa method (1994) with the finite element program.

The research with the topic "Analysis of High-Level Building Pole Needs with Finite Element Spring Analysis on Sandy Soft Soil" is expected to be a consideration in modeling the construction of under construction of buildings.

1.2. Purpose

Based on the background described in section 1.1, the purpose of this paper was to obtain the need for piles that can withstand axial and elastic forces using the Nakazawa method using spring finite elements and manual calculations as validation. The detailed objectives of the research are described as follows:

1. Getting efficient pile needs
2. Obtain displacemen

2. Literature Review

2.1. Modeling a High Rise Building Structure

This high-rise building structure modeling is carried out with Sap2000 structure analysis program. This modeling is in the form of a 10-storey building that functions as an office building with a calculation of the load according to the Indonesian National Standard (SNI). This modeling is intended to get the reaction of the base in the building that will be used in foundation planning.

2.2. Calculation of Soil Capacity

By using the N-SPT data at the location of the high-rise building, the carrying bearing capacity (QL) is calculated. to calculate bearing capacity the pile method is used by **Luciano Decourt** (1996): [2]

$$Q_u = Q_p + Q_S = K.A_p.N_p.\alpha + \left(\frac{Ns}{3} + 1\right).A_S.\beta \quad (1)$$

$$Q_{ijin} 1\ tiang = \frac{Q_u}{SF} \quad (2)$$

Information :

Qu = The carrying capacity of the ultimate pole
K = soil characteristic coefficient
- sand soil: 40 T / m^2
- sandy silt soil: 25 T / m^2
- clay silt soil: 20 T / m^2
- clay: 12 T / m^2

Ap = base area of pole (m^2)
Np = Average value of SPT (N) along 4D above to 4D below the end of the pole
α = Correction factor at the end of the pile = 1
β = Correction factor on pile blanket = 1
Ns = Mean value (N) along the pole with a value of $3 \leq N \leq 50$
As = Pile blanket value (perimeter x pole length)
SF = Safety factor

Factor for piles efficiency in groups, based on Converse-Labarre: [5]

$$E_{LA} = 1 - \theta\frac{m(n-1)+n(m-1)}{90mn} \quad (3)$$

Information :

E_{LA} = Efficiency of pile groups
m = Number of pile lines
n = Number of piles in 1 row
θ = arc tg d / s
s = Distance of the center of the pole to the center
d = Pole diameter

2.3. Control of Pile Resistance to Horizontal Force

The foundation of the pile also needs to be controlled for horizontal forces. Deflection that occurs is not allowed to exceed the specified limits. The following is a formula to determine the deflection that occurs at the stake. [3]

$$\delta_P = F_\delta\left(\frac{PT^3}{EI}\right) < 2{,}54\ cm \quad (4)$$

$$T = \left(\frac{E\ x\ l}{f}\right)^{1/5}$$

2.4. Estimated K Coefficient From Horizontal Force

The following is the equation for estimating the coefficient of soil reaction in a vertical force based on technical standards in Japan. [4]

$$k = k_0 \,.\, y^{-\frac{1}{2}} \tag{5}$$
$$k_0 = 0{,}2 \,.\, E_0 \,.\, D^{-3/4}$$

Information :

k_0 = Value k if the shift on the surface is made 1 cm (kg / cm^2)
y = The amount of shift to be searched (cm)
E_0 = Modulus of foundation soil deformation, usually estimated from E0 = 28 N using N prices from standard penetration experiments (standard penetration test)
D = Piles diameter (cm)

2.5. Estimated K_V Coefficient From The Vertical Force

Estimated K_V (spring constant) in the vertical force of the pile is an elastic constant which is expressed as a force in the vertical direction and can cause displacement on the mast head. K_V is used in the calculation of the size of the pole head reaction or elastic decline that occurs. Here is the empirical formula for estimating K_V constants.

$$K_V = \alpha \,.\, \frac{A_P.E_P}{l} \tag{6}$$

For a steel pipes

$$\alpha = 0{,}027\ (l/D) + 0{,}2$$

For prestressed concrete

$$\alpha = 0{,}041\ (l/D) - 0{,}27$$

For Cast in Situ

$$\alpha = 0{,}022\ (l/D) - 0{,}05$$

nformation :
AP = Net cross-sectional area of pole (cm^2)
EP = Modulus of elasticity of the pole (cm^2)
l = Pole length (cm)
D = pole diameter (cm)
The formula above is used when $l/D \geq 10$.

3. Research Methodology

The following is an explanation at the step of this research.
What was done for the first time was collecting land data at the construction site of a high-rise building in Surabaya. The selected high-rise building is a building with office functions. Then modeling high-rise building structures using SAP2000. From the structural modeling, the placement reaction at the bottom of the building can be used to analyze the needs of the number of piles using the spring finite element method based on the type of soil obtained from secondary data.

By using the Nakazawa method (1994) the number of effective piles will be produced, with validation using manual calculation method. In addition, from this study, found piles that are resistant to axial forces and elastic forces.

4. Data and Analysis Result

4.1. Structure Planning Data

The building structure data analyzed in this paper are as follows:

1. Location = Surabaya

2. Building function	= Office
3. Building height	= 10 floors + 1 roof top floor
4. Material	= Concrete
5. Concrete quality	= 30 MPa (plates and beams)
	= 35 MPa (Column)
6. Column dimensions	= 1000 x 1000 mm
7. Dimensions of the beam	= 50 x 70 mm
8. Distance of column	= 8000 mm

4.2. Modeling of Structure Building with SAP2000

In this study high-rise buildings were modeled with the properties of the materials used in accordance with the discussion in Section 4.1. By using SAP2000, the building structure is modeled and given the loads that work on the structure (Figure 2). Structure modeling can be seen in Figure 1.

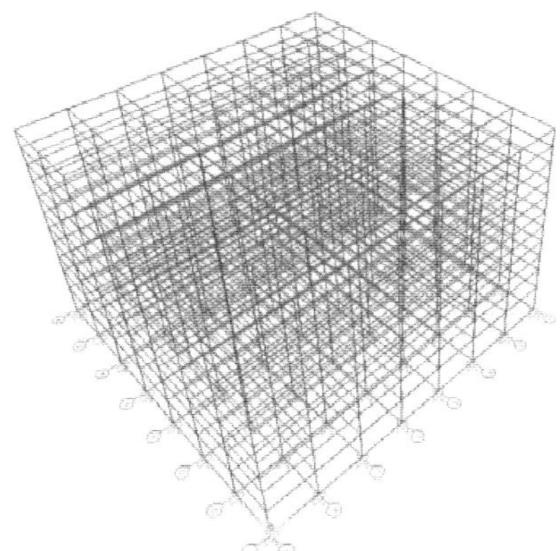

Figure 1. Show undeformed shape.

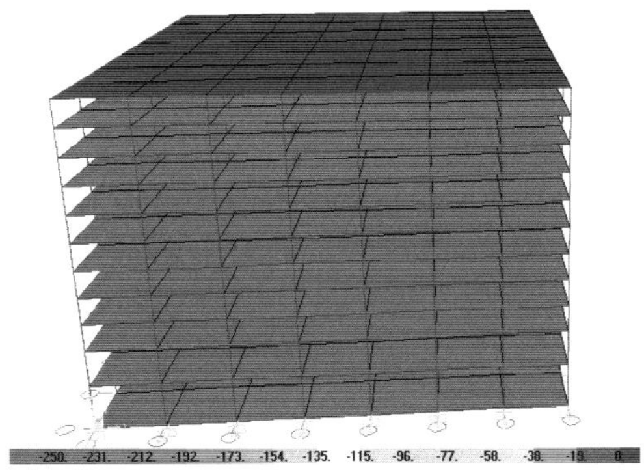

Figure 2. Show Area loads, Uniform load to frame contours.

To analyze the need for number of piles, resistance to axial forces and elastic forces, as well as displacements that occur in the horizontal direction, analysis is needed with spring analysis modeling. For modeling with the spring finite element method as shown in Figure 3.

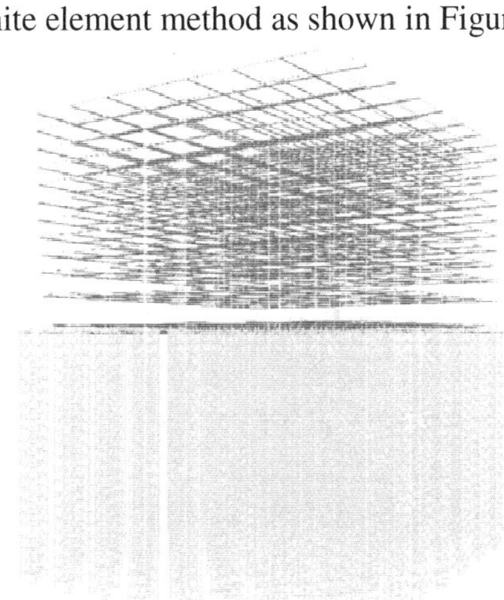

Figure 3. Show undeformed shape.

4.3. Calculation of Soil Capacity
The calculation of the bearing capacity of the soil against the pole includes the calculation of the endbearing carrying capacity and the carrying capacity of the positive and negative skin friction blankets. Before calculating the capacity, the NSPT value is first corrected to water level and to soil pressure overburden. [6]

4.4. Control Strenght material of Piles
The pile is planned with the tip pinched by a flat foundation against the subgrade. So that the material strength control is carried out on the axial force and moment force. The location of the pile pinch point from the ground surface (Zf) is 3.0 m.
Axial force (Pmax) is obtained at 130.09 Ton. So that this axial force must be smaller than the Pall pile based on specifications. The pile is planned to use dimensions ø 50 cm class C with Pall which is 169 Ton at a depth of 41 m.

4.5. Pile Needs Analysis
Analysis of the pile using the Nakazawa method, namely the Spring Fimite Element analysis will be validated using the Luciano Decourt (1996) method, can be seen in Table 1.

Table 1. Pile Needs Analysis

Methode	Pile need	Pmax
Luciano Decourt	320	644.00
Nakazawa	244	130.09

From the table above, it is found that the needs generated from the Nakazawa method are spring analysis which results in a smaller pole requirement compared to the Luciano Decourt method. Due to the considerable differences in styles of the two methods above.

5. Conclusion
The conclusion of this study is that the need for piles is more effective with the Nakazawa method, because the number of pile requirements is much smaller than the results of manual calculations in the same case study.

6. Acknowledgement
The author would like to thank PT. ITSC which has allowed the use of secondary data in this study.

7. Bibliography
[1] Pardoman Silitongan, Ary dan Iskandar Rudy *Perbandingan Daya Dukung Pondasi Tiang Pancang Metode Analitis dan Loading Test Terhadap Metode Elemen Hingga Menggunakan Model Mohr-Coulomb dan Soft Soil pada Bore-Hole I (Studi Kasus Pembangunan Rusunawa Jatinegara Barat Jakarta Timur)* (Sumatra Utara : Universitas Sumatera Utara)
[2] Ath Thaariq, Brilliant Joko Wahyu Adi, Tri dan Trihanyndio rendy Satrya 2013 *Analisa Pemilihan Alternatif Sistem Pondasi pada Gedung Kampus ABC Balikpapan – KALTIM Ditinjau dari Aspek Teknis, Biaya dan Waktu* (Surabaya : Institut Teknologi Sepuluh Nopember (ITS))
[3] Ilham Gumilang S, M B. Mochtar, Indrasurya, dan Lastiasih, Yudhi 2016 *Perencanaan Pondasi Tiang Pancang dengan Memperhitungkan Pengaruh Likuifaksi pada Proyek Pembangunan Hotel di Lombok* (Surabaya : Institut Teknologi Sepuluh Nopember (ITS))
[4] Nakazawa, Kazuto, et al 1994 *Mekanika Tanah & Teknik Pondasi* Kazuto Nakazawa, et al, Suyono Sosrodarsono (Jakarta : PT Pradnya Paramita)

[5] Anaya Nabilla Antari Terenggana, S 2014 *Analisa Perhitungan Pile – Raft Foundation Pada Proyek The 18 Office Park Jakarta* (Sumatera Selatan : Universitas Sriwijaya)
[6] Kurniawan, Ali B. Mochtar, Indrasurya dan Lastiasih Yudhi 2016 *Alternatif Perencanaan Pondasi Tiang untuk Gedung Tinggi di Atas Tanah Lunak pada Proyek Pembangunan Kota Baru Summarecon-Bandung* (Surabaya : Institut Teknologi Sepuluh Nopember (ITS))

IOP Conf. Series: Materials Science and Engineering **462** (2019) 012046 doi:10.1088/1757-899X/462/1/012046

Characteristics of Flat-Wall Impinging Spray Flame and Its Heat Transfer under Small Diesel Engine-Like Condition. 3th Report: Effect of Oxygen Concentration

R Mahmud[1], T Kurisu[1], K Nishida[1], Y Ogata[1] and O Akgol[1]
[1]Department of Mechanical System Engineering, University of Hiroshima, 1-4-1 Kagamiyama, Higashi-Hiroshima, Japan

Abstract. To investigated heat transfer phenomena inside the combustion chamber wall in diesel engines, we investigated the effects of spray/flame impingement on transient heat flux to the wall. By using a constant volume vessel under engine-like condition, surface heat flux of the wall at spray/flame impingement was measured with three thin film thermocouple heat flux sensors. Fuel was injected using a single-hole injector with a 0.133 mm diameter nozzle. Under these conditions, spray evaporates, then burns near the wall. In order to investigate the relation between diesel flame and wall heat loss, two colour-method was applied to observe local flame temperature distribution. The effect of oxygen concentration on the heat transfer was investigated parametrically. The results showed that the ratio of oxygen concentration influenced the combustion behaviour, which has significant effect on flame temperature and flame contact area that gave rise to further heat loss on the wall.

1. Introduction

Heat transfer losses are one of the largest factor loss of engine thermal efficiency, especially in small-size diesel engines. More than half of total heat energy in the internal-combustion engine is lost by cooling through the cylinder wall to the atmosphere and the exhaust gas. Heat loss through combustion chamber wall largely determines cooling loss. For this reason, fully understanding of heat transfer mechanism from combustion gas to cylinder wall is required for maximizing the thermal efficiency.

In general, heat loss (q) from the combustion gas to cylinder wall in internal combustion engine as shown in formula (1). Considering the decreasing value of surface area (A), temporal changes in the heat transfer coefficient (h) and temperature difference between gas temperature and wall temperature (T_{gas}-T_{wall}) are required for decreasing heat loss (q).

$$q = A.h.\left(T_{gas} - T_{wall}\right) \quad (1)$$

Some studies showed that minimizing temperature difference between in-cylinder gas and combustion chamber wall surface was possible by applying insulation material such as ceramic coating on the surface [1-3]. As a result, the heat loss was reduced. Thus, both exhaust energy and piston work increased which, in turn, led to improved thermal efficiency.

Another simulation study, applying the Exhaust Gas Recirculation (EGR) rate also become considering to decreasing flame temperature distribution on heat loss. Comparing the single injection results at different EGR rate, it was confirmed that the heat flux was decrease at higher EGR rate [4]. With increasing EGR rate, the oxygen concentration in the fuel/gas mixture decreases and the specific heat of the cylinder increases and leads reaction rate decrease. Further, in-cylinder pressure and combustion decrease [5].

As described by the previous researchers, oxygen concentration has an effect on combustion behavior, which closely relates with lower temperature combustion. Therefore, it is important to clarify an effect of oxygen concentration on decreasing heat loss. For this reason, this paper aims to investigate the oxygen concentration effect on the transient local heat flux of combustion. In order to investigate the relation between diesel flame and wall heat loss, two colour-method [4][6-7] was applied to observe local flame temperature distribution. Instantaneous temperature was detected by three Thin Film Thermocouple Heat Flux Sensors (TFTHFSs) which were mounted on the wall surface. By using the measured temperatures as boundary condition, we computed non-steady heat flux through the surface with finite difference method.

2. Experimental Apparatus and Method

2.1 High-Pressure and High-Temperature Chamber Vessel

The experiment was previously conducted by the authors [8] using high-pressure and high-temperature chamber vessel. Pressurized hot air was supplied through an electric muffle furnace and heated with an electric heater mounted at the bottom of the chamber in order to simulate an engine-like thermodynamic environment. The chamber pressure was manually operated, whereas the chamber temperature was controlled by a volt slider for the heater. Four K-type thermocouples were installed around the combustion area. Diesel fuel from a common rail system was injected into the chamber when temperature difference of the thermocouples was within range of 5 K. Injection quantity and timing was controlled by a delay pulse generator (Stanford Research Systems DG645). Injection pulse had been calibrated by the injection rate measurement previously.

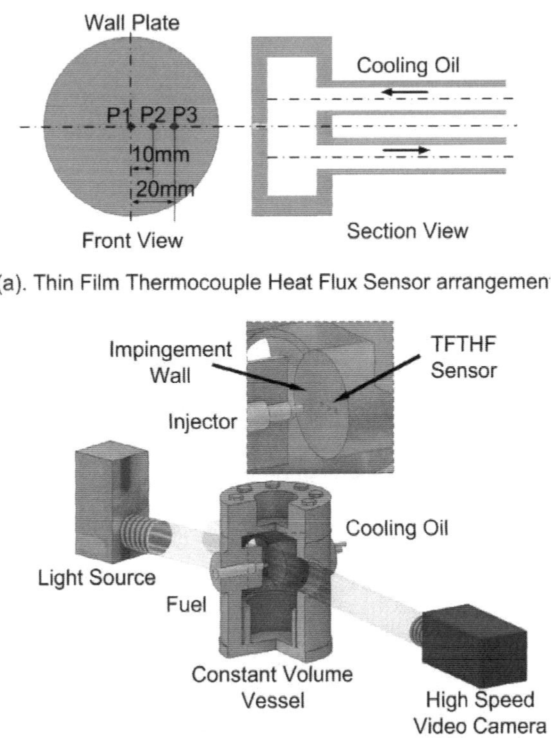

(a). Thin Film Thermocouple Heat Flux Sensor arrangemen

(b). Optical set-up of high-speed video camera observation

Figure 1. Thin Film Thermocouple Heat Flux Sensor arrangement and optical set-up of high-speed video camera observation for spray/flame

2.2 Heat Flux Measurement and High-Speed Video Camera Photography
Figure 1 (a) shows Thin Film Thermocouple Heat Flux Sensor (TFTHFSs) arrangement. Three Alumel/Chromel TFTHFSs (Medtherm 10702B) were applied to obtain the instantaneous surface heat flux on the wall. The diameter of the sensors is ø 1.55 mm, and distance between their thermal contacts (surface and inside) is 3.30 mm. The sensors were radially located on the wall surface with a 10 mm distance each other. Positions of the sensors are designated as Position1 (centre of wall impingement), Position2 (10 mm from centre) and Position3 (20 mm from centre). The flat wall was made of stainless steel which has a thickness of 7.3 mm. The back side of the wall plate was cooled by oil in order to form one-directional heat flow across the wall. The cooling oil temperature was kept constant by a thermostatic oil bath.

Instantaneous local heat flux was computed by applying one-dimensional non-steady heat conduction equation using two measured temperatures by the identical sensor as boundary conditions. The equation was explained in the previous report [8]. Total heat loss was calculated by integrating the heat flux with concentric area and time. In area integration, heat flux affected areas were defined as concentric circles of 0-5 mm for Position1, 5-15 mm for Position2, and 15-25 mm for Position3, respectively.

Figure 1 (b) shows optical set-up of high-speed video camera observation. Spray and flame behaviour were recorded by high speed video camera (Nac Image Technology Inc, HX-3) with an imaging speed of 20,000 fps and resolution of 320 x 448 pixels. To observe the spray behaviour, Mie scattering method was adopted to support the spray impingement illumination photography. Light from Xenon lamp illuminated the spray in the chamber through the sideward transparent quartz window to provide observation at non-evaporating and evaporating conditions. Observation of flame was conducted with natural flame luminosity. In order to acquire high quality spray images, image processing was employed in this study. To clarify the flame temperature distribution, two-colour method was applied. We used a two-colour pyrometry software named "Thermias" of Nac Company for analysing the natural flame colour images. A standard light was used as an illumination source for calibration.

3. Test Conditions
Measurement conditions are listed in Table 1. It was decided based on actual operation of small size diesel engines. Injection quantity was 5 mm^3 with injection pressure 120 MPa using a nozzle of 0.133 mm hole diameter. The impinging distance between the nozzle tip to the impingement wall was set at 40 mm as shown in Figure 2. Oxygen concentrations were conducted at 21% and 16% as experimental parameters to investigate the effect of the heat transfer characteristics on the impinging flame on the wall. The conditions were "non-evaporating spray" at room temperature with N_2 gas, "evaporating spray" at high temperature with N_2 gas, and "combusting spray" at high temperature with air. To keep the gas density same for all of the conditions, which is 16 kg/m^3, ambient pressures was set at 1.4 MPa for non-evaporation and 4.1 MPa for others.

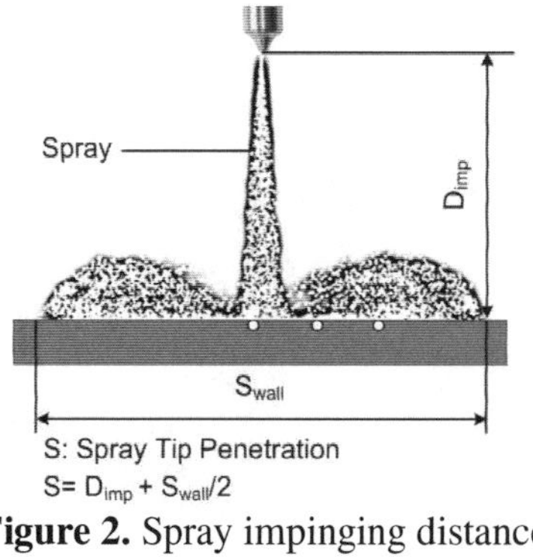

Figure 2. Spray impinging distance

Table 1. Experimental conditions

	Non-Evaporation	Evaporation	Combustion
Fuel			
Fuel	Diesel fuel	←	←
Ambient Condition			
Ambient gas	N_2	←	Air (N_2:79%,O_2:21%) Air (N_2:84%,O_2:16%)
Ambient pressure (MPa)	1.4	4.1	←
Ambient temperature (K)	300	873	←
Ambient density (kg/m³)	16	←	←
Injector and Injection Condition			
Injector type	Piezo actuator type	←	←
Number of nozzle holes	1	←	←
Injection quantity (mm³)	5	←	←
Diameter of nozzle holes (mm)	0.133	←	←
Injection pressure (MPa)	120	←	←
Impingement Wall			
Impingement wall	Flat plate, Stainless steel	←	←
Impingement distance (mm)	40	←	←
Wall temperature (K)	300	460 ±10	←
Cooling method	-	Oil cooling	←

4. Results and Discussion

4.1. Injection Rate

According to Wakisaka and Azetsu [9], shapes of fuel injection rate influences the combustion and emission formation. As Bower and Foster [10] performed and confirmed that the Zeuch method is more accurate for measuring injected volume. Injected fuel quantity was calibrated by Zeuch method injection ratemeter (Ono Sokki Co. Ltd, FJ-7000). Figure 3 shows an averaged result of a ten times injection rate measurement with a single hole injector at 120 MPa. In all experiments, injected quantity was kept at 5mm^3, with 1.2 ms injection duration.

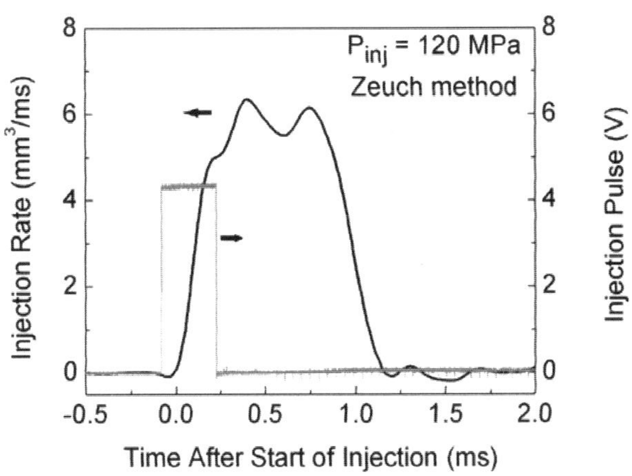

Figure 3. Injection rates

4.2 Spray Behavior at Non-evaporation and Evaporation

In this section, the impinging spray behaviour under non-evaporation and evaporation conditions was studied.

4.2.1 Spray Behaviors. Figure 4 shows comparison between non-evaporating and evaporating wall impinging spray images. Spray travel from the top to the bottom at each photo and impinges on the wall. Non-evaporating spray images were observed at room temperature (300 K), while the evaporating spray at high temperature (873 K). In the case of evaporating spray, intensity of white colour shows liquid spray concentration. The images revealed that non-evaporating spray led to shorter impingement timing and larger spray spreading to the circumferential area of the flat wall. On the other hand, in the case of evaporation, the spray almost evaporated completely before impinging to the wall. As it can be seen at 0.9 ms After Start of Injection (ASOI), there is some fuel droplets spread on the wall.

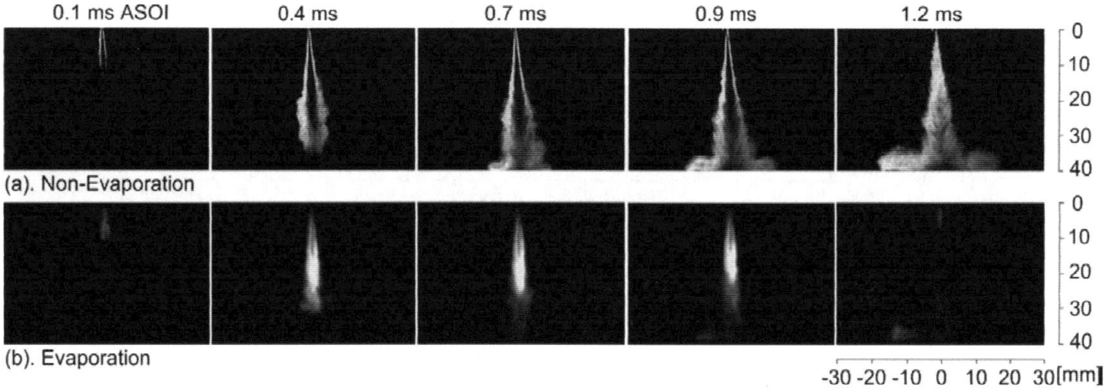

Figure 4. High-speed video camera images of wall impinging spray at non-evaporating and evaporating

Spray tip penetration lengths, which are obtained from the non-evaporating and liquid lengths from evaporation images are plotted in the Figure 5. Spray tip penetration length is defined as a summation of axial and radial penetration distances as shown in Figure 2. The liquid length, which was defined similar to the non-evaporating spray, were obtained by the image processing as the maximum liquid penetration during the injection using a pre-determined threshold intensity. We used a threshold intensity of 10% on the 0 to 255 intensity scale of the image.

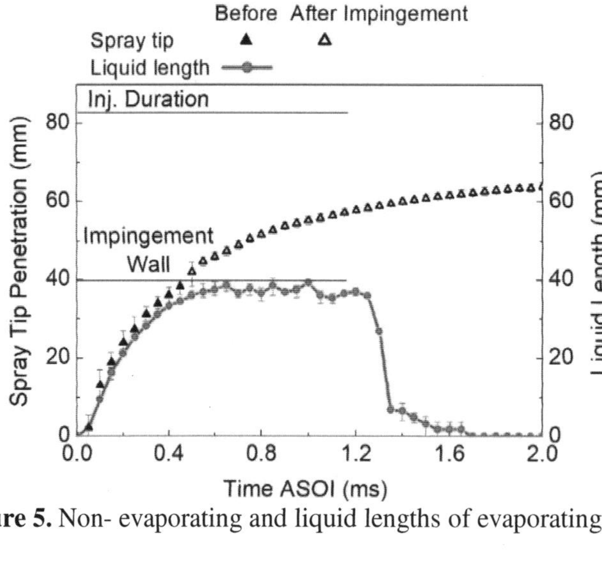

Figure 5. Non- evaporating and liquid lengths of evaporating spray

Closed and open triangles show before and after the spray impingement on the wall, respectively. Figure shows that the length of evaporating was rapidly declined at ASOI 1.2 ms and reached zero, while the length of non-evaporating spray steadily grew. As can be seen in the figure, liquid penetration length rapidly decreases after the end of injection compared with non-evaporation state.

4.3 Combustion Flame Behavior and Wall Heat Transfer under Oxygen Concentrations

In this section, impinging flame behaviour and combustion effect on the wall heat transfer under oxygen concentration are studied. Effects of the oxygen concentrations, which were 21 and 16% were investigated at impingement distance of 40 mm.

4.3.1 Combustion Flame Behavior. Figure 6 shows images of impinging flame at different oxygen concentrations (O_2). Comparing with non-evaporating spray photographs in Figure 4 (a), luminous flames occur a few deci-seconds after the spray impingement. In the Figure 6, the luminous flame appeared earlier with O_2 of 21%. It had started at 0.9 ms ASOI before the end of the fuel injection and at 1.25 ms ASOI just after the end of the fuel injection for O_2 of 16%. It is found that the reduction of the oxygen concentration in the enclosure mainly affects the reactions in the gas phase and results in changes of the combustion behaviour such as ignition timing and flame development. Furthermore, comparing the photographs of evaporation in Figure 4 (b) and combustion in Figure 6, it can be seen that the luminous part, which is product of combustion, existed in the vapor area.

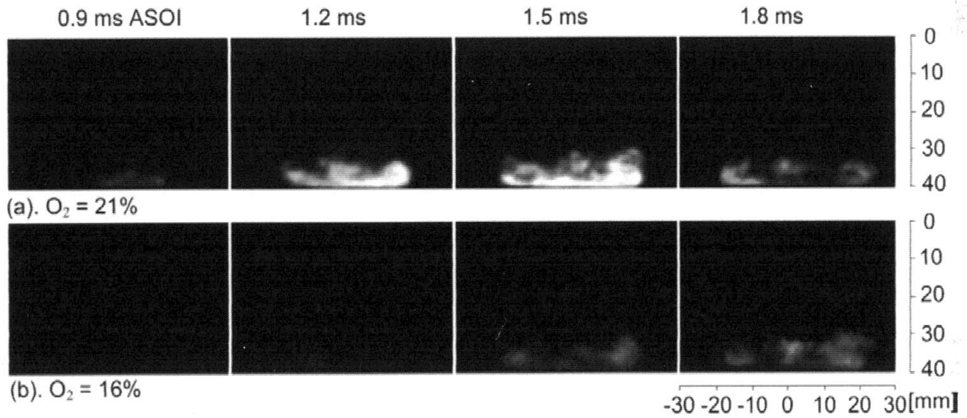

Figure 6. High-speed video camera images of impinging flame at different oxygen concentration

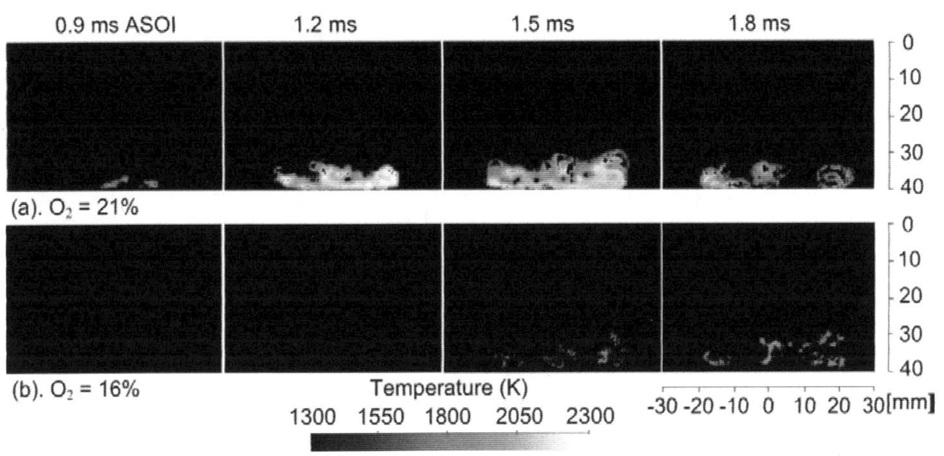

Figure 7. Flame temperature distribution at different oxygen concentration

Figure 7 shows flame temperature distributions at different oxygen concentrations obtained from two-colour method analysis, which was performed with flame natural luminosity images. The temperature distribution shapes were similar to those of flame, because they were obtained from the flame luminosity. The colour distribution scale varying from dark to bright indicates temperature.

By comparing the temperature distributions with different oxygen concentrations, it can be seen that the O_2 of 16% had the lowest temperature. It can be inferred from Figure 8 that burned gas temperature decreases with decreasing volume fraction of oxygen because the heat capacity of ambient gas increase with decreasing the volume oxygen concentration [11].

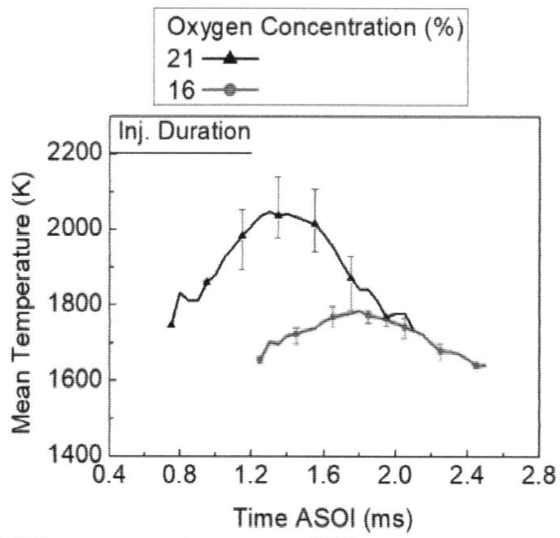

Figure 8. Mean temperature at different oxygen concentration

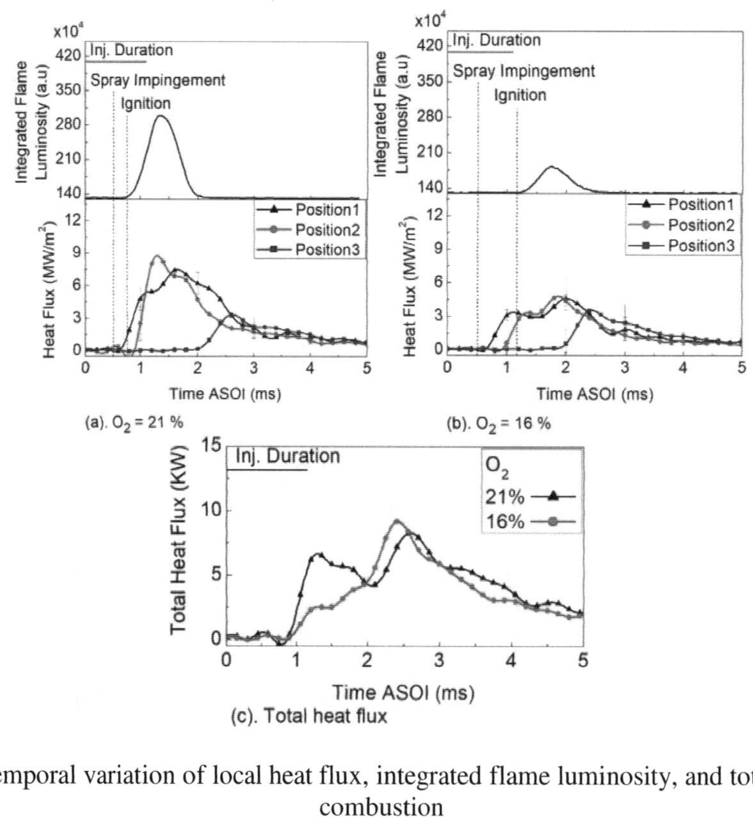

Figure 9. Temporal variation of local heat flux, integrated flame luminosity, and total heat flux at combustion

4.3.2 Heat Flux on the Wall Surface. Figure 9 (a-b) show non-steady local heat fluxes with flame impinging on the wall and integrated luminosity from flame images at different oxygen concentration. Timings of the spray impinging and ignition, which were obtained from the video images, are also plotted. The luminosity increased from near-the ignition timings at all oxygen concentrations. Local heat flux at Position1 started to increase earlier than Position2 and 3 at each oxygen concentration. This increase in local heat flux at Position1 was followed by ignition at O_2 of 21% and after Position2 for O_2 of 16%, which had the minimum value. The low temperature due to decreasing oxygen concentration causes unsuitable equivalence ratio for ignition. As a result, ignition takes time to start after spray impinging on the wall.

Regarding the effect of oxygen concentration on the heat flux, it could be found from the graphs as follows: (1) earlier start of local heat flux increase was observed at all positions with the O_2 of 21%, except Position3 was similar (2) higher peak value was observed at Position1 and 2 with the O_2 of 21%.

Figure 9 (c) shows temporal variation of the total heat flux at different oxygen concentrations. It can be seen in the graph, O_2 of 21% had two peaks waves i.e. there were two stages for the local heat transfer. At first, the local heat flux was increased by turbulence of impingement spray. Then, heat transfer by flow along the wall, which started after impingement, took place. While O_2 of 16% accumulated at the second stage for local heat transfer due to significant ignition delay.

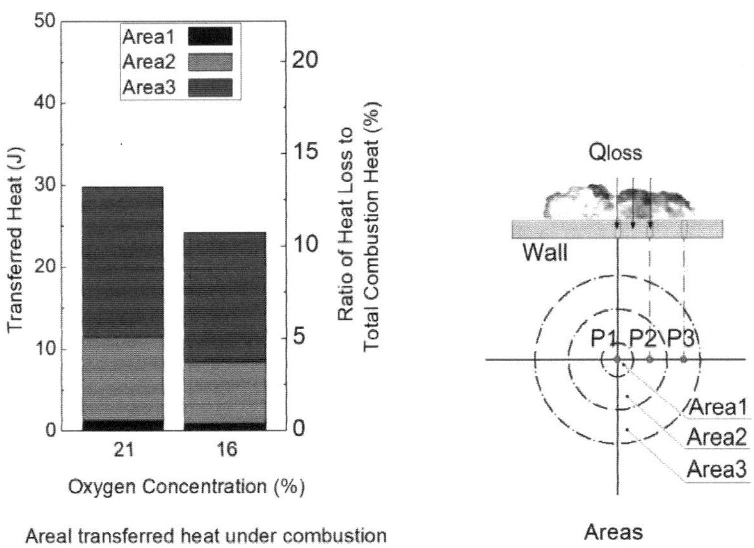

Figure 10. Comparison of the areal transferred heat and ratio of heat loss to total combustion heat at different oxygen concentration

A comparison of the transferred heat at each area under different oxygen concentrations are shown in Figure 10. Transferred heat value was decreased with lower oxygen concentration. This effect is mainly due to the decrease in the burned gas temperature with decreasing the volume fraction of oxygen concentration [11]. Ratios of heat lost through the wall to total combustion heat are also shown in the graph. It is about 11 to 14 % of total combustion energy.

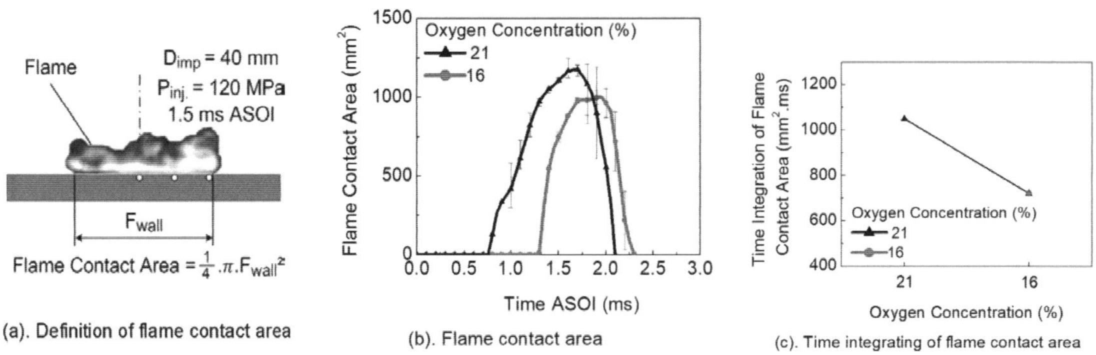

Figure 11. Comparison of flame contact area and time integrating of flame contact area at different oxygen concentration

Effect of flame contact to the wall at different oxygen concentrations as shown in Figure 11. Figure 11 (a) is definition of flame contact area. Figure 11 (b) shows that O_2 of 16% had smaller contact area than that 21%. It means with oxygen concentration of 21%, leads to flame contact significantly to spread wide into circumferential area of flat wall. This flame contact area contributes to the heat transfer by conduction. Time integration of the flame contact area can be seen in Figure 11 (c), which shows that flame traveling distance was shorter as the oxygen concentration is decreased.

5. Conclusion

Effects of oxygen concentration on the combustion wall heat transfer under diesel engine-like conditions were investigated. The main conclusions are summarized as follows. By using the two-colour pyrometry analysis, we found that the combustion distribution was low temperature at oxygen concentration of 16%. The low temperature due to decreasing oxygen concentration delays start of ignition. As a result, total heat flux accumulated at the peak of second waveform stage. Comparing different oxygen concentration, O_2 of 16% showed the lower transferred heat to the wall. This was caused by decrease in burned gas temperature and small flame contact area.

Nomenclatures

q : Heat Loss
A : Surface Area
h : Heat Transfer Coefficient
EGR : Exhaust Gas Circulation
TFTHFS : Thin Film Thermocouple Heat Flux Sensor
ASOI : After Start of Injection
O_2 : Oxygen Concentration
a.u. : arbitrary units

6. References

[1] Kogo T, Hamamura Y, Nakatani K, et al. High-Efficiency Diesel Engine with Low Heat Loss Combustion Concept - Toyota ' s Inline 4-Cylinder. SAE paper, 2016, 2016-01-0658.
[2] Wakisaka Y, Inayoshi M, Fukui K, et al. Reduction of Heat Loss and Improvement of Thermal Efficiency by Application of "Temperature Swing" Insulation to Direct-Injection Diesel Engines. SAE Technical Paper, 2016, 2016-01-0661.
[3] Fukui K, Wakisaka Y, Nishikawa K, et al. Development of Instantaneous Temperature Measurement Technique for Combustion Chamber Surface and Verification of Temperature Swing Concept. SAE Technical Paper, 2016, 2016-01-0675.
[4] Osada H, Uchida N, Shimada K, et al. Reexamination of Multiple Fuel Injections for Improving

the Thermal Efficiency of a Heavy-Duty Diesel Engine. SAE paper 2013, 2013-01-0909.
[5] Cheng X, Chen L, Yan F, et al. Investigations of Split Injection Strategies for the Improvement of Combustion and Soot Emissions Chraracteristics Baesed on the Two-Color Method in a Heavy-Duty Diesel Engine. SAE Technical Paper 2013, 2013-01-2523.
[6] Cheng X, Chen L, Yan F at al. Investigation of Split Injection Strategies for the Improvement of Combustion and Soot Emissions Characteristics based on the Two-Color Method in Heavy-Duty Diesel Engine. SAE Technical Paper 2013, 2013-01-2523.
[7] Li K, Ido M, Ogata Y et al. Effect of Spray/Wall Interaction on Diesel Combustion and Soot Formation in Two-Dimensional Piston Cavity. SAE Technical Paper 2013, 2013-32-9021.
[8] Mahmud R, Kurisu T, Nishida K, et al. Experimental Study on Flat-wall Impinging Spray Flame and Its Heat Flux on Wall under Diesel Engine–like Condition (First Report: Effect of Impingement Distance). 2018, Journal of Automobile Engineering, IMechE, Part D.
[9] Wakisaka Y and Azetsu A. Effect of Fuel Injection Rate Shaping and Injection Pressure on Intermittent Spray Combustion, SAE Technical Paper 2000, 2000-01-2793.
[10] Bower G and Foster D. A comparison of the Bosch and Zuech rate of injection meters. SAE Technical Paper 1991, 910724.
[11] Kuboyama T, Moriyoshi Y, and Kosaka H. Heat Transfer Analysis in a Diesel Engine Based on a Heat Flux Measurement using a Rapid Compression and Expansion Machine. SAE Technical Paper, 2017, 2107-32-0115.

Acknowledgements
We would like to thank Mazda Motor Corporation for their support with measurement in this study.

IOP Conf. Series: Materials Science and Engineering **462** (2019) 012047 doi:10.1088/1757-899X/462/1/012047

Pico-hydro as A Renewable Energy: Local Natural Resources and Equipment Availability in Efforts to Generate Electricity

A Khomsah[1], Sudjito[2], Wijono[3], A S Laksono[4]

[1]Mechanical Engineering, Adhi Tama Surabaya Institut of Technology, Surabaya, Indonesia
[2]Mechanical Engineering, Brawijaya University, Malang, Indonesia
[3]Electrical Engineering, Brawijaya University, Malang, Indonesia
[4]Environmental Resource Management and Development, Brawijaya University, Malang, Indonesia

alikhomsah@itats.ac.id

Abstract. Pico hydropower (PHP) is a small power plant system to generate electricity that is suitable for implementation in the rural area. The potential of local natural resources such as bamboo and the existing of many induction motors with small capacity, both of them respectively can be used as turbine material and as an induction generator. Bamboo as a natural and bio-composite material there are many in Indonesia and Southeast Asia. The main objective of the present research is how local natural resources bamboo can be used as turbine material and induction motor using as an induction generator, with the aim to finding cheap and ease equipment for pico hydro implementing. Moreover, through this research, for the first time, pico hydropower (PHP) was applied with a using bamboo turbine and Induction Motor As Generator (IMAG). The results obtained are a cross-flow turbine with dimension diameter of 125 mm, turbine width of 50 mm and with 20 blades. From laboratory test result at water pressure, 0.55 kilogram per square centimetre and flow rate 4 litres per second, power system capable of load 114 Watt LED lamps, maximum voltage and current respectively 100 Volts and 0.38 Amperes. Pico-hydro with cross-flow turbines made from bamboo material and generator IMAG has been implemented on demonstration site in Jembul village, local natural resources material and equipment availability which is cheap and ease of implementing. The pico hydro exhibits a good electrical output result that can be improved for a good performance for the next.

1. Introduction

The Quran is an ancient text which is the word of God and is a guide for Muslims, which there is knowledge about the environment in it. That knowledge needs to be explored and published because in the future will provide benefits for environmental sustainability. Nature will be balanced if all forms of human behaviour or activities do not cause damage to the natural environment, because basically, environmental damage will cause a natural imbalance. Quran is science and science from an Islamic outlook corner must show the interrelatedness of all parts of the universe [1]. *"He sends down from the sky, rain, and valleys flow according to their capacity, and the torrent carries a rising foam. And from that [ore] which they heat in the fire, desiring adornments and utensils, is a foam like it. Thus Allah presents [the example of] truth and falsehood. As for the foam, it vanishes, [being] cast off; but as for that which benefits the people, it remains on the earth. Thus does Allah present examples."* (Quran 13:17), the Quran surah talks about the cycle of water which expresses about renewable energy. With so many mountains in Indonesia it is an advantage because according to the Quran, the mountain is a stake. Moreover, as a place where water is stored which then flows on the valley naturally according to their capacity, the term *'...flows on valley naturally according to their capacity'* express the kinetic energy of water. In the Quran, it is stated that the mount firmly stands because it functions as a stake of the earth which consists of large sturdy rocks and basically, rock is the mountain building material. *"Have we not made the earth a resting place? And the mountains as stakes?"* (Quran 78: 6-7). Moreover, in the mountain that is the place where the rain falls from the sky and the place where springs come from, that can be used as water potential energy. Based on the contents of the verse in the Quran, it is necessary to utilise the potential of water energy by developing and implementing turbines and generators of the potential that exists in the surrounding environment.

As an archipelagic nation and mountainous region, Indonesia has abundant resources which can be exploited as the sources of energy to sustain life. However, over time, the natural availability is continuously depleting and need to anticipate it, New and Renewable Energy (NRE) becomes the best alternative solution, because environmentally friendly. What is Renewable Energy? Renewable energy is a source of energy from nature that is continually produced and recharge by nature —the sun, the wind, water, the earth's heat, and plants [2]. Renewable energy sources are those resources which can be used to produce energy, again and again, e.g. solar energy, wind energy, biomass energy, geothermal energy, or others are also often called alternative sources of energy. Renewable energy technologies turn these fuels into usable forms of energy—most often electricity, but also heat, chemicals, or mechanical power. The main topic of this paper concern about water energy because it is still available for use in Indonesia, especially about pico hydro (the power under 5 kW), especially in the mountainous region in the rural communities. Until now, need to know that the Indonesian government in its regulation only includes micro-hydro, while the pico hydro has not been mentioned up to adopt. Indonesia country has a huge amount of hydropower resources. The potential of hydropower is estimated about 75,000 MW which makes Indonesia ranking fourth in Asia after China, the former Russian Federation and India. However, only 34,000 MW of 75,000 MW is exploitable [3].

Renewable energy resources and intensive explorations of different alternatives energy are currently being conducted worldwide. Pico-hydro power (PHP) is at the forefront of these options because it is considered the most cost-effective renewable energy option to provide electricity for rural areas. Energy can enable to be derived from extremely low head and flow streams of 1 m and 1 L/s, respectively. The future of the pico-hydro market looks prosperous because there is a substantial availability of low head and flow hydroelectric sites in less developed countries. In the future, technology can play a crucial role in the lighting of houses in remote communities, with the energy source derived from a domestic water supply [4]. The wasted energy of daily used water and rainwater also can be converted into the electrical energy by the hydroelectric mechanism through several energy conversion processes [5]. Pico hydro project is implemented with low cost using irrigation water and generated power and can be utilized to household appliances in rural areas [6].

The baseline concept for multi-year system development of a pico-hydroelectric power system in Nepal. How green energy solutions for rural applications should be highly contextualised to ensure maximum impact and long-term sustainment. Another result is a viable baseline concept for a green electric power solution in remote rural locations [7]. Stand-alone power systems can be a solution to the problem of the lack of electricity, which highlights challenges such as modules installation costs as well as the sufficient knowledge required for installing, maintaining and operating such systems. With this fact in mind, a 250 W Low Head turbine (propeller) system was developed and optimised to have the advantage of the simplicity of installation and construction from readily available materials [8]. Pico hydropower (PHP) offers an efficient, reliable and cost effective of alternative power sources in developing nations, especially in Malaysia. By and large, the objective is to provide essential strategic implications for Malaysia and other Southeast Asian countries [9].

Independence of villagers must be encouraged about the use of water energy to generate electricity, namely by introducing easy and simple technology. Bamboo which grows a lot in Indonesia and the ease of access to getting induction motors with small capacity, which both of them respectively can be used as turbine materials and made induction generators. The objective of this paper is how to make a turbine with material from local natural wealth and make a generator from an existing electrical motor. Turbine, generator, and auxiliary arrangement as a pico hydro equipment system. The cheap and easy manufacturing of turbines and the ease of getting generators by utilising electrical motors is the aim. The hope that the pico hydropower plant easy to make and inexpensive. So it can spur villagers to be able to implement it, in a rural area, in mountainous. An experiment about Induction motor as the generator was done by [10] to found cheaper and easy generator.

2. Material and Method

2.1. Turbine

Bamboo is a fast-growing, renewable, widespread, low cost, environmental enhancing resource with great potential to improve poverty alleviation and economic development [11]. Bamboos occur and grow well in tropical forests in Indonesia [12], and bamboo is an important plant that is needed to be used for various purposes. More than 90 per cent of bamboos are in village lands and homesteads, mostly cultivated by people and is local natural wealth. Indonesia is a tropical country; there is much variety of bamboo and Indonesia becoming the biggest exporter of bamboo in the world after China. Indonesia has around 154 types from 1,250 to 1,500 type of world bamboo diversity. Bamboo is found in many places, either grow naturally or cultivated deliberately [13].

(a) (b)

Figure 1. Bamboo turbine: (a) Runner with shaft; (b) Parts of turbine.

Turbines are the main equipment for the Pico hydro power plant, turbine innovation needs to be carried out technically and economically, with cheap orientation and ease to implementation. Some proper turbines are expensive so in the Xiagnabouli province, Lao People's Democratic Republic for cost-effective solution for rural electrification, choice pump as turbine to implementation [14]. In this research, crossflow turbine is used by using the type of ori bamboo (*Bambusa arundinacea*), petung bamboo (*Dendrocalamus asper*) and bamboo of Japan (*Pseudosasa japonica*) which are used as material. The dimensions include diameter, turbine width and number of base blade calculations, referring to Mockmore article, with reference to water flow capacity of 5 liters per second and a nominal turbine rotation of 750 rpm.

2.2. Generator

For a small-scale generation, the induction generator converts potential energy to kinetic energy then electric energy. The rotor spins above synchronous speed and develops a counter-torque that opposes this over-speed, same effect as a brake.

The generators rotor returns the power as electrical energy instead of dissipating it as heat, commonly referred to as asynchronous generation. Then kinetic energy is converted into electrical energy and the generator delivers active power to the electrical grid. However to operate successfully the electrical utility grid must also provide reactive power to create the stator's rotating magnetic field.

That is an induction generator receives its excitation or magnetizing current directly from the utility grid. Motors can usually function as generators, and vice versa, and since the induction generator is actually an induction motor being driven by a turbine called the prime mover. An experiment has been done for 200-watt lamps load by [15] with water wheel turbine as prime over and induction motor as induction generator with specification1.5 kW. three phase, 230 volts, 4.5 A, 1440 rpm. Experiments carried out by Katre did not clearly exhibit how much total efficiency was produced.

(a) (b)

Figure 2. (a) Motor as generator; (b) Circuit generator 3 phase output 1 phase.

When an induction machine is driven by an external prime mover at a speed greater than the synchronous speed (negative slip) the direction of induced torque is reversed and it starts working as an induction generator. Induction motor can be operated as a motor or as a generator, in this research, three-phase induction motor (motor specification as a nameplate) was functioned as a one-phase induction generator, generator set up and circuit as seen on Figure 2 (a) and (b). As a reactive load, in generator dipasang 2 buah capacitor seharga 15 μF dan 12 μF.

2.3. Laboratorium Experiment and Field Implementation

Experimental set up for laboratory test show on Figure 4, water pumping for flow simulation to force and rotating the turbine, and the turbine drive induction generator, besides that's the light emitting diode (LED) lamp as an electrical load. On this experiment measuring water flow, pressure, voltage, and electrical current to determine pico-hydro performance.

Figure 3. Laboratorium experiment set up.

(a)

(b)

Figure 4. (a) Under construction; (b) Trial.

Pico-hydropower is becoming one of the most important renewable energy sources in the world for remote or rural locations day by day. It does not encounter the problem of population displacement and is not as expensive as solar or wind energy. However, PHP units are usually off grid and isolated from the grid network; thus, they require user friendly self control to maintain a constant frequency [16].

Figure 5. Electrification to villager house.

3. Result and Conclussion

The paper presents the potential of pico-hydro plant. The envisaged scheme is well suited in remote rural areas where transmission of power proves uneconomical. Pico-hydro plants can be installed at such places to power one or few homes. The power requirement at such location is minimal during off periods which can be utilized for charging batteries and other electronic gadgets. The pico hydro plants can be installed at much lower financial requirements compared with solar plants and wind mills. The prominent impacts include changes in flow regimes and water quality, barriers to fish migration, loss of biological diversity and population displacement but can also provide multiple benefits beyond energy supply. Technological innovation and material research can further improve environmental performance and reduce operational cost. With a changing climate, the resource potential could change due to changes in river flow particularly in precipitation and temperature in catchment area. This may lead to changes in runoff volume, variability of flow and seasonality of the flow (e.g. by changing from spring/summer high flow to more winter flow), directly affecting the resource potential for pico- hydropower generation. Changes in extreme events (floods and droughts) may increase the cost and risk for the such hydropower projects. Increased sediment load could also fill up reservoirs faster and decrease the live storage, reducing the degree of regulation and decreasing storage services [17].

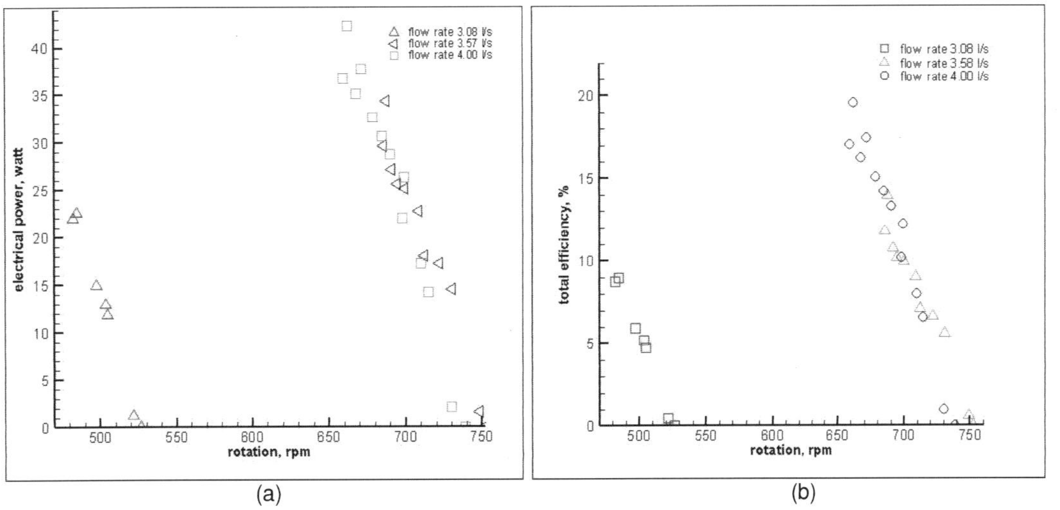

(a) (b)

Figure 6. (a) rpm vs electrical power; (b) rpm vs total efficiency.

The result of experiment plotting on the graphic, it is seen in Figure 6, the greater the flow rate is, the greater the power produced (Figure 6a) and also the greater the flow rates increase the total efficiency (Figure 6b), the largest total efficiency of 19.57% occurs at flow rate 4 liters per second at a pressure of 0.55 kgf / cm2 (head 5.5 meters), the total efficiency including turbine efficiency, transmission efficiency and generator efficiency. Another experiment that has been done before by the author has achieved a total efficiency of around 30%, with a water flow rate of 7 liters per second. Whereas from the research information by [18] using a pump as a turbine at an 18-meter head and a flow rate of 28 liters per second carried out in Kenya at 45% efficiency. The total efficiency obtained in bamboo turbines with IMAG generators is around 30%, which is smaller than the total efficiency of the pico hydro generator in Kenya, by 45%. This is because the smaller the turbine capacity, the smaller the efficiency, according to what [19] stated.

Combination of bamboo ori, bamboo petung and teak wood which is a natural material that many found in Indonesia, can be used to manufacture cross-flow turbine. The cross-flow turbine utilized to drive an IMAG generator that produces electric power (small-scale hydropower, pico hydro), and the result of laboratory tests shows the electric power output which is significant enough to be utilized as electric power generating that is with the load of 114 watt of LED lights at 100 Volt voltages, as well as the result of field implementation tests capable for lighting the LED lamps at a voltage of 167 Volt. The pico hydro exhibits a good electrical output result that can be improved for a good performance for the next. Pico hydro has still needed although rural areas have been received electrification because in the rainy season the light often extinguished or the existing of pico-hydro can help to reduce high electricity cost, so the author hopes that PHP with bamboo turbine and IMAG generator especially can be widely implemented in Indonesia as an alternative choice. It is proven that the materials that are around us in the form of bamboo from several types and induction motors, can be used and used to make pico hydro power generation equipment.

4. References

[1] Faruqi, Y.M. 2007. Islamic view of nature and values: Could these be the answer to building bridges between modern science and Islamic science. International Education Journal, 2007, **8**(2), 461-469. ISSN 1443-1475 © 2007 Shannon Research Press. http://iej.com.au.

[2] Panwar, N.L, Kaushik, S.C., Kothari, S. 2011. Role of renewable energy sources in environmental protection: A review. Renewable and Sustainable Energy Reviews – ELSEVIER **15** (2011) 1513–1524.

[3] Hasan, M.H., Mahlia, T.M.I., Nur, H. 2012. A review on energy scenario and sustainable energy in Indonesia. Renewable and Sustainable Energy Reviews **16** (2012) 2316–2328. journal h om ep age: www.elsevier.com/locate/rser

[4] Lahimer , A.A., Alghoul , M.A., Sopian, K., Amin, N., Asim, N., Fadhel, M.I. 2012. Research and development aspects of pico-hydro power. Renewable and Sustainable Energy Reviews **16** (2012) 5861–5878. journal homepage: www.elsevier.com/locate/rser

[5] Al Amin, R., Talukder, S.H. 2014. Introducing Pico Hydro From Daily Used Water And Rain Water. Int. Journal of Engineering Research and Applications www.ijera.com ISSN : 2248-9622, Vol. **4**, Issue 1(Version 2), January 2014, pp.382-385

[6] Hunachal, S.M., Kumar, M., Oak, S., Burate, A., Kanekar, R., Shirke, P. 2016. DESIGN AND DEVELOPMENT OF PICO HYDRO POWER SYSTEM BY IRRIGATION WATER. International Research Journal of Engineering and Technology (IRJET) e-ISSN: 2395 -0056 Volume: **03** Issue: 08. Pp 1904-1910. www.irjet.net

[7] Sturdivant, R., Yeh, J., Stambaugh, M., Zahnd, A., Edwin K.P. Chong, Fellow. 2012. Pico-Hydro Electric Power In The Nepal Himalayas. 2017 Ninth Annual IEEE Green Technologies Conference. 232-237

[8] Hofmeister, J., Krebs, S., Schickhuber, G., Scharfenberg, G. 2015. Design and Development of a Pico Hydro Turbine System for the use in Developing Countries. 5th International Youth Conference on Energy (IYCE)-IEEE Published. Pisa, Italy. 1-7. **ISBN:** 978-1-4673-7171-1

[9] Kadier, A., Kalila, MS., Pudukudya, M., Hasana, HA., Mohamed, A., Hamid, AA. (2018). Pico hydropower (PHP) development in Malaysia: Potential, present status, barriers and future perspectives. *Renewable and Sustainable Energy Reviews*-ELSEVIER, **81**(2018) 2796–2805.

[10] Adhikary, B., Bhandari, V., Dahal, R. 2010. Design of an Automatic Synchronizer for Connecting Induction Motors as Generators in Mini-Grid Systems. IEEE ICSET 2010 6-9 Dec 2010, Kandy, Sri Lanka. 1-4

[11] Xuhe, C.2003. Promotion of bamboo for poverty alleviation and economic development. J. Bamboo and Rattan, Vol. **2**, No. 4, pp. 345– 350 (2003).

[12] Hakim, L., Nakagoshi, N, Isagi, Y. 2002. Conservation Ecology of Gigantochloa Manggong: an Endemic Bamboo at Java, Indonesia. Journal of International Development and Cooperation, Vol.9, No.1, 2002, pp. 1–16

[13] Abdullah, A H D, Karlina, N., Rahmatiya , W., Mudaim,S., Patimah, Fajrin, A.R. 2017. Physical and mechanical properties of five Indonesian bamboos. IOP Conf. Series: Earth and Environmental Science **60** (2017). 1-5.

[14] Arriaga, M. 2010. Pump as turbine – A pico-hydro alternative in Lao People's Democratic Republic. Renewable Energy **35** (2010) – ELSEVIER. 1109–1115

[15] Katre, S.S., Bapat, V.N. 2015. Induction generator for Pico- hydro generation as a renewable energy source. 2015 International Conference on Energy Systems and Applications (ICESA 2015)-IEEE. Dr. D. Y. Patil Institute of Engineering and Technology, Pune, India 30 Oct - 01 Nov, 2015. 130-134.

[16] Adhikary, P., Kundu, S., Roy, P.K., Mazumdar, A. 2013. Fuzzy Logic based user friendly Pico-Hydro Power generation for decentralized rural electrification. International Journal of Engineering Trends and Technology- Volume4Issue4- 2013. ISSN: 2231-5381 http://www.internationaljournalssrg.org. 507-511

[17] Nimje, A.A., Dhanjode, G. 2015. Pico-Hydro-Plant for Small Scale Power Generation in Remote Villages. IOSR Journal of Environmental Science, Toxicology and Food Technology (IOSR-JESTFT) e-ISSN: 2319-2402,p- ISSN: 2319-2399.Volume 9, Issue 1 Ver. III (Jan. 2015), PP 59-67 www.iosrjournals.org

[18] Maher, P., Smith, N.P.A., Williams, A.A. 2003. Assessment of pico hydro as an option for off-grid electrification in Kenya. Renewable Energy **28** (2003) 1357–1369. www.elsevier.com/locate/renene

[19] Mockmore, C.A, Merryfield, F. (1949). *The Banki Water Turbine*. Oregon State System Of Higher Education, Oregon State College, Corvallis.

Acknowledgement
This work was financed by Indonesia Ministry of Higher Education from scholarship program and this research was funded by Research and Technology Ministry for doctoral scholarship in Doctoral Program Environmental Science Brawijaya University-Indonesia. Special thanks to Mr. Prayitno head of Jembul Village have provided place for pico hydro power plant implementation of experiment research.

IOP Conf. Series: Materials Science and Engineering **462** (2019) 012048 doi:10.1088/1757-899X/462/1/012048

Design And Simulation Of Electric Center Distribution Panel Based On Photovoltaic System

T Wati[1], A Sahrin[2], T Suheta[1] and I Masfufiah[1]

[1]Institut Teknologi Adhi Tama Surabaya, Indonesia
[2]Politeknik Energi dan Mineral Akamigas Cepu, Imdonesia

trisnaw@itats.ac.id

Abstract. The solar energy as an alternative energy to generate electrical energy using photovoltaic (PV) connected to the grid of batteries that work all night widely developed in Indonesia. The ability to use PV for power conditioning and power factor compensation has been proven by researchers. However, most of the applications described in the engineering literature are used on a three phase system using techniques developed for active filters. Inverter can supply requirements in the vessel control room Baruna Jaya 8. PV systems are designed to generate 250 kW. From the simulation and test results using PSIM software it is found that the PV system can generate 250 kW power to supply the distribution panel as alternate of fuel on the ship Baruna Jaya 8.

1. Introduction

Indonesia is one of tropical country that solar energy is widely used to generate electricity. Many photovoltaic (PV) that have been developed among the community both for industrial and research purposes. In this paper, the PV system is used to supply electrical energy at the Distribution Panel Control of the Baruna Jaya vessel. PV is installed on the ship due to solar energy opportunities in the sea to produce large amounts of electrical energy. Solar Power (PLTS) are electricity generator that use solar energy as a source. The main tool is a catcher, modifier and producer of electricity called photovoltaic or Solar Cell Panel (module) [1]. Solar energy is converted into electricity through the process of negative electron flows, then becomes a DC electric current that will directly charge the Battery / Accumulator based on the required voltage and current. The average of marketed module product produces 12 to 18 VDC and 0.5 to 7 Ampere. The module has a variety of capacities, ranging from 10 watt peaks to 200 watt peaks. The module consists of monocrystal and polycrystal cell types [2]. The core components of this solar power are solar cell modules, regulators/controllers, battery, accumulators, DC to AC inverters and loaders. In utilizing PV as an electrical energy source, planning for the installation process is needed to obtain maximum results and reduce wasted energy [3][4].

2. Literature Review

In the recent research solar energy using Arduino UNO controller, its compare with static parabolic reflector and without parabolic reflector. Solar energy using Arduino UNO with parabolic reflector has a best performance with average output voltage 7.8 volts and 6,14 mA. The system can be efficient of energy for many application especially it use to operate motor DC [5]. Numerical method is present to validate optimization solar panels position for Beirut [6]. Where the method is for determine of solar irradiance for Beirut, Lebanon. The method can applied to any region and the result has a constant tilt angle, a monthly adjustable tilt angle, and solar tracking system. However this method is applied for the normal direct radiation. Mirror reflected solar panel (MRSP) was advance from parabolic reflector with automatic cooling and tracking [7]. Tracking system target is sun light approximately in perpendicular direction. The tracking with cooling reflection was higher power than tracking without reflection cooling. But the reflection cooling has appetite because the output power downward during noon to afternoon. In this paper the solar energy system for the needs of electricity power is designed based on the load power requirements of the Distribution Control Panel. The solar power system design can be seen in Figure 1. The initial step in designing a PLTS system for the Control Panel Distribution is the power capacity of the distribution panel of 30 KVA, 230V-50 Hz where the panel supplies and is connected as a controller of research equipment.

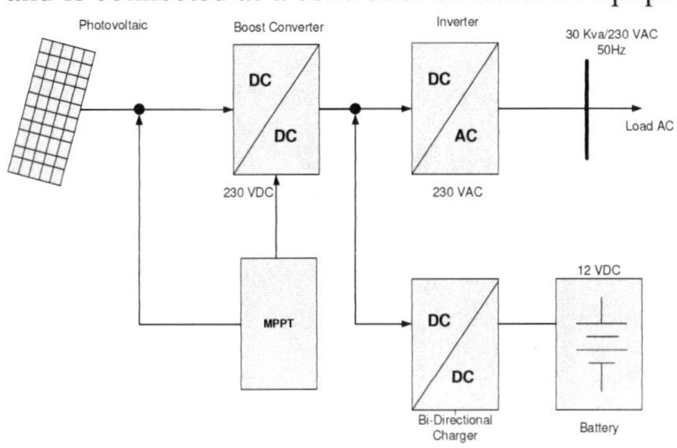

Figure 1. Block of planning system diagram

2.1 Load Estimation

In the design of the PLTS system for the distribution panel of the Baruna Jaya vessel, determines the total daily load, from the determination of that total daily load, we will obtain an electric load curve [2]. The total daily load is the amount of energy needed by the electricity load every day, and can be shown in Table 1.

Table 1. Load Estimation of the Baruna Jaya Ship Distribution Panel

No	Load	Power (W)	Jml	Total power (W)	Length of Use (Jam)(H)	Energy (WH)
1	Fan	100	4	400	10	4000
2	Computer	700	8	5600	10	56000
3	LCD Proyector	255	2	510	3	1530
4	Printer	30	4	120	3	360
5	TL Lamp	25	10	250	10	2500
6	DC power supply	160	8	1280	10	1280
7	Gyro Compas	25	1	25	10	250
8	DGPS	5	1	5	10	50
9	Others	200	1	200	8	160
			Total			66130

Note: Data refers to Oceanography of LIPI RV Baruna Jaya 8

2.2 Calculation Of Plts Load Energy Equipment Specifications

The amount of load energy that will be supplied by the PLTS on the Baruna Jaya vessel is 8390 W. The power capacity of solar cell modules can be calculated by several factors, those are the required system energy, solar insulation, and adjustment factor. The power capacity of the solar modules produced is:

$$\text{Module Power Capacity} = \frac{66130\, wH}{3{,}91}$$

$$= 16913{,}04W$$

Where 3,91 is average sun isolations.

2.3 Number of Modules

The number of modules are determined by:

$$\frac{16913{,}04}{7\, x\, 250} = 9{,}66\ modul$$

Rounded up, so the number of modules are 10 units with the capacity of each module is 250 Wp.

2.4 Battery Capacity

Battery capacity is calculated based on the amount of power produced by 10 solar modules for 10 hours with a 12VDC battery voltage, namely:

$$AH = \frac{10\, x\, 250\, w\, x\, 10\, jam}{12} = 2083{,}33\, AH$$

The battery capacity used is 100 Ah with a terminal voltage of 12 Volt DC, so the battery needed is 20 batteries. The load on the solar cell system takes energy from the Battery Charger Regulator (BCR). The current capacity that flows in BCR can be determined by knowing the maximum load installed. The maximum load occurs is approximately 1500 Watts with the maximum load of the system voltage is 12 volts, so the current capacity that flows in BCR is:

$$Imaks = \frac{1500}{12} = 125\, A$$

So the BCR capacity used must be greater than 125 A.

2.5 Equipment Specifications

2.5.1 Solar cell or photovoltaic (PV). Solar cells or more often called photovoltaic (PV) function to capture and convert solar energy into electricity through a negative electron flow process. Electricity generated is DC electricity. This study uses 100 units of PV modules with specifications:

Model SL ET-P660250

Maximum Power (Pm)	: 250 W
Open Circuit Voltage (Voc)	: 37,58 V
Short Circuit Current (Ioc)	: 8,98 A
Maximum Power Voltage (Vmp)	: 30,02 V
Maximum Power Current (Imp)	: 8,33 A
Working Temperature	: 45,3±2°C
Tolerance	: ± 3%

2.5.2 *Battery Control Regulator (BCR).* Battery Control Regulator (BCR) is an equipment used for setting the charging system from the Solar Panel. This study uses BCR that can be enabled under Normal or Auto Load position conditions. This study uses 3 units of BCR with specifications:

Output Voltage	: 12 VDC
Power Consumption	: 10 mA (standby)
Turn of Charge Battery	:14.50 +/- 0.01 V
Recharge Batteray position	: 13.50 +/- 0.10 V
load shutdown	: 11.20 +/- 0.01 V
Reset the load voltage	: 12.60 +/- 0.01V

Low battery warning sound :11.60 +/- 0.10 V

2.5.3 Inverter. The inverter function is to change the DC voltage produced by PV (stored in a battery) into the AC phase. This is done because most electrical equipment is designed with an AC voltage source. The inverter in this study was able to produce voltage from the original condition that is 12 VDC to 220 VAC. This study uses 1 inverter unit with specifications:

Model	: SDA-1000
Type	: Modified Wave
Input Voltage	:12 VDC
Output Voltage	: 220 VAC-240 VAC
Rated power	: 1000 W
Peak power	: 2000 W
Efficiency	: > 70%

3. Metode

3.1 Simple Equivalent Series

Solar cell consists of p-n semiconductor junction where when it exposed to light, the proportional currents in the sollar will generate radiation. The PV circuit model can be illustrated in Figure 2.

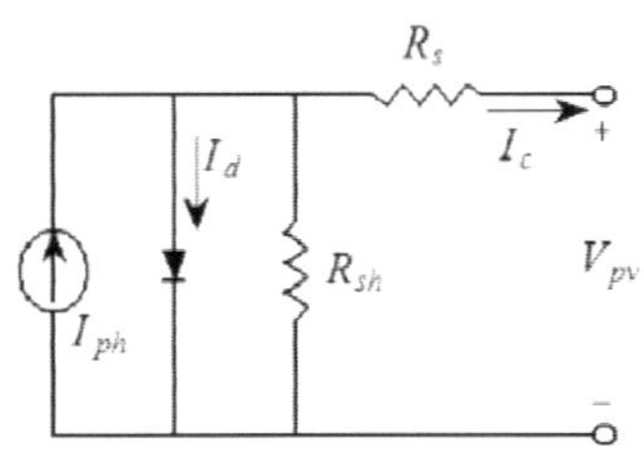

Figure 2. Potovoltaic model with single Battery

3.2 Mathematical Model Theory

From Figure 1, the characteristics of solar cells in the equivalent circuit can be formulated into:

$$I = I_{ph} - I_d(e^{\frac{qV_{PV}}{kT}} - 1) \quad (1)$$

3.3 Practical Model Of PV Series

Photovoltaic arrays can be connected in series-parallel batteries. The photovoltaic array output will be shown in the following equation:

$$Vtot = N_s U_{cell}$$
$$Itot = N_p I_{cell}$$
$$Ptot = N_s N_p I_{cell} \quad (2)$$

We assume using two series of circuits, the photovoltaic array circuit model is shown in Figure 3.

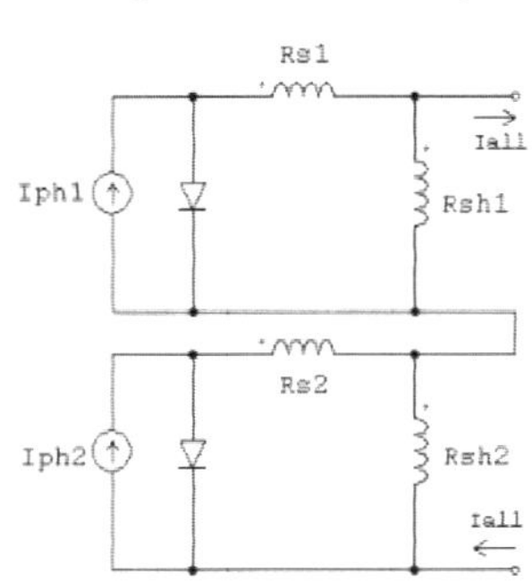

Figure 3. Model series of PV array

3.4 PV Simulation Model on PSIM

Referring to the PV SL ET-P660250 parameter that one module has a maximum power of 250 watts. In PV integration modeling, Figure 4 is 10.

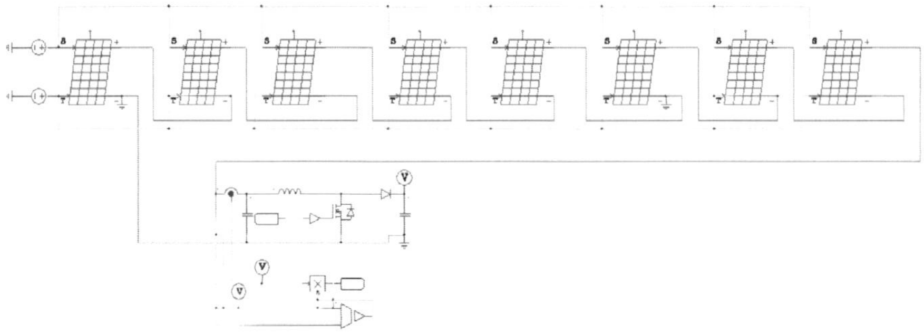

Figure 4. PV simulation on PSIM

3.5 MPPT Design with Modified Buck-Boost ConverterKonverter

Buck-Boost is used to increase the voltage level of the PV output. Figure 5 is a modification of the Buck-Boost converter circuit.

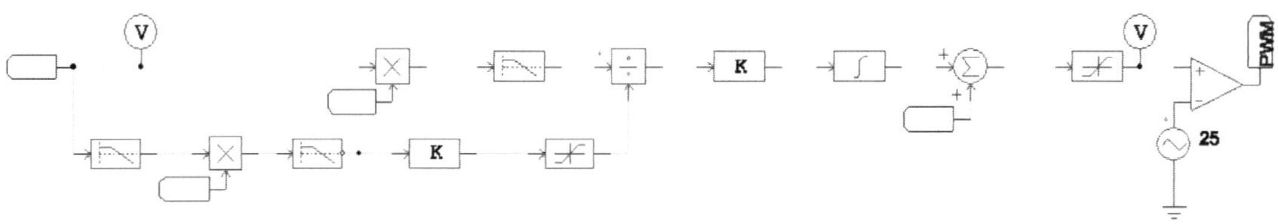

Figure 5. Modification series of Buck-Boost converter

3.6 Battery Modeling

Battery modeling in Figure 6 is based on the equation:

$$E = E_0 - K\frac{Q}{Q - \int i_b dt} + A\exp(-B.\int i_b dt) \quad (3)$$

Battery modeling consists of voltages that can be controlled by Vin which is connected in series with the resistance R, while the equation is implemented with control blocks whose output values can regulate Vin voltage.

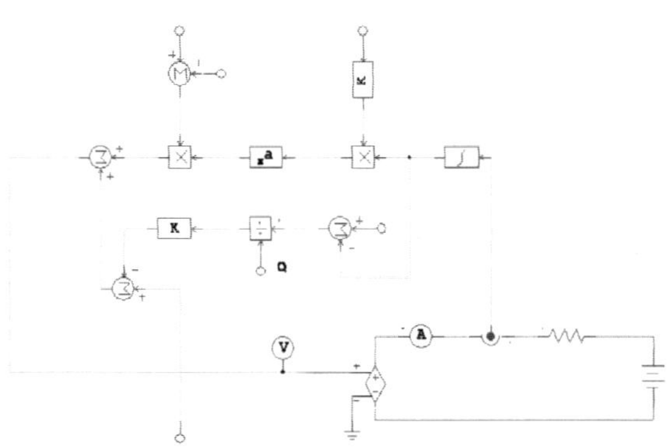

Figure 6. Equivalen series of battery

3.7 One Phase Inverter Design

This inverter used is the ful-bridge inverter type. The inverter modeling circuit consists of Vdc, S1, S2, S3, S4, and R loads. While L is the output filter of the inverter circuit that is shown in Figure 7.

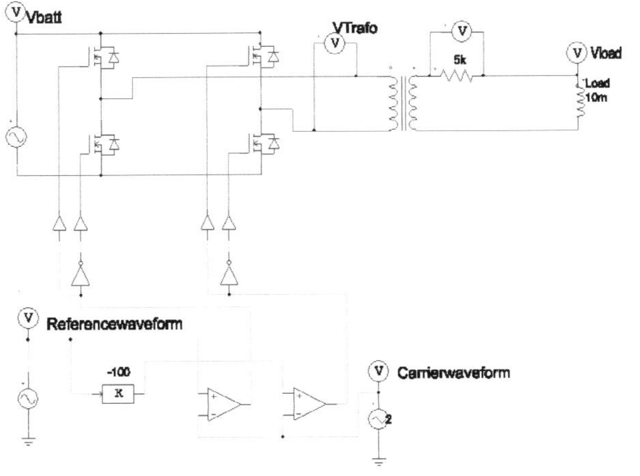

Figure 7. One Phase Inverter circuit

4. Simulation Result

PV output can be seen in Figure 8. We can see that the output is a 230 volt DC signal

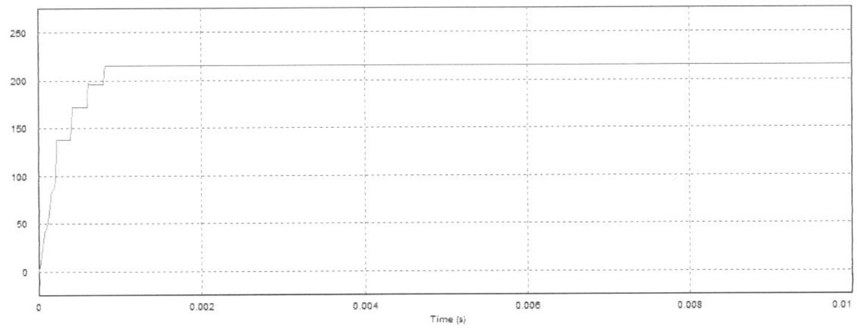

Figure 8. Output of PV Voltage

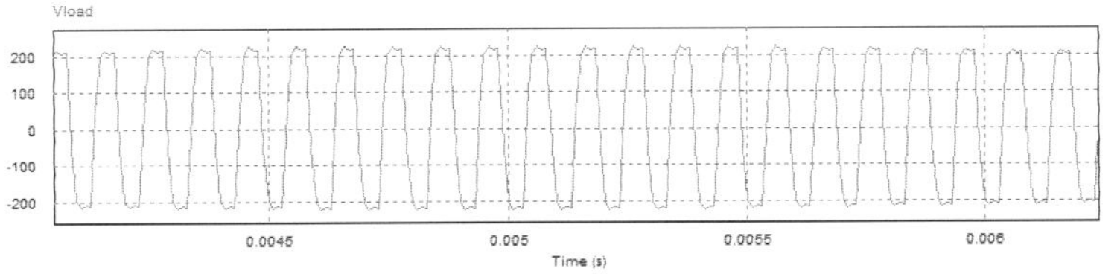

Figure 9. Output of Inverter Voltage

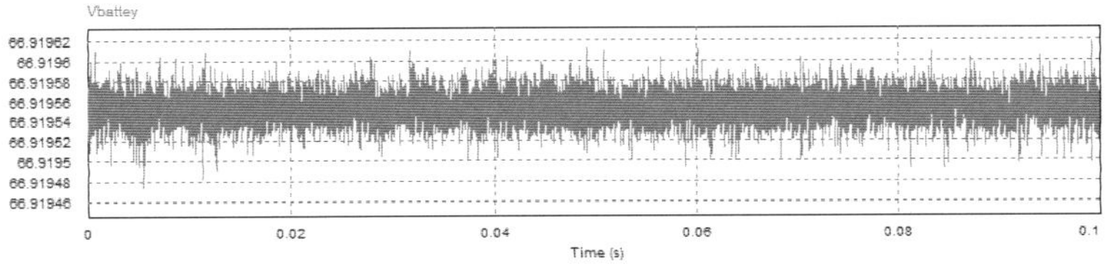

Figurer 10. Charging and Discharging Battery

5. Conclusion

From the simulation results it can be concluded that the PV output voltage of 230 Volts DC with an output of 230 volts is expected to supply distribution panels on the Baruna Jaya 8 vessel with a capacity of 25000 watts. The modules needed to meet a power of 25,000 watts require100 units modules with a capacity of module is 250 watts per unit. From the simulation results, it is expected to be implemented in real-time to supply the Baruna Jaya 8 Vessel panel as a substitute for fossil energy.

6. References

[1] Abidin, Zainal dan Custer, Johny, 2013 Desain Sistem Energi Alternatif Sebagai Sumber Energi Listrik Laboratorium Listrik Dasar Politeknik Negeri Bengkalis Vol 3 Nomor 2 hlm 98-102
[2] Imawan, Pramudya Santoso dan Aria I Ketut Pria Utama 2012 Studi Pengembangan Kapal Ikan Katamaran dengan Penggerak Hybrid Institut Teknologi Sepuluh Nopember Seminar Nasional Kedaulatan Pangan dan Energi
[3] Li, Ji., Chen, Yu., Liu, Yunda 2012 Research on a Stand-alone Photovoltaic System with a Supercapasitor as the Energy Storage Device School of Information Engineeering, Inner Mongolia University of Science and Technology Baontu Science Direct Energy Procedia
[4] Salman, Adi., Williams, Arthur., Amjad, Hanya., Kamran,M., Saad, M 2015 Simplied Modeling of PV Panel by Using PSIM and its Comparison with Laboratory Test Results Dublin Institut of Technology (DIT) Global Humanitarian Technology Conference
[5] Maninder Singh, Jaspreet Singh, Anshula Garg, and Ekambir Sidhu 2016 Efficient Autonomous Solar Energy Harvesting System Utilizing Dynamic Offset Feed Mirored Parabolic Dish Integrated Solar Panel IEEE WISPNET Conference
[6] Emilio Sassine, 2016 Optimal Solar Panels Positioning for Beirut International Renewable Energy Congress (IREC),
[7] Sheikh Md. Shain Alam, Mizanur Rahman 2016 Performance Comparison of Mirror Reflected Solar Panel with Tracking and Cooling International Conference on the Development in the in Renewable Energy Technology (ICDRET)
[8] Suresh E. Puthanveettil, M.Sankaran and S.B. Gupta 2014 Mitigation Techniques for arcing on Space Solar Panles-Results From ISRO Mumbai, Indida XXVI International Symposium on Discharges and Electrical Insulation in Vacuum.
[9] Vasudha Sharma, Vijay Kumar Tayal 2017 Hardware Implementation of Sun Tracking Solar Panel Using 8051 Micro-Controller amity University Uttar Paadesh, Noida, India International Conference on Reliability, Infocom Technologies and Optimization (ICRITO)
[10] Usha G, Vasudevan R, Krishna Priya G, Uma.B.R and Sankaran M 2014 Establishing Suitability of Standard Multijunction Solar Cells For Mars Orbiter Mission Photovoltaic Specialist Conference

IOP Conf. Series: Materials Science and Engineering **462** (2019) 012049 doi:10.1088/1757-899X/462/1/012049

Implementation of Waste Reduction at Operational Division with Lean Manufacturing Concept

Lukmandono[1], N L P Hariastuti[1], Suparto[1] and D I Saputra[1]

[1]Industrial Engineering Department, Faculty of Industrial Technology, Institut Teknologi Adhi Tama Surabaya, Jl. Arief Rahman Hakim 100, Surabaya, Indonesia

lukmandono@itats.ac.id

Abstract. PT. Semen Gresik Packaging Industry (IKSG) Tuban is a subsidiary established by PT. Semen Gresik (Persero) Tbk which is tasked to fulfil the needs of packaging bags for cement production. Problems that appear often occur in a company, especially PT. IKSG is a waste that often occurs in the production flow, such as the large number of defective production bags, production machines that often stop in the middle of the production process and other problems. From these problems, this research aims to find a solution using the lean manufacturing approach in handling and reducing waste that occurs in the existing production processes in the company. Three types of waste often occur in the company, namely, defect, waiting, and overproduction, as well as recommendations for improvement based on the selected mapping from VALSAT analysis in the form of elimination of non-value, added activities on PAM mapping with the results of manufacturing cycle effectiveness (MCE). It is increased by 3.18%, the application of the master production schedule (MPS) method on SCRM mapping with a more optimal amount of production results, and the implementation of maintenance and policies in the production process to reduce defects in the QFM mapping.

1. Introduction

Making cement bags in bag factories is done to meet the needs of the cement production process. Demand for cement bags is not only for the needs of cement production in the factory, but also to meet the demand and needs in several cement industries and other industries. To meet the growing demand for cement bags, a more considerable cost is needed, and higher profits are expected.

PT. Semen Gresik (Persero) Tbk as one of the companies engaged in cement production in Indonesia established a subsidiary that will serve as a supplier of cement bags for cement production that the company does. PT. Semen Gresik Packaging Industry (IKSG) Tuban is a company engaged in the manufacture of cement bags. Problems that appear often occur in a company, especially PT. IKSG is a waste that often occurs in the production flow as there are still many pockets of production that are defective due to printing results, glueing or folding bags that are not optimal, production machines that often stop amid the production process and other problems that can cause waste. The existence of waste in the production process of a company influences income and produces a waste of time and production processes.

From these problems, the purpose of this research is to find a solution in handling and reducing the waste that occurs in the existing production processes in the company [5]. So that with the discovery of an answer about handling and handling waste, it is expected that the waste that previously often occurs can be minimized to increase efficiency and increase profits from the company itself. One method that can be done to eliminate waste is lean manufacturing. Lean is a continuous effort to reduce waste and improve value added of products (goods/services) to provide value to customers (customer value) [1].

2. Literature Review

2.1 Lean Manufacturing
Lean manufacturing is a systematic approach to identify and eliminate waste through continuous improvement, to create a smooth flow of production processes with fast lead times and little waste. The production process that can be categorized as lean is a process that transforms input into output with activities that add value to a very low level of waste, so that the customer gets a product that is worth the expected [4]. Waste Toyota has identified a type of waste that does not add value to a business or manufacturing process. The following is a type of waste [2]: (1) Excessive procedure (overproduction); (2) Delays/ waiting time; (3) Transportation that is not necessary; (4) Excessive processing (inappropriate processes); (5) Excess inventory; (6) Excessive movement; (7) Defective products.

2.2 Value Stream Analysis Tools (VALSAT)
In principle, VALSAT is used as a tool to map specific value streams that focus on the value-adding process [3]. There are seven kinds of detail mapping tools that are most commonly used, namely:

1. Process Activity Mapping (PAM) PAM is a technical approach used in activities on the production floor. However, the expansion of this tool can be used to identify lead times and productivity both physical production flow and information flow.
2. Supply Chain Matrix Response It is a graph that illustrates the relationship between inventory and lead time on distribution channels so that there can be an increase or decrease in inventory levels and distribution time in each area in the supply chain (material area, production area, and finished product area).
3. Production Variety Funnel A visual mapping technique that attempts to map the number of product variations in each stage of the manufacturing process. This tool can be used to identify where a generic product point is processed into a specific product. In addition, this tool can also be used to show the bottleneck area in the process design.
4. Quality Filter Mapping Is a tool used to identify the location of the problem of quality defects in the existing supply chain. Evaluation of quality loss that is often done for short-term development. This tool is able to describe three different types of quality defects such as product defects, scrap defects, and service defects.
5. Demand Amplification Mapping The tools used to visualise demand changes along the supply chain. This phenomenon adheres to the law of industrial dynamics, where the demand transmitted throughout the supply chain through a series of order and inventory policies will experience ever increasing variations in each movement from downstream to upstream.
6. Decision Point Analysis Tools that show a variety of different production system options, with a trade-off between the lead times of each choice with the level of inventory needed during the lead time process.
7. Physical Structure Is a tool used to understand the condition of the supply chain at the production level. This is needed to know how the industrial operating conditions are and direct attention to areas that may not have received enough attention for development.

2.3 Value Stream Mapping
Value Stream Mapping (VSM) is a tool to identify value-added activities and non-value activities added to the manufacturing industry, to make it easier to find the root of the problem in a process [3]. This tool is able to show errors through an overview of the current state system and is used as a basis for making ideal conditions.

3. Methods

The research method is a description of the steps taken in solving the problems that exist in this study. These stages are prepared as a reference for researchers in the implementation and resolution of these problems.

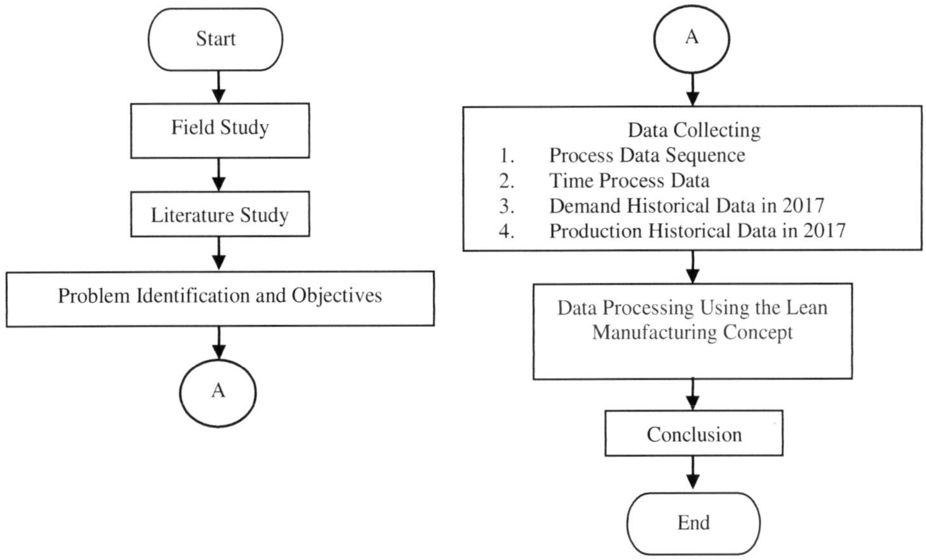

Figure 1. Flowchart of Research Methodology

4. Result and Discussion

4.1 Current Value Stream Mapping

Current value stream mapping is an illustration of the current production process conditions.

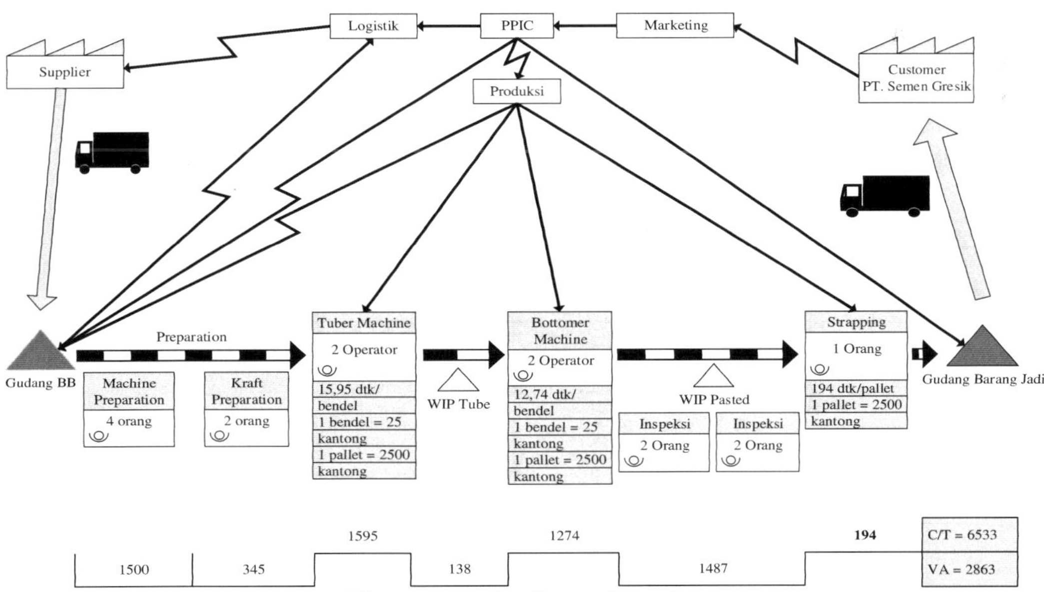

Figure 2. Value Stream Mapping

4.2 VALSAT

After knowing the description of the production process at this time using the current value stream mapping, the next step is to prepare a VALSAT matrix to find out the detailed mapping that will be used in the process of identifying waste that occurs based on the conditions in the company.

Table 1. *Value Stream Analysis Tools* (VALSAT)

Waste	*Weight*	*Process Activity Mapping*	*Supply Chain Response Matrix*	*Product Variety Funnel*	*Quality Filter Mapping*	*Demand Amplification Mapping*	*Decision Point Analysis*	*Physical Structure*
Over production	2,50	2,50	7,50	0,00	2,50	7,50	7,50	0,00
Waiting	3,17	28,53	28,53	3,17	0,00	9,51	9,51	0,00
Transport	0,00	0,00	0,00	0,00	0,00	0,00	0,00	0,00
Inappropriate Process	0,33	2,97	0,00	0,99	0,33	0,00	0,33	0,00
Inventory	1,00	3,00	9,00	3,00	0,00	9,00	3,00	1,00
Unnecessary Motion	0,33	2,97	0,33	0,00	0,00	0,00	0,00	0,00
Defect	4,00	4,00	0,00	0,00	27,00	0,00	0,00	0,00
Total		43,97	45,36	7,16	29,83	26,21	20,34	1,00
Rank		2	1	6	3	4	5	7

4.3 Quality Filter Mapping (QFM)

Quality filter mapping is used to analyse waste type defects that occur in the process of producing pasted kraft cement bags. The following is the data defect in the production process of pasted kraft cement bags from January 2017 until December 2017.

Table 2. Percentage *Defect*

Month	*Kraft* Usage (kg)	Afval (kg)	Percentage (%)
January	3.024.330	46.272	1,53
February	1.907.310	36.569	1,92
March	2.073.827	43.870	2,12
April	1.983.212	35.284	1,78
May	1.473.722	34.407	2,33
June	2.028.698	35.125	1,73
July	2.438.144	54.755	2,25
August	2.238.455	48.789	2,18
September	2.782.767	54.658	1,96
October	2.740.091	86.011	3,14
November	2.563.824	61.096	2,38
December	2.454.078	87.957	3,58
Total	27.708.458	624.791	2,25
Average	2.309.038	52.066	2,24

Table 3. Master Production Schedule bag's *Pasted Kraft* Type 40 kg

	Years in 2017 (Unit)											
	Jan	Fe	March	April	May	June	July	August	Sep	Oct	Nov	Dec
Actual demand	0	14.326.000	14.020.000	11.163.000	13.000.000	12.264.000	12.290.000	14.999.000	16.779.000	15.726.000	11.977.500	13.744.000
Actual Stock	18.580.000	4.254.000	3.518.500	5.640.000	5.924.500	6.945.000	7.939.500	6.225.000	2.730.500	289.000	1.596.000	1.136.500
MPS		1.328.4500	1.3284.500	13.284.500	13.284.500	13.284.500	13.284.500	13.284.500	13.284.500	13.284.500	13.284.500	13.284.500

Information :
Actual Stock = 18.580.000; *Lot Sizing* = 13.284.500

Table 4. Master Production Schedule of bag's Pasted Kraft Type 50 kg

	Years in 2017 (Unit)											
	Jan	Fe	March	April	May	June	July	August	Sep	Oct	Nov	Dec
Actual demand	90.000	3.155.000	2.833.000	2.637.000	3.113.000	1.335.500	2.153.500	4.025.984	2.838.000	2.871.000	2.494.811	3.794.811
Actual Stock	2.660.000	3.007.410	174.410	1.039.820	1.429.230	93.730	1.442.640	919.066	1.583.476	2.214.886	3.222.485	2.930.084
MPS		3.502.410		3.502.410	3.502.410		3.502.410	3.502.410	3.502.410	3.502.410	3.502.410	3.502.410

Information :
Actual Stock = 2.660.000; *Lot Sizing* = 3.502.410

Table 5. Comparative of Corporate Policy and MPS

Comparative	*Pasted Kraft* 40kg (unit)	*Pasted Kraft* 50kg (unit)
Demand	150.288.500	31.341.606
Corporate Policy	163.387.750	31.729.000
MPS	146 129.500	31.521.690
Margin	17.258.250	207.310

Table 6. Comparison of Afval Amounts Before and After Repair Recommendations

Month	Afval (kg)	
	Before	After
January	46.272	41.477
February	36.569	31.774
March	43.870	39.075
April	35.284	30.489
May	34.407	29.612
June	35.125	30.330
July	54.755	49.960
August	48.789	43.994
September	54.658	49.863
October	86.011	81.216
November	61.096	56.301
December	87.957	83.162
Total	624.791	567.253
Average	52.066	47.271

5. Conclusion

From the results of the research that has been done, the findings obtained are as follows: Based on the value stream analysis tools (VALSAT) identification, there are three wastes with the highest value, namely the defect of 4.00, waiting for 3.17 and overproduction of 2.50. Also, VALSAT identification also obtained 3 mapping with the highest value, namely supply chain response matrix (SCRM) of 45.36, process activity mapping (PAM) of 43.97, and quality filter mapping (QFM) of 29.83.

From the analysis of each selected mapping, the following results are obtained: From the supply chain response matrix (SCRM) analysis using a master production schedule (MPS), the optimal production amount was 17,258,250 units compared to the company's policy for 40kg pasted kraft bags. Whereas for 50kg pasted kraft bags 207,310 more optimal results were obtained from company policy in 2017. From the analysis of process activity mapping (PAM), the results of manufacturing cycle effectiveness (MCE) were 47.10%, up 3.18% from the previous 43.92%. From the quality filter mapping (QFM) analysis, the results of an increase in defects that are expected to occur is 567,253 kg during 2017, a decrease of 9.21% from the actual condition of 624,791 kg.

6. Reference

[1] Gasperz Vincent. 2007. *Lean Six Sigma for Manufacturing and Service Industries.* Jakarta. PT Gramedia Pustaka Utama

[2] Gasperz, Vincent dan Avanti Fontana. 2011. *Lean Six Sigma for Manufacturing and Service Industries.* Bogor.Vinchristo Publication

[3] Hinnes P., and N. Rich. 1997. The Seven Value Stream Mapping Tools. *International Journal of Operationsand Production Management*

[4] Likert, J. K. 2004. *The Toyota Way.* Jakarta. Penerbit Erlangga

[5] Prabowo, Rony. 2016. Penerapan Konsep Line Balancing untuk Mencapai Efisiensi Kerja yang Optimal PadaSetiap Stasiun Kerja Pada PT. HM. Sampoerna Tbk. *Jurnal IPTEK*, vol. 20 (2). Pp 9-20. Institut Teknologi Adhi Tama Surabaya

Plastic debris in sediments from the east coast of Surabaya

A C Ni'am[1,2], S J You[1], Y F Wang[1], and J J Jiang[1]
[1]Department of Environmental Engineering, Chung Yuan Christian University
200 Zhongbei Road, Chung Li 320, Taiwan
[2]Deparment of Environmental Engineering, Adhi Tama Institute of Technology Surabaya
Arief Rahman Hakim Road, 100 Surabaya

Abstract. Surabaya is the capital city of East Java and it has a coastal area. This coastal area has lot of communities and tourism activities. As a result, this will have an impact of increasing plastic waste in the environment. This circumstance will have an impact on increasing plastic waste in the environment. The accumulation of plastic waste on the east coast of Surabaya is due to human activities. The present study has identified the polymer of plastic debris collected from the east coast of Surabaya. Bulak was the major district having high plastic debris (89.41%). The highest of three polymer types were identified viz. polypropylene (PP), high-density polyethylene (HDPE), and low-density polyethylene (LDPE). Polypropylene was the most polymer plastic found on the east coast of Surabaya (67.86%), particularly in Bulak followed by Kenjeran, Gunung Anyar, and Rungkut.

Keywords: coastal area, plastic debris, sediments, Surabaya

1. Introduction

Production of plastics has risen significantly since the improvement of synthetic polymers in the intermediate of the 20th century in the world [1]. In 2014, the global production of plastics estimated at approximately 311 million tonnes and still probability to rising [2]. Plastics are largely used because they are simple, light, durable and cheap [3]. The highest of a plastic polymer was polypropylene (19.2%) for packaging [2]. The accumulation of plastics in the environment caused by poor management rates and inappropriate exile of plastic waste. This base probability leads to rising levels of plastic debris worldwide, especially in the marine environment.

Currently, accumulation and fragmentation of plastic debris in terrestrial environments, in the ocean, on shorelines of even the most secluded islands and in the deep sea. Furthermore, the marine debris has attributed to land-based and ocean-based sources such as tourism activities, fishing activities, and human activities around coastal areas [4]. The marine environments are acknowledged to be useful to the fragmentation of plastic debris [5]. Previous studies on plastic debris accumulation in the marine environments [1, 3-6]. A lot of plastic marine debris accumulating in coastal areas before moving to expanses of the ocean [3].

Indonesia has more than 100 cities that are located in coastal areas. The amount is growing because of a tendency of urbanization rate in the urban region. Therefore, the human population living around coastal areas will consequently develop [7]. Indonesia produces 187.2 million tonnes of plastic waste each year, it means Indonesia is the second highest marine polluter in the world following China [8]. In fact, 43.4% of plastic in Indonesia is used as packaging [9]. Surabaya City as the second largest city has been developing rapidly. Surabaya is a large city that has the coastal area like Jakarta as the capital city of Indonesia. Surabaya city has been produced 400 tonnes of plastic waste [9].

The east coast of Surabaya is susceptible to plastic waste accumulation on the coastline from terrestrial source owing to human activities around coastal areas, marine sources owing to fishing and tourism activities, as well as river discharges. The aim of the present study is to identify the type of plastic debris in sediments on the east coast of Surabaya.

2. Material and methods

2.1 Study area and sampling location

The east coast of Surabaya was chosen because of its accessibility and the huge number of living communities. Samples were collected at four districts on the east coast of Surabaya viz. Rungkut district, Gunung Anyar district, Bulak district, Kenjeran district. The sediments sampling was conducted using a random station in each district. Subsequently, each district has four stations. The detail of the station is shown in Table 1. Sampling at each station performed was occurred repetitions by using Global Positioning System (GPS).

2.2 Sample processing and analysis

The focus of items in the present study was plastic debris in sediments viz. mesoplastic (> 5-20 mm) and macroplastics (> 20 mm). The sediments were taken by *Ekman grab*. The sediments sample were placed directly in a plastic bag. The sediments sample was dried at room temperature overnight. Subsequently, sediments were dried at 60°C in an oven for 24 hours.

The visible plastic debris were separated from sediments with tweezers. In addition, the plastic debris were sorted from non-plastic materials based on the physical and morphological characteristics. Furthermore, the sediments were sieved using commercial mesh with distinct mesh sizes to separate plastic items with distinct sizes and other particles. Most of the plastic debris were identified with the naked eye.

Figure 1. (a) station location in mangrove area (b) Sediments sampling using an Ekman grab.

3. Results and discussions

The sites of data collection located on the east coast of Surabaya. The sampling sites divided two areas viz. mangrove areas (Rungkut district and Gunung Anyar district) and urban areas (Bulak district and Kenjeran district) (Table.1). The plastic debris in sediments was taken when the east monsoon in Surabaya with the sea temperature approximately 28-29°C.

Table 1. Sampling location study

District	Characteristics area	Characteristics Sediment	Stations	Coordinates
Rungkut	Mangrove area	Clay	1	S 07° 19' 07.38" E 112° 50' 03.99"
		Clay	2	S 07° 19' 15.73" E 112° 50' 12.00"
		Clay	3	S 07° 18' 30.20" E 112° 50' 46.41"
		Clay	4	S 07° 19' 16.85" E 112° 50' 25.18"
Gunung Anyar	Mangrove area	Clay	1	S 07° 20' 05.12" E 112° 49' 49.39"
		Clay	2	S 07° 20' 06.71" E 112° 49' 48.50"
		Clay	3	S 07° 20' 01.26" E 112° 49' 42.27"
		Clay	4	S 07° 19' 56.79" E 112° 49' 35.67"
Bulak	Urban area	Clay	1	S 07° 13' 33.43" E 112° 47' 23.00"
		Clay and Silt	2	S 07° 13' 28.32" E 112° 47' 18.88"
		Silt and Sand	3	S 07° 13' 02.56" E 112° 47' 06.20'
		Silt and Sand	4	S 07° 13' 11.89" E 112° 47' 15.14"
Kenjeran	Urban area	Clay	1	S 07° 14' 11.42" E 112° 47' 49.28"
		Clay	2	S 07° 14' 17.66" E112° 47' 51.81"
		Clay and Silt	3	S 07° 13' 57.94" E 112° 47' 42.10"
		Silt and Sand	4	S 07° 13' 49.43" E 112° 47' 38.20"

3.1 The quantities of plastic debris

The total of 529 plastic debris was taken successfully in four districts along the east coast of Surabaya (Table 2). The high percentages of plastic debris have obtained in Bulak District was 89.41% followed by Kenjeran 9.07%, Gunung Anyar 0.95% Rungkut 0.57%. This result indicates that Bulak district (urban area) was the highest district have a source of plastic waste. Bulak district has the majority of human activities. In addition, Bulak district has the fishing settlement that very close with the coastal areas, domestic activities, traditional markets [10, 11]. Therefore, allows if scraps of human activities directly discard to marine or carried away the wave currents to marine.

Table 2. The number of plastic debris collected from east coast of Surabaya

District	Plastic number
Gunung Anyar	5
Rungkut	3
Bulak	473
Kenjeran	48

The plastic debris was inputted in the sediments due to activities from ocean-based sources. Fishing activities include nets, lines, ropes, gillnet or trawl floats has contributed plastic debris in marine areas [12, 13]. As far as the present study, Bulak district was the fishing settlement. Therefore, a huge of fishing boats leaning on the shore areas. These conditions will contribute plastic scraps for the marine environment.

3.2 The composition of plastic debris

The composition of plastic debris refers to a polymer of plastic debris. The amount of plastic debris based on polymer observed at the distinct station in the east coast of Surabaya is shown in Figure 1. In total 529 plastic debris items were found in the following viz. Polypropylene (67.86%), High Density Polyethylene (11.53%), Low Density Polyethylene (11.15%), Polyethylene terephthalate (5.1%), Polyvinyl Chloride (3.21%), other (1%) and Polystyrene (0.19%).

IOP Conf. Series: Materials Science and Engineering **462** (2019) 012050 doi:10.1088/1757-899X/462/1/012050

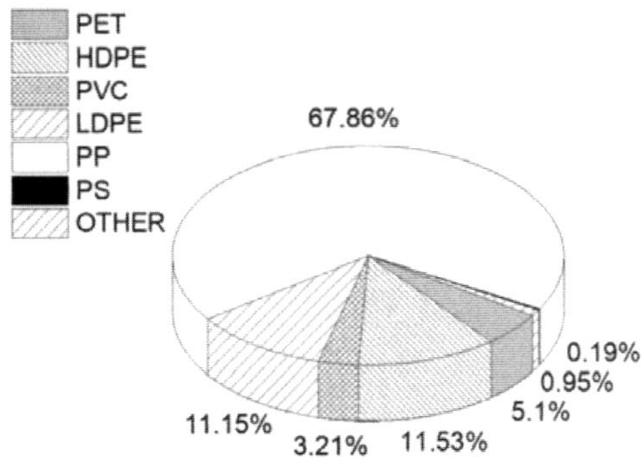

Figure 2. The total of plastic debris on the east coast of Surabaya

This study found that the majority of plastic debris on the east coast of Surabaya derives from terrestrial-based sources followed by ocean-based sources. The majority of plastic debris polymer was from polypropylene (67.86%). Based on the field observation, most of plastic debris were fragments from fishing activities and disposal from domestic activities. The source of fragments can be from fishing tools such as line fibres and nets [6, 13]. Based on our investigation, the source of HDPE plastic debris is fragmentation of plastic bottles and plastic bags. Fragmentation in the environment is mainly compelled by photo-oxidative degradation, followed by thermal and biodegradation [14]. Polypropylene fragments are more slowly and more point to mechanical degradation [15].

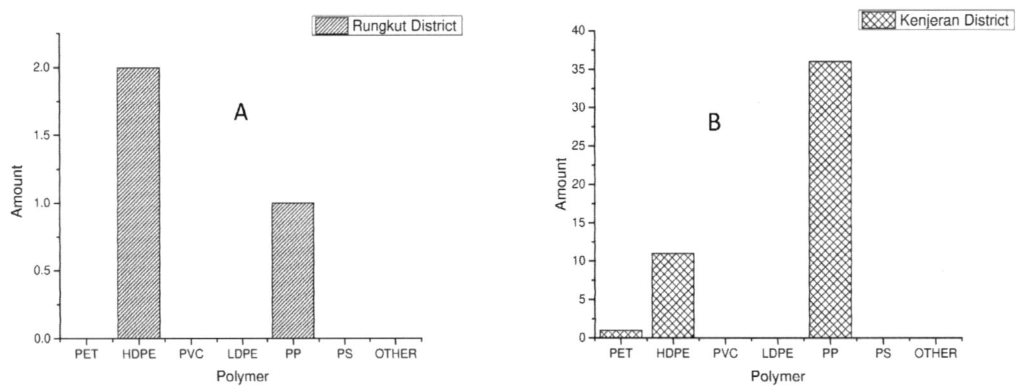

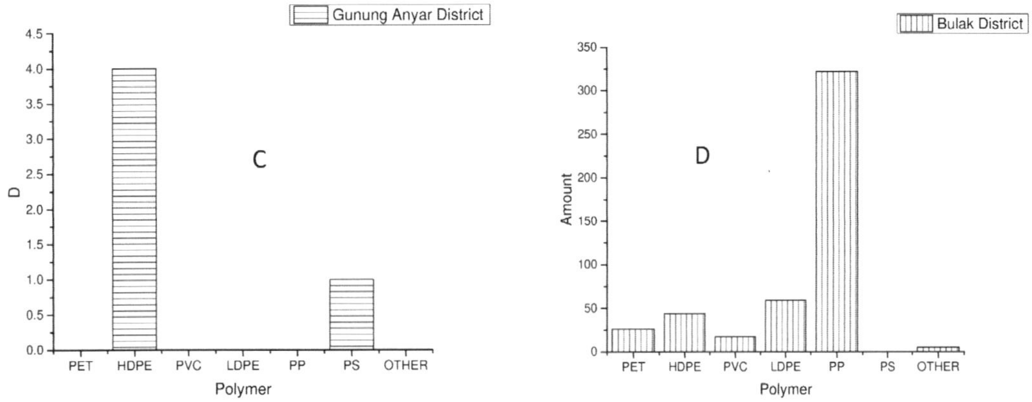

Figure 3. Summary of plastic debris based on its polymer in a) Rungkut district b) Kenjeran district c) Gunung Anyar district d) Bulak district

The amount of plastic debris in sediments based on the polymer in four districts was shown in figure 2. The highest number of plastic debris was obtained in Bulak district followed by Kenjeran district. On the other hand, Polypropylene (PP) was the dominant polymer plastic debris on the east coast of Surabaya. Bulak and Kenjeran were the districts with huge polypropylene. The previous paper reported fishing nets and line fibers include polypropylene strands [11]. Based on the observation, most of the plastic debris originated from a fishing net followed by plastic bottles scraps, and scraps of crackle bag. This indicates that plastic debris is from land-based and ocean-based sources.

The plastic debris found originated from fishing activities viz. line fibers and nets especially in Bulak district. Bulak and Kenjeran are districts whose majority population's livelihood depends on fishing. In addition, a lot of fishing boats were moored on the Bulak district. The present study has similar information to the previous report that Bulak district produces huge waste especially plastics, diapers and various other types of waste in the marine environment [16]. The present study showed Gunung Anyar and Rungkut districts have plastic debris from HDPE plastics originated from bottles of mineral water and crackle bags. This indicates that the plastic debris are from communities' activities discarded by the people into the marine environment. The finding of present result similar to previous studies of plastic debris in which polypropylene and HDPE were the among the most of polymer types found in marine environments

4. Conclusions

The plastic debris found at all the four districts sampled in the east coast of Surabaya. The plastic debris in sediments mainly consisted of polypropylene, where Bulak district had a majority of plastic debris. Overall, it can be concluded that plastic debris is commonplace in the east coast of Surabaya. Surabaya requires a strength and good management related to plastic waste to reduce the amount of plastic debris in the environment. The present study is a beginning for further study related to marine debris or plastic debris in the marine environment, especially in Surabaya.

5. References

[1] A. L. Andrady, "Microplastics in the marine environment," *Mar Pollut Bull,* vol. 62, no. 8, pp. 1596-605, Aug 2011.

[2] P. Europe, "Plastics – the Facts 2015," Belgium2015, Available: http://www.corepla.it/documenti/5f2fa32a-7081-416f-8bac-2efff3ff2fbd/Plastics+TheFacts+2015.pdf.

[3] N. H. Nor and J. P. Obbard, "Microplastics in Singapore's coastal mangrove ecosystems," *Mar Pollut Bull,* vol. 79, no. 1-2, pp. 278-83, Feb 15 2014.

[4] S. P. Wilson and K. M. Verlis, "The ugly face of tourism: Marine debris pollution linked to visitation in the southern Great Barrier Reef, Australia," *Mar Pollut Bull,* vol. 117, no. 1-2, pp. 239-246, Apr 15 2017.

[5] L. Fok, P. K. Cheung, G. Tang, and W. C. Li, "Size distribution of stranded small plastic debris on the coast of Guangdong, South China," *Environ Pollut,* vol. 220, no. Pt A, pp. 407-412, Jan 2017.

[6] D. K. Barnes, F. Galgani, R. C. Thompson, and M. Barlaz, "Accumulation and fragmentation of plastic debris in global environments," *Philos Trans R Soc Lond B Biol Sci,* vol. 364, no. 1526, pp. 1985-98, Jul 27 2009.

[7] A. Rahmat, N. Syadiah, and B. Subur, "Smart Coastal City: Sea Pollution Awareness for People in Surabaya Waterfront City," *Procedia - Social and Behavioral Sciences,* vol. 227, pp. 770-777, 2016.

[8] A. C. Staff, "In Indonesia, commuters pay for the bus with plastic waste," in *Asian Correspondent* vol. 2018, ed, 2018.

[9] S. Cahyono, "Surabaya Hasilkan Sampah Plastik 400 Ton per Hari," in *Jawapos*, ed. Surabaya: Jawapos, 2018.

[10] A. D. Hardianto, "The survival mechanisms of fisherman families (study of fisherman in urban village of kenjeran, bulak sub-district, Surabaya city)," *Journal Universitas Airlangga,* vol. 5, no. 2, pp. 1-15, 2016.

[11] Surabaya City. (2016). *Medium Term Development Plan Surabaya City 2016-2021.* Available: http://surabaya.go.id/uploads/attachments/2016/11/16408/bab_2.pdf

[12] A. W. Michelle Allsopp, David Santillo, and Paul Johnston, "Plastic Debris in the World's Oceans," Greenpeace Netherland.

[13] V. Hidalgo-Ruz, L. Gutow, R. C. Thompson, and M. Thiel, "Microplastics in the marine environment: a review of the methods used for identification and quantification," *Environ Sci Technol,* vol. 46, no. 6, pp. 3060-75, Mar 20 2012.

[14] P. Vermeiren, C. C. Munoz, and K. Ikejima, "Sources and sinks of plastic debris in estuaries: A conceptual model integrating biological, physical and chemical distribution mechanisms," *Mar Pollut Bull,* vol. 113, no. 1-2, pp. 7-16, Dec 15 2016.

[15] D. A. Cooper and P. L. Corcoran, "Effects of mechanical and chemical processes on the degradation of plastic beach debris on the island of Kauai, Hawaii," *Mar Pollut Bull,* vol. 60, no. 5, pp. 650-4, May 2010.

[16] N. Wijayanto, "Perairan Bulak Kenjeran Makin Banyak Sampah, Nelayan: Dapat Ikan 30 kg Sekarang Sudah Banyak," A. F. Nurani, Ed., ed. Surabaya: Tribun Jatim, 2017.

IOP Conf. Series: Materials Science and Engineering **462** (2019) 012051 doi:10.1088/1757-899X/462/1/012051

Policy Implication For Economic Losses Reduction Due To Earthquake Disaster In Bantul City, Indonesia

M F N Aulady[1,2] and T Fujimi[1]

[1]Department of Architectural and Civil Engineering, Kumamoto University, Japan
[2]Department of Civil Engineering, Institut Teknologi Adhi Tama Surabaya, Indonesia

170d9410@st.kumamoto-u.ac.jp

Abstract. One of the worst earthquakes occurred in Indonesia is in Bantul city, Yogyakarta province in 2006. It suffered huge amount of economic losses and many casualties. This issue become more significant, because residential buildings in Bantul city is impoverished. Moreover, Bantul city has high density of population. Here, we estimate the economic losses due to the earthquake in the future based on the vulnerability of dwellings in Bantul area using risk curve. Furthermore, we simulate a hypothetical mitigation policy of risk reduction for future earthquakes. After policy implication with various subsidy scenario, the economic losses can be reduced up to 648,017.48 USD for 0.5 % probability of exceedance in 50 years. This study indicates that the government can reduce the risk of economics significantly as the beginning in reducing risk with various factors.

1. Introduction

Natural disaster is a disaster as a result of an event series that is caused by nature, such as: earthquake, tsunami, volcanic eruption, flood, drought, hurricane and landslide. Indonesia is one of the countries that is prone to natural disaster, especially earthquake. This is due to Indonesia is surrounded by three tectonic plates, the Indo-Australian plate, Eurasia, and the Pacific plate. Earthquakes are not directly related to climate change, but the risk of natural disasters is ever increasing because of population growth, increasing urbanization and continuous unregulated construction of buildings. More importantly, many existing buildings may not withstand future earthquakes. Due to continuous changes in seismic codes caused by increasing human knowledge and experience about earthquakes, seismic rehabilitation of existing structures and components is inevitable [1].

One of the worst earthquakes occurred in Indonesia is at Bantul Regency, Yogyakarta province in 2006. It suffered many casualties; the reported fatalities are 5,778 people dead [2]. National and local government attempted to recovery the Regency. Many developments have been done, but nescience about earthquake resistant building might make people experience the same event. Moreover, vulnerability in this Regency might increase.

Based on the experience of the natural disasters in the past, a building which has a fairly high vulnerability becomes a major cause of fatalities in an earthquake disaster. This is in accordance with the saying goes: "Earthquake does not kill people, but buildings do". Some earthquakes that occurred recently, such as the earthquake in Yogyakarta Special Region and in other areas have been many casualties and property losses, including the destruction of building houses, as well as damage to public facilities such as schools, hospitals and office buildings insufficiently established to anticipate the danger.

IOP Conf. Series: Materials Science and Engineering **462** (2019) 012051 doi:10.1088/1757-899X/462/1/012051

This issue become more significant, because residential buildings in Bantul Regency is impoverished [3]. Furthermore, Bantul Regency has high density of population [4]. According to areas occupied by residential dwellings and other facilities, it shows that urban development in Regency Bantul does not follow logical process, even the population growth has increased [5].

Due to the number of losses, lives as well as properties, it needs a proper strategy. Disaster management focuses on implementing strategies that may lead to eliminating the risk of disasters [6]. This research becomes important because by evaluating earthquake risk in Bantul Regency based on scientific data, the public will be aware of the importance of the earthquake impact and will be able to reduce the vulnerability of earthquakes.

A related research concerning to earthquake hazard analysis in Indonesia developed hazard curves for major cities of Indonesia including Yogyakarta Special Region for 10 % probability of exceedance in 50 years [7]. Figure 1 shows Peak Ground Acceleration (PGA) of Indonesia at 2% probability of exceedance in 50 years. Seismic hazard map of Indonesia has been established by residential research and development center under Ministry of public works and residence [8]. As depicted in the figure, the color of Bantul city is peach or around 0.5-0.6g which is equivalent to intensity of VIII. Consequently, there would be a great shaking and heavy damage for unreinforced masonry buildings.

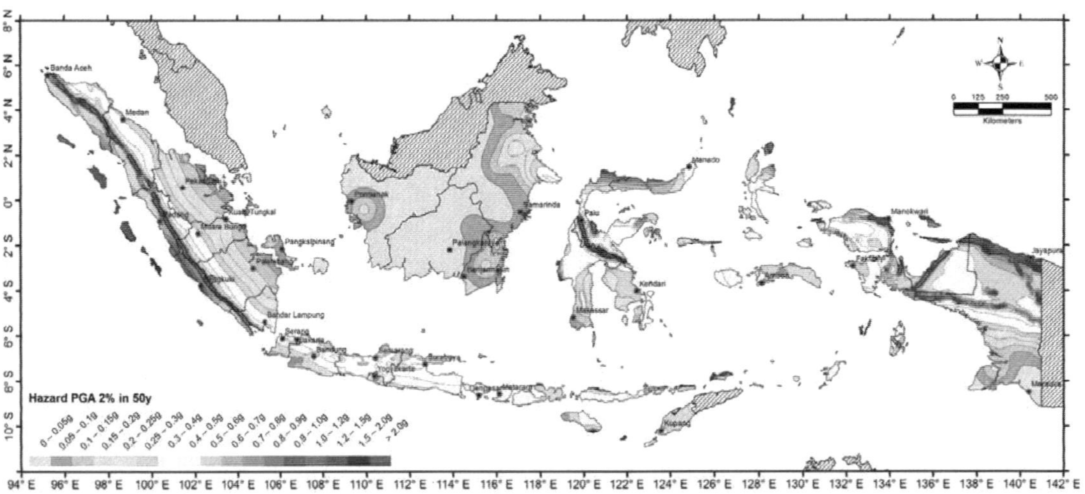

Figure 1. Indonesia hazard map, ground motion for all modeled sources for peak ground acceleration at 2% probability of exceedance in 50 years [9]

The main objective of this research is to evaluate earthquake risk in terms of economic loss in Bantul, regarding current situation of Bantul by using risk curve. The risk curve is used to estimate the surviving casualties and fatalities that might be caused by a potential earthquake in the future. Furthermore, it applied the hypothetical policy to reduce the risk. Since, most developments in Bantul are incompatible with building code, homeowners usually provide home construction work directly to the foreman or workers without planning process by an expert. It is intended to given an illustration about simple home that resistant to earthquake. Furthermore, the survey about the ability of homeowner to retrofitting based on the building code shall be conducted. If the ability of homeowners for retrofitting is low, it is expected that the government will give subsidy by the applicable terms and conditions or give the consultation services to the public. Hence, both government and homeowners can participate in reducing the risk of earthquakes.

2. Target Area

2.1. Indonesia

Indonesia is an Archipelago country that covers a total area of 1.905 billion square kilometers. 4.85 % of the Indonesia area is water territory. The population density is approximately 124.66/Km2. It is administratively divided into 34 provinces. Each province which has a capital city is governed by a provincial governor and provincial council. Each province consists of some regencies headed by regional head. One of the densest islands in Indonesia as well as in the world is Java island. More than 50% of Indonesian population number is in Java island. The south area of Indonesia, especially in the middle and south of Java Island, is prone to earthquake disaster. It is due to subdural zone activity which is formed from collision between Indo-Australia plate and Eurasia plate in the south of Java

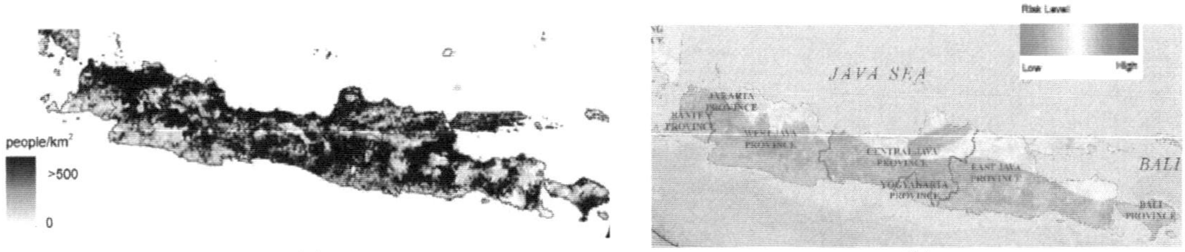

(a) (b)

Figure 2. Correlation between Population density map of Java Island (a) [10] and Earthquake hazard map of Java Island (b) [11]

Figure 2 how density map in Java island. From the figure, we can conclude that Yogyakarta is special province with population density is reach more than 500 people/Km2, it has earthquake risk level that is approximately red color. It will increase the possibility of fatalities and economic loss due to earthquake significantly.

2.2. Yogyakarta Special Region

Located in south of Java, Yogyakarta is an area with long history. Yogyakarta become a special region due to that long history. It has free procedure in Yogyakarta regarding the head regional position, duties, governor and vice governor authorities, the local government institutions of YSR; culture; land; and layout. YSR has an area of 3,185.80 km^2 with a total population of more than 3 million people. YSR consists of 6 Regions which Bantul is one of region occupying the third position in population density which is reaching 1884 people/km^2.

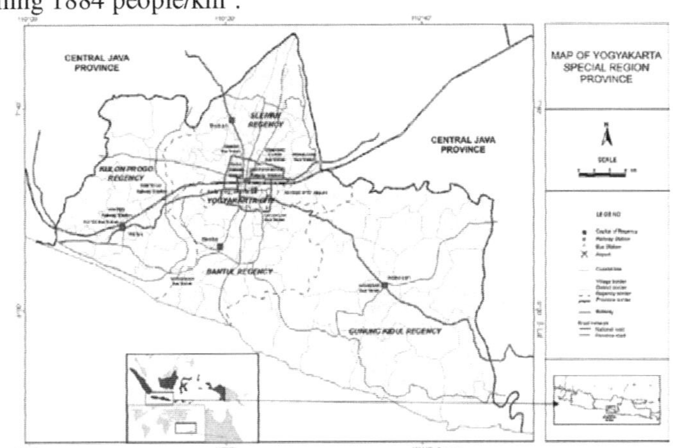

Figure 3. Yogyakarta Special Region (YSR) administration map [12]

2.3. Bantul City

Bantul City which is a target area of this research is located on south YSR. The southern part of this district is a limestone mountain, the western tip of the Sewu Mountains. The large rivers flowing along Bantul city are Progo River (limiting this district to Kulon Progo Regency), Opak River, Tapus River and its tributaries. Bantul is directly adjacent to the Indian Ocean. It is divided into 17 sub-regions to ease administrative matters.

According to official website of Bantul government, the population of Bantul city in 2012 is 955,952 people, with an area of 506.85 km^2. The number of live able houses in 2013 reached 146,628 units or about 65.91% of the total houses in Bantul. There are several health and school facilities to support the population life. Most of the facilities including trade centers, hospitals and educational centers have been developed along the main road so that this situation has put the burden on households who live far from this main road for supplying their essential goods. Road conditions in Bantul district can also be categorized well from the total length of road 872,125 km, with 652,775 km is a paved road.

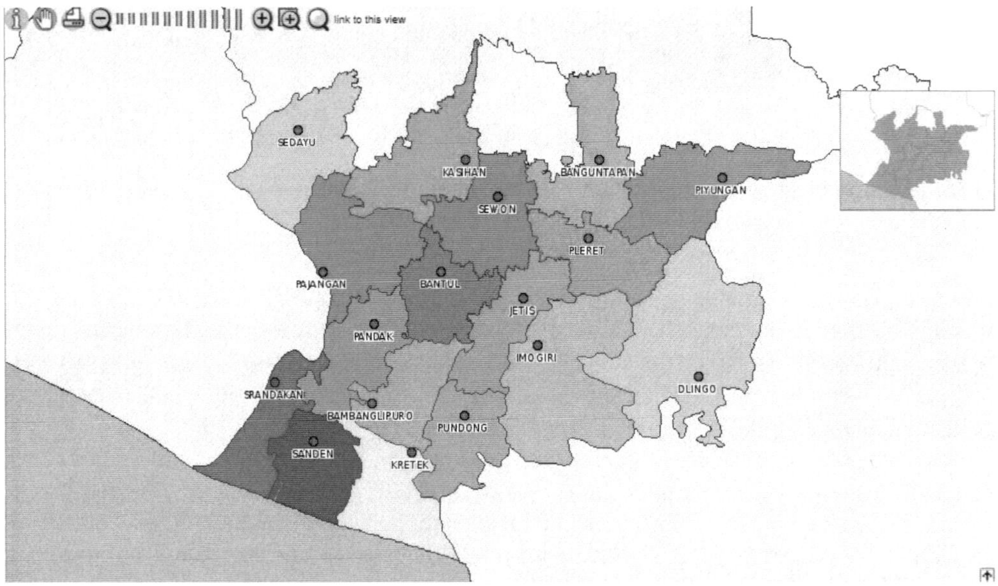

Figure 4. Administration Map of Bantul City [13]

3. Methodology

3.1. Risk Curve

Risk curve is a two-dimensional plot of real or projected financial risk which shows a display of correlation between probabilities of exceedance represented in vertical axis versus real or projected financial reward (horizontal axis). The risk curves show how often the occurrence of an event such as earthquake may have consequences such as economic loss indicating level of loss with different return period. The area underneath the curve represent the expected losses in in a specific probability of exceedance in 50 years which calculated considering accident scenarios due to the earthquake.

3.2. Filed Survey

To determine the impact of the hazard, the first step is analyzing and reconstructing the environment that can be affected. In general, the data identify the different types of physical ground entities, including built assets, infrastructure, agricultural land and people. Based on the number of residential dwellings and population living in Bantul which were obtained through the literatures and public statistics, a sampling in the target area and performing questionnaire survey distinguishes the percentage of various building typologies, the number of inhabitants in each type dwellings, cost value of the house, and other necessary data for loss estimation [14].

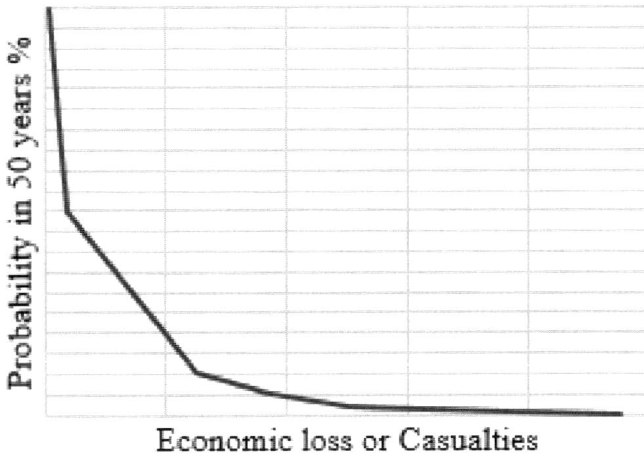

Figure 5. Conceptual Risk Curve

Table 1. Typology of Residential Building in Bantul city

No	Building Type	Symbol	Description
1	Unreinforced Masonry	UM	Constructed from brick, with no columns or beams, roofs made of tiles or asbestos
2	RC Low Rise Frame with Masonry infill walls	RCLRFM	The building has a frame (beams and columns) constructed from concrete. The frame is built first, then it is continued for the wall
3	Confined Masonry	CM	An improved version of MU, widely used by Indonesian people. The construction between walls and columns is done almost simultaneously
4	RC Medium Rise Frame with Masonry infill walls	RCMRFM	Constructed from frame concrete for buildings with medium height
5	Timber Frame with Structure infill	TFS	Building structure are from timber, and filled by rock wall and concrete
6	Unreinforced Masonry River Rock Walls	UMRRW	Building are not from concrete but using River rock wall
7	Timber Frame Residential	TFR	Constructed from timber and using for residential
8	Timber Frame Non-Residential	TFNR	Constructed from timber and not using for residential

Figure 6. Sample of Unreinforced Masonry

Figure 7. Sample of RC Low Rise Frame with Masonry infill walls

4. Result

Bantul is small city with strong heredity of culture and most of the population are farmer. As of it, more than 80% of buildings in Bantul City are residential, therefore most of economic loss are related to these buildings. Figure 8 summarizes the result of risk calculation in terms of economic loss for residential buildings in Bantul City, respectively.

Figure 8 depicts risk curve in terms of economic loss for all residential building in Bantul city. The curve was obtained by calculating the selling price of building and its assets, followed by multiplying the damage ratio and the building number of each type. Then, the result is summed according to probability of exceedance in 50 years.

The building typologies unreinforced masonry, confined masonry, RC medium rise frame with masonry infill walls, timber frame with structure infill, unreinforced masonry river rock walls, timber frame residential and timber frame non-residential are assumed to be in category of masonry, RC low rise frame with masonry infill walls and RC medium rise frame with masonry infill walls are assumed to be in the category of timber for adjusting parameters.

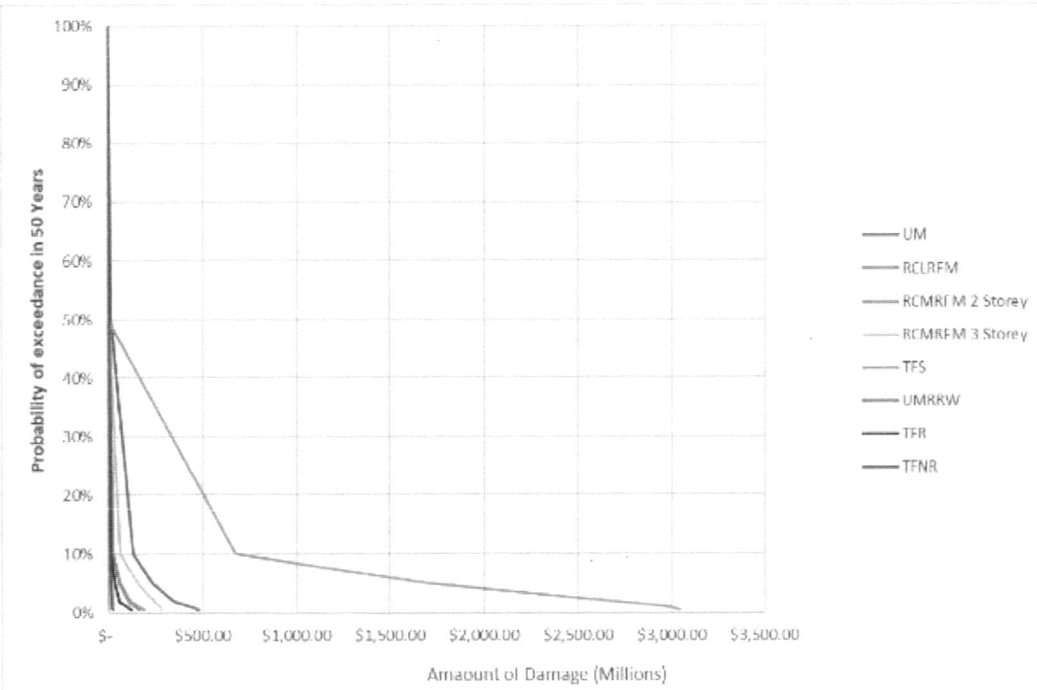

Figure 8. Calculation of risk in terms of economic loss for each building typologies of residential buildings in Bantul city

5. Discussion

The estimation result of the loss exceedance probability (EP) curves in term of economic loss for each building typology is shown in Figure 8. The concaved-shaped curve observed in all curves describe that the smaller probability exceedance, the more increase of damage. Figure 8 indicates that 10% probability exceedance in 50 years is equal to 90% of risk value or 90% probability of damage value which is amounted to be 676,357.63 million USD for RCMRFM 2 Story. Namely, the minimum cost that need to be pay out by related organization for recovery RCMRFM 2 Story of 90% probability is 676.357.63 million USD. In the other hands, 10% probability exceedance in 50 years of UM building is approximately 128,854,77 million USD. The economic loss of 95% and 98% value at risk would be 234,736.47 and 346,736.476 million USD for UM residential building, respectively. And for RCMRFM 2 Story are 1,704,984.86 and 2,677,248.96 million USD

The worst-case scenario of earthquake is at 0.5% probability of exceedance in 50 years which is equivalent to earthquake magnitude of M7.15. In this case, the economic loss for RCMRFM 2 Story and UM are at the highest value, which are approximately to 3,000 and 500 million USD. The economic loss for RCMRFM 3 Story is approximately to 276 million USD.

6. Policy Implication

In this section. We simulate a hypothetical mitigation policy of risk reduction for future earthquakes. As indicated in the previous section, the number of RCMRFM 2 story and UM plays a crucial role to increase the risk. It is because more than 90 % house in target area is inadequate. In this section, the policy applied is to replace the UM building type with the standard building code. The policy applied is only for UM building because RCMRFM 2 story building is for capable family, so that they can apply the building code without government support. Based on budget plan (APBD) of Bantul regency data in 2014, the budget plan of Public Work Department is 14,021,886.46 USD and it is realized to 13,431,298.50 USD. This means that it is still 590,587.96 USD remaining in 2014. We assume that the remaining budget can be used to build concrete house with earthquake national standard. Then it will have 95 units new house that can be built.

Table 2. The number of concrete houses from various subsidy scenario

Procerntage Subsidy		Price		Number of New House
Government	Home Owner	Government	Home Owner	
100%	0%	$ 6,223.23	$ -	95
90%	10%	$ 5,600.91	$ 622.32	105
80%	20%	$ 4,978.58	$ 1,244.65	119
70%	30%	$ 4,356.26	$ 1,866.97	136
60%	40%	$ 3,733.94	$ 2,489.29	158
50%	50%	$ 3,111.62	$ 3,111.62	190
40%	60%	$ 2,489.29	$ 3,733.94	237
30%	70%	$ 1,866.97	$ 4,356.26	316
20%	80%	$ 1,244.65	$ 4,978.58	475
10%	90%	$ 622.32	$ 5,600.91	949

It can be seen from Table 2 the number of concrete houses from various subsidy scenario , from the rest budget, the government can build up to 949 new houses with a subsidy of 10% per house, and it will then decrease the economic losses until 648,017.48 USD for 0.5% Probability of Exceedance in 50 years. Due to lack of data (Government website only publish the data on 2014). We assume that every year government has the same amount of remaining budget. So that, in three years government can subsidy up to 2,847 of concrete house. It will then reduce the economic losses for 0.5 Probability of Exceedance in 50 years up to 1,944,052.44 USD.

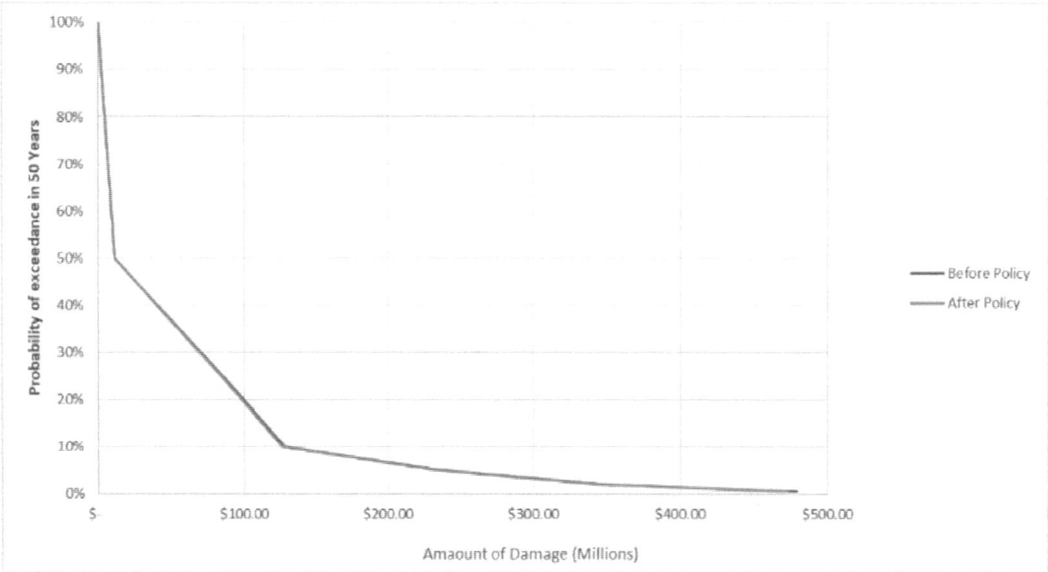

Figure 9. Calculation of risk in terms of economic loss after policy implication for UM residential building

7. Conclusion

The estimation of economic losses due to Bantul earthquake in 2006 has been done. The Residential building, especially UM building type is exposed highest value of economic losses, it is approximately to 500 million USD. After policy implication with various subsidy scenario, the economic losses can be reduced up to 648,017.48 USD for 0.5 % probability of exceedance in 50 years. Thus, the Bantul government can consider this policy properly. It can also reduce the vulnerability of Bantul city due to earthquake disaster. Certainly, it still has many parameters to consider in this policy, such as the ability of homeowners to spend the costs, and other unexpected factors. However, this study indicates that the government can reduce the risk of economics significantly as the beginning in reducing risk with various factors.

8. References

[1] "National Disaster Prevention Plan 2015-2019," The National Board for Disaster Countermeasure Abbreviated, Jakarta, Nov. 2014.

[2] S. Husein, S. Pramumijoyo, and D. Karnawati, "Micro Vulnerability Zone Mapping Vulnerability Province D.I. Yogyakarta," presented at the Geology Symposium Yogyakarta, 2010, pp. 1–10.

[3] "Percentage of Poor People (Percent)," *Statistic of D.I. Yogyakarta*, 28-Feb-2018. [Online]. Available: https://yogyakarta.bps.go.id/dynamictable/2017/08/02/31/kepadatan-penduduk-menurut-kabupaten-kota-di-d-i-yogyakarta.html. [Accessed: 28-Feb-2018].

[4] "Population Density by Regency / City in D.I.Yogyakarta," *Statistic of D.I. Yogyakarta*, 28-Feb-2018. [Online]. Available: https://yogyakarta.bps.go.id/dynamictable/2017/08/02/31/kepadatan-penduduk-menurut-kabupaten-kota-di-d-i-yogyakarta.html. [Accessed: 28-Feb-2018].

[5] R. Handayani, "Evaluation of Land Usage in Bantul District Based on Spatial Planning," Universitas Gadjah Mada, Yogyakarta, 2013.

[6] N. Etinay, C. Egbu, and V. Murray, "Building Urban Resilience for Disaster Risk Management and Disaster Risk Reduction," *Procedia Eng.*, vol. 212, pp. 575–582, 2018.

[7] J. Nugraha, G. Pasau, B. Sunardi, and S. Widiyantoro, "Hazard Analysis of Earthquake and Isoseismal for Java-Bali-NTB Region," *J. Meteorol. Geophys.*, vol. 15, no. 1, pp. 1–11, 2014.

[8] "Source and Earthquake Hazard Map 2017," Center of Research and Development of Housing and Settlements.

[9] "Source and Earthquake Hazard Map 2017," Center of Research and Development of Housing and Settlements, 28-Feb-2018.

[10] J. K. Baird, "Asia-Pacific malaria is singular, pervasive, diverse and invisible," *Int. J. Parasitol.*, vol. 47, no. 7, pp. 371–377, Jun. 2017.

[11] "Earthquake Risk Map Indonesia 2011," National Agency For Disaster Management, 16-Oct-2012.

[12] A. Wismadi, M. Brussel, M. Zuidgeest, H. Sutomo, L. E. Nugroho, and M. van Maarseveen, "Effect of neighbouring village conditions and infrastructure interdependency on economic opportunity: A case study of the Yogyakarta region, Indonesia," *Comput. Environ. Urban Syst.*, vol. 36, no. 5, pp. 371–385, Sep. 2012.

[13] "SIM Kewilayahan Berbasis WebGIS." [Online]. Available: http://kewilayahan.bantulkab.go.id/rtrw.php?mod=1. [Accessed: 28-May-2018].

[14] I. . Sengara, M. Suarjana, D. Beetham, N. Corby, M. Edwards, and Wehner, "The 30th September 2009 West Sumatra earthquake: Padang Region damage survey," Geoscience Australia, Canberra, 2010.

The Development of LIDI: A Web-Based Car Rent Marketplace Application in Sidoarjo, Indonesia

N F Rozi[1], M Ruswiansari[1], A Rachman[1], S R Wardhana[1], L Istiyanto[1]

[1] Department of Informatics, Institut Teknologi Adhi Tama Surabaya, Jl. Arief Rachman Hakim 100, Surabaya, East Java, Indonesia

Abstract. The common problem encountered by a car rental company is the transaction inefficiency. Only to know the car's availability, customers need to call by phone or coming directly to the rental company one-by-one. So, it is considered time-consuming. Whereas, the company could make use of the internet to make the transaction simpler. In this paper, a car rent marketplace application which is called LIDI developed to face that problem. To gain the high accessibility, the development of LIDI uses the responsive web principle so that users can access the application, both in a desktop or mobile environment. Its development phase consisting of requirement analysis, system design, implementation, and testing. The requirement analysis is carried out by collecting the rental data and analyzing the user needs. The system design will model the requirements into the diagrams. The implementation will transform the design into runnable codes utilizing an MVC framework. The opportunity is given to the rental companies and the customers in Sidoarjo, Indonesia to try the application as a part of the testing phase. The acceptance test's result shows that the application is feasible to be generally released.

1. Introduction

Sidoarjo is one of the districts of East Java in Indonesia. The total population of Sidoarjo is 2,199,171. It consists of 50.44% male and 49.56% female [1]. There are intangible business sectors or services in Sidoarjo such as a car rental. However, based on the survey and collected information, the car rental services in Sidoarjo are few compared to the one in Surabaya.

The survey's result on the car rental entrepreneurs in Sidoarjo concludes there are obstacles in doing such business. The customer usually does not know the location of the rental owner. The availability of the car is still checked by calling the company again and again, especially for the recurring customer transaction. Moreover, the first option users would do is to come directly to the rental company to check the car condition.

The existence of a system which could face the problem is needed, such as software. Software, or specifically called an application, is the key success of a computerized system [2]. It can be used to do special work and solve a specific problem [3]. For example, a most common application for doing word processing would be Microsoft Word [4]. Anyone in the world has used it as it can help someone in creating the report or other word processing tasks. Another advanced example would be the use of statistical software by the researchers in doing statistical analysis or simulating the model [5]. The development of such an application could also be used to solve the rental company problems.

The possible application platform would be the web platform. The consideration is the rapid growth of the internet users. By December 2017, there are approximately 2 billion people, or 48.7% of all internet users over the world are residing in Asia with the basis of 4,156,932,140 internet users [6].

IOP Conf. Series: Materials Science and Engineering **462** (2019) 012052 doi:10.1088/1757-899X/462/1/012052

It is the biggest one compared to the other region. The full report of the internet users can be looked at Figure 1.

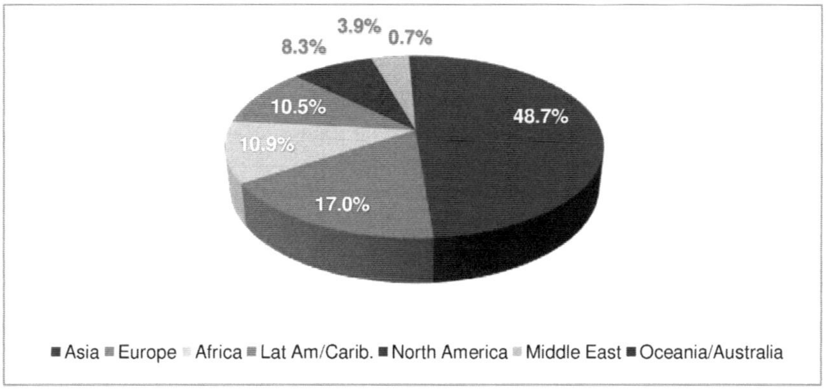

Figure 1. The number of internet users in the world by regions [7].

Following the data on Figure 1, the data centralization concept also considered as a big reason for choosing the web platform for developing the application. The participation level of business could be monitored easily.

Therefore, this paper will show the development process of the car rental marketplace application called LIDI. It will be used to resolve the current car rental company problems, especially in Sidoarjo, Indonesia. The application makes use of the web platform so that it can be in line with the internet users' growth. The similar application would be *BukaLapak*, *Lazada*, and *Tokopedia*, which is also using the marketplace concept. But the existing applications are targeted more on the electronic commerce for buying and selling goods. A Model-View-Controller (MVC) framework also used for standardizing the implementation process as also done by [8]. In the end, this paper will show the testing phase result to know the feasibility of the application to be released.

2. Development Model

The application development model used is a plan-driven model which also stated in [9]. It consists of some activities, such as requirement analysis, system design, implementation, and testing. The model also can be seen in Figure 2.

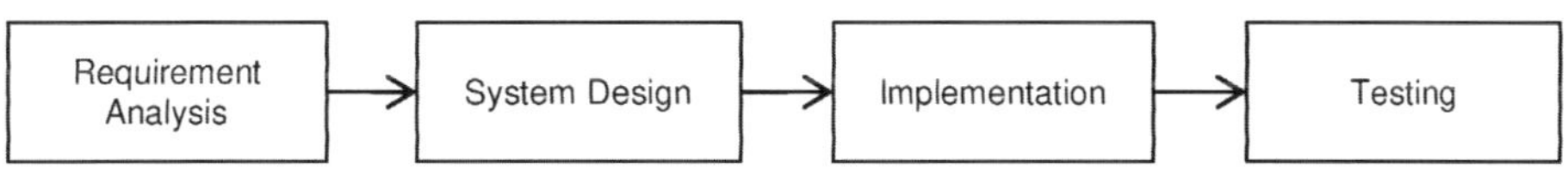

Figure 2. The development model.

Requirement analysis process is done by surveying the several car rentals in Sidoarjo, Indonesia. The existing business process of the car rental is asked. In this process, the user needs can be determined. The system design will utilize the UML to model the requirements into the standardized form. The implementation phase will transform the model into runnable code using responsive web design pattern to make the application accessible in the desktop and mobile environment without losing the user experience side. An MVC framework also implied. It can force the developer to separate the tasks into the three parts so that the application can be developed in more aligned architecture and managed easier rather than the one that does not utilize it. The testing process will be taken by five car rental companies (entrepreneurs) and 30 potential customers. The application will be run and used. The survey then conducted. The elements measured are the application functionalities and information quality in the web. The statement list confirmed could be checked in Table 1. The options are categorized into 4 points Likert scale (strongly disagree, disagree, agree, and strongly agree). The 4 points scale chosen to know also the sentiment of users about the application. The first two options are interpreted as negative, and the last two are interpreted as positive.

Table 1. Statement list to be confirmed by users.

No.	Statements
1	Application Functionalities
1.1	The application menu matches its function
1.2	The application functionalities matched to the user needs
1.3	The application run without error
2	Information Quality
2.1	The chosen font style is good
2.2	The user guides are clearly defined
2.3	The text on the web can be read clearly (without zooming)
2.4	The main menu can be understood
2.5	The car pictures can be seen clearly
2.6	The button and other visual elements are sized correctly

3. Result and Discussion

The discovered results in the preliminary phase of the requirement analysis related to the current car rental business process as follows.

1) The order process is still done by phone by inquiring the car availability before.
2) Customer must come and check the existence and condition of the desired car manually.
3) Rental owners are still using the conventional promotion strategy such as word-of-mouth marketing as a part of the promotional process.
4) Customers could only use manual process or cash to pay the rental service; the bank transfer is currently not available.

In this process, a few things to help car rental owners in promoting and renting a car finally determined. The following are the specifications related to the user needs.

1) Dynamic web applications that accessible via desktop and mobile environment are needed.
2) The application should be able to load text and images for promotional use.
3) The developed application should be able to receive payment by bank transfer.
4) The application should be able to provide information to the customers about the company profile, the car's availability, as well as other information.

The possible users of the application would be the *customer*, *car rental company*, and *application administrator*. The administrator added to the application with the privilege of managing and giving support to the stakeholders.

Modelling the requirements into the UML diagram is done in the system design phase. The LIDI marketplace application features which are modelled using the use case diagram could be seen in Figure 3. The application administrator will have a privilege to manage the web data. Every transaction should be done by login first to the application. The car company and customers will be able to register to the application by themself. The car company will have privileges to manage its company profile and car data. Customers will be then able to rent the car from any company in the marketplace. The application wireframes also created to help the preliminary requirements validation. This sample wireframe could be seen in Figure 4. The wireframe in Figure 4 is the prototype of the main page. The main page will consist of LIDI logo, main menu, search box, car category list, company header, and the main view which will be used to show the car list.

In the third phase, the specifications then transformed into runnable code. The user interface will employ in Bahasa Indonesia as language as it is targeted for the car companies which are resided in Sidoarjo, Indonesia. The coding process emphasizes the use of an MVC framework. The use of such a framework will force in separating the codes into the controller, model, and view. This architecture will make the code maintenance easier than the one which does not implement.

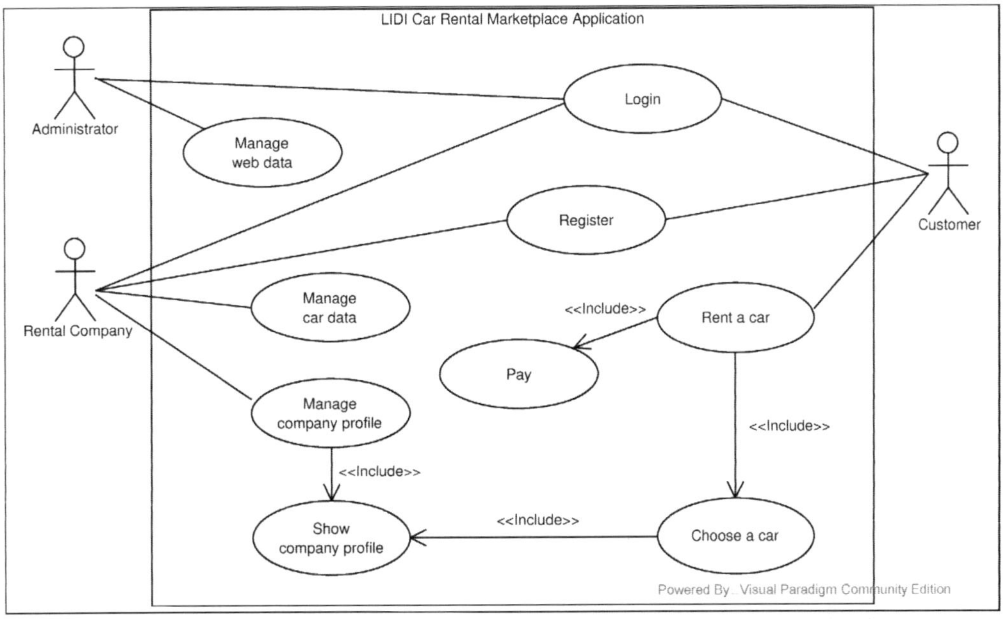

Figure 3. The use case diagram of LIDI car rental marketplace application.

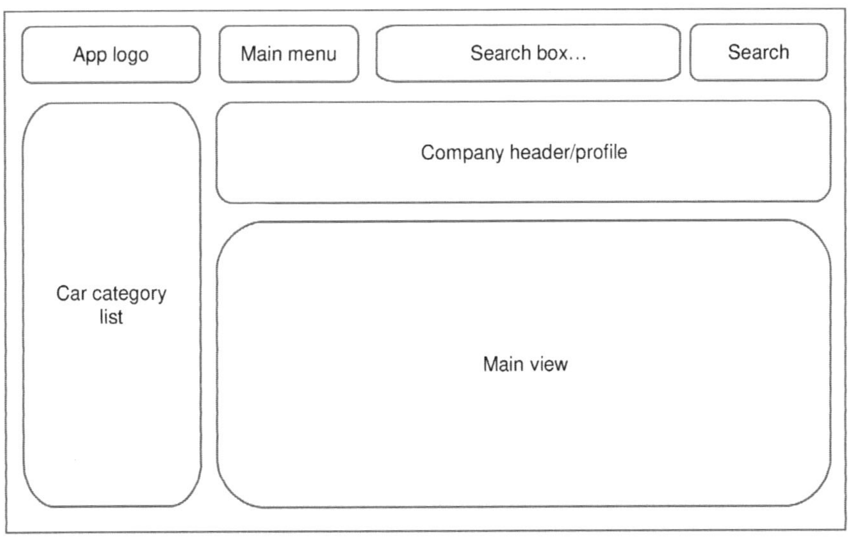

Figure 4. Wireframe of LIDI marketplace main page.

The result of developed application screenshots could be seen in Figure 5. The application could show the list of cars rented by a company as seen in Figure 5. Customers will be able to get the car's availability by just looking at the car status in the car detail page as seen in Figure 6. The page will also inform the price tag of the rent service related to the shown car. If there is a discounted price, the customers will also be able to check in that page. The application also provides the cashless payment mechanism. The customers could pay its transaction by bank transfer as can be seen in Figure 7. Customers could finish its transaction by confirming that the payment has been transferred to the provided bank account number.

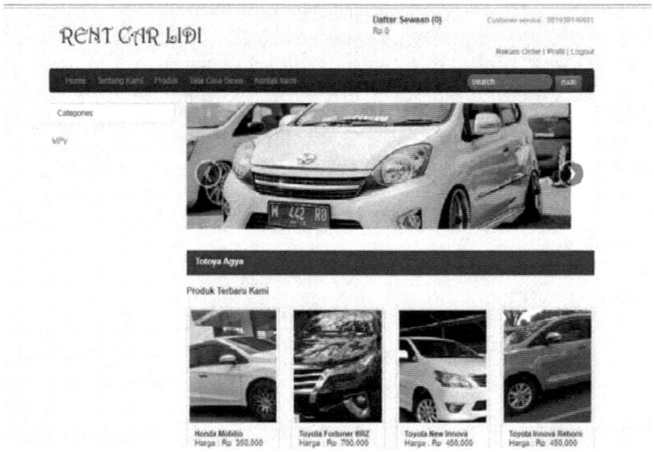

Figure 5. A car for rent listed in the LIDI marketplace.

Figure 6. A car detail page with the price tag.

Figure 7. Payment confirmation page.

The last phase is the feasibility test of the LIDI marketplace application by surveying the users based on the statements listed in Table 1. The result could be checked in Figure 8 and Figure 9. From the Figure 8, it can be seen that only one person who is stating disagree with every statement and no one stating strongly disagree. It can be stated that the application functionalities are in a good acceptance rate. As can be seen in Figure 9, users who choose to select the disagree option are very few with the maximum of 5 respondents compared to the users who choose to select the agree and strongly agree on options. Based on this condition, it can also be stated that the information quality provided by the application is in a good acceptance rate. From the results, it can be concluded that the application is feasible to be used and generally released.

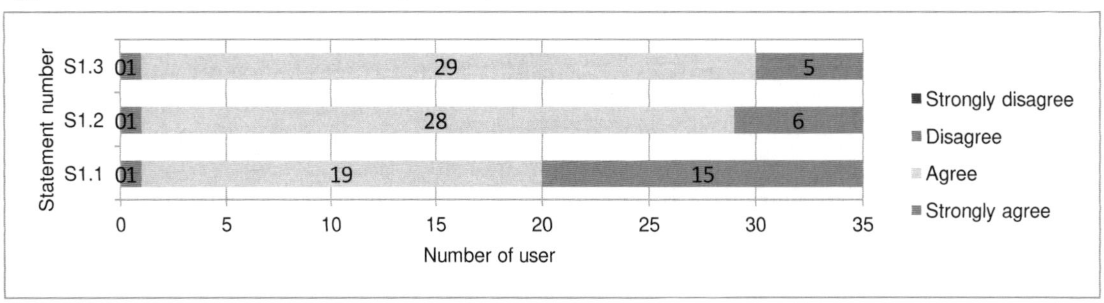

Figure 8. Application functionalities check result.

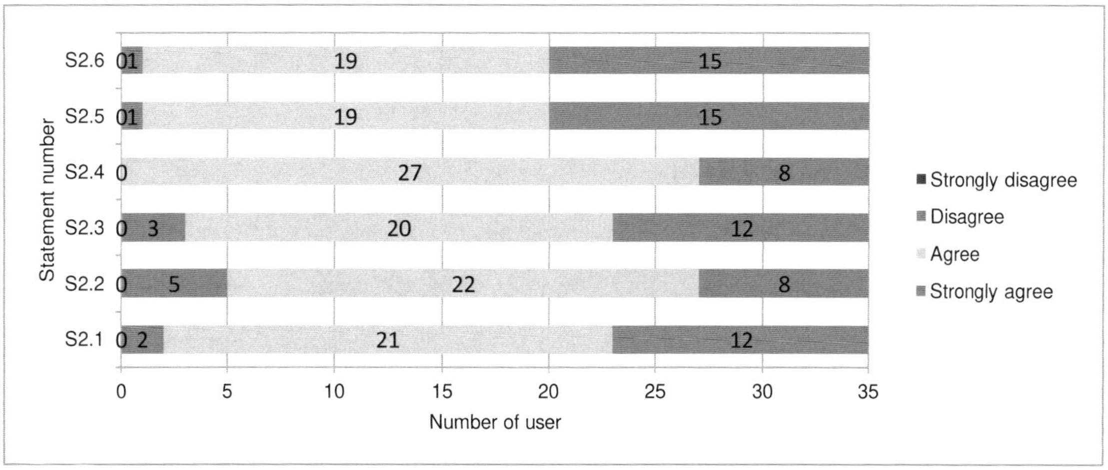

Figure 9. Information quality check result.

4. Conclusions

Based on the good acceptance rate of the application, it can be concluded that the car rental company problems in Sidoarjo, Indonesia could be solved by utilizing the LIDI marketplace for its business. The future work could be improving the two tests result by looking more at the requirements analyzed.

5. References

[1] Disdukcapil Kabupaten Sidoarjo 2016 Data Kependudukan *Dinas Kependud. Dan Pencatatan Sipil Kabupaten Sidoarjo*

[2] Rachman A, Rochimah S and Sunaryono D 2016 Komentar Semi Otomatis untuk Memudahkan Pemahaman pada Bahasa Pemrograman Java *J. Ilm. Nero* **2** 145–152

[3] Anees A and Agrwal A 2017 Software process improvement models and their comparison *Int. J. Adv. Res. Comput. Sci.* **8**

[4] Florell D 2011 Using Advancing Technologies in the Practice of School Psychology *A Practical Guide to Building Professional Competencies in School Psychology* (Springer, Boston, MA) pp 227–44

[5] Nunes C A, Alvarenga V O, de Souza Sant'Ana A, Santos J S and Granato D 2015 The use of statistical software in food science and technology: Advantages, limitations and misuses *Food Res. Int.* **75** 270–80

[6] Miniwatts Marketing Group I W S 2018 *Internet User in the World by regions - December 31, 2017*

[7] Perez M R L, Lagman A C and Adao R T 2017 Event Management Solution Using Web Application Platform *Proceedings of the 2017 International Conference on Information Technology* ICIT 2017 (New York, NY, USA: ACM) pp 206–211

[8] Sendiang M, Alelo M and Ponggawa V 2017 The Development of Accreditation App with MVC Architectural Pattern *Int. J. Comput. Appl.* **176** 28–32

[9] Kaya M, Conley S, Othman Z S and Varol A 2018 Effective software refactoring process *2018 6th International Symposium on Digital Forensic and Security (ISDFS)* 2018 6th International Symposium on Digital Forensic and Security (ISDFS) pp 1–6

Acknowledgment

We would like to thank Direktorat Riset dan Pengabdian Masyarakat (DRPM) from Ministry of Research, Technology and Higher Education (Kemenristekdikti) for funding this research through the Penelitian Dosen Pemula (PDP) research grant.